CISM International Centre for Mechanical Sciences

Courses and Lectures

Volume 570

The series presents lecture notes, monographs, edited works and proceedings in the field of Mechanics, Engineering, Computer Science and Applied Mathematics. Purpose of the series is to make known in the international scientific and technical community results obtained in some of the activities organized by CISM, the International Centre for Mechanical Sciences.

More information about this series at http://www.springer.com/series/76

Friedrich Pfeiffer · Hartmut Bremer
Editors

The Art of Modeling Mechanical Systems

Springer

Editors
Friedrich Pfeiffer
Institute for Applied Mechanics
Technical University of Munich
Garching, Bayern
Germany

Hartmut Bremer
Institute of Robotics
Johannes Kepler University Linz
Linz
Austria

ISSN 0254-1971 ISSN 2309-3706 (electronic)
CISM International Centre for Mechanical Sciences
ISBN 978-3-319-40255-0 ISBN 978-3-319-40256-7 (eBook)
DOI 10.1007/978-3-319-40256-7

Library of Congress Control Number: 2016942798

Printed on acid-free paper

This Springer imprint is published by Springer Nature
The registered company is Springer International Publishing AG Switzerland

Preface

Engineering and Physics cannot be thought of without models; models, which represent the real world to the best of our knowledge. And before we start with any mathematical description, with any mathematical model, we have to establish something like a phenomenological picture, a symbolic map of the real-world structures with elements like masses, springs, dampers, fluid system, thermodynamic elements and so forth and, not to forget, with elements of interconnections frequently more complex than the elements themselves.

This very first step of physical or mechanical modeling is mostly underestimated, but it decides very substantially about the success of all following activities like mathematical modeling, numerical algorithms, and finally computer codes. Therefore, it is worth looking at that more systematically, in spite of the fact that there does not exist any systematic approach to these problems. It is still more an art than a science.

Good modeling requires a deep insight into the performance of the real-world artifact, may it be a machine, an airplane, or human walking. We must understand how it works, in terms of operations, functions, dynamics, kinematics, stability and deformation, noise and wear, and even costs. But this is only one important precondition. Other aspects are the goals and requirements connected with such models.

First, there are the simple models, which nevertheless represent the main features of a problem, for example of a vibration problem, in such a good way, that they can be used to give some analytical insight into that problem with regard to dynamics but also with regard to parameter influences. Establishing such models is an art for a very few number of experts. It requires a perfect knowledge of the specific problem under consideration, and it affords intuition and intelligence to reduce such a system to a few parameters. But we often can learn from such models in a couple of days much more than by long-lasting computer simulations.

Second, we may establish models by considering as many details as possible. Such models are large and costly regarding computing times. And even in this case we have to investigate very carefully all physical effects for doing the correct neglects without endangering realistic results. Done in a skillful way such models are the basis for physical understanding and for improving design concepts. These two types of models aim at generating some results, which are as realistically as possible related to our real-world problem.

Third, if we leave that requirement, we may find models with similar features as our real-world case, but only in a more or less qualitative sense. This might help sometimes, but usually it is too far away from practice. Anyway, establishing models includes very strong phenomenological issues. This is mostly underestimated, because only good models in a mechanical sense, at this stage not in a mathematical sense, give access to good solution algorithms and finally to good results. Models should be as simple as possible and so complex as necessary, not more and not less.

As a rule we understand the word model as a theoretical construct. But model and modeling applies in the same way to experimental setups. Lack of thought very often identifies experiments with the dogmatic truth of practice, which is only sometimes true. To design and to establish good experiments really related to the practical system under consideration is a difficult task. And it is also a difficult task to find the correct interpretations of measured data. Therefore, comparing theory and measurement requires very much care on both sides, on the side of theory and that of experiments.

From all this we know, that modeling mechanical, and generally physical systems, requires insight and intuition, which usually is connected with long and broad professional experience. The course concerned with such a topic aims at presenting some rules for mechanical models in a more general systematic way always in combination with small and large examples able to illustrate the most important features of modeling. It will be not a course presenting mathematical solution algorithms, but discussing more advantages and disadvantages of potentially well-suited mathematical branches. It is a course with a strong focus on the art of modeling.

The following lectures have been given:

- Hartmut Bremer, Fascination of Making Models
- Friedrich Pfeiffer, Model Objectives and Realization
- Michel Raous, The Art of Modeling in Contact Mechanics
- Ahmed Shabana, Flexible Multibody System Dynamics
- Steven Shaw, Modeling for Nonlinear Behavior in Dynamics Systems
- Peter Wriggers, The Art of Modeling in Computational Mechanics

These lectures cover aspects of dynamics and also, to a certain extent, of continuum mechanics. They demonstrate, that the modeling problems have very much in common with respect to various fields, but of course with differences from the structural point of view. In the following these lectures will be presented in text-form.

Garching, Germany — Friedrich Pfeiffer
Linz, Austria — Hartmut Bremer

Contents

Modeling Objectives and Realization

Friedrich Pfeiffer
‡

Lehrstuhl fuer Angewandte MechanikTU-Muenchen
Boltzmannstrasse 15, D-85748 GARCHING, Germany ‡

Abstract Engineering and Physics cannot be thought of without models; models, which represent the real world to the best of our knowledge. And, before starting with any mathematical description, we must establish something like a phenomenological picture, a symbolic map of the real world's structures with elements like masses, springs, dampers, fluid system, thermodynamic elements and so forth and, most important, with elements of interconnections.

This first step of mechanical modeling is mostly underestimated, but it decides very substantially about the success of all following activities like mathematical modeling, numerical algorithms and finally computer codes. Therefore, it is worth looking at that more systematically, in spite of the fact that there do not exist systematic approaches to these problems. Establishing models includes very strong phenomenological issues. Models should be as simple as possible and so complex as necessary, not more and not less. And all this is still more an art than a science.

1 Preface

Good modeling requires a deep insight into the performance of the real world's objects, may it be a machine, a building, an airplane or human walking. We must understand how it works, in terms of operations, functions, dynamics, kinematics, stability and deformation, noise and wear under given loading conditions. But this is only one important precondition. Other aspects are the objectives and requirements of models.

Firstly, very simplified models might nevertheless represent the main features of a problem in such a way, that they provide some physical insight especially with regard to parameter influences. Secondly, we may establish models by considering as many details as possible. Such models are large, costly and sometimes leading to cloudy results. But done in a skillful way such models are also the basis for physical understanding and for improving design. Thirdly, we may find models

F. Pfeiffer and H. Bremer (eds.), *The Art of Modeling Mechanical Systems*,
CISM International Centre for Mechanical Sciences 570,
DOI 10.1007/978-3-319-40256-7_1

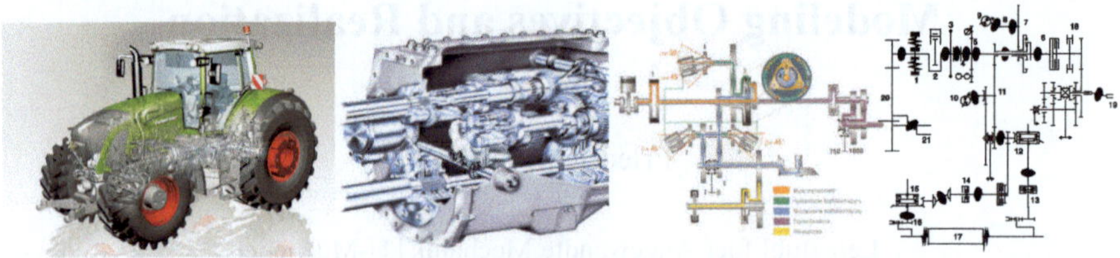

Figure 1: *The process of modeling shown by a tractor: the starting position for this example has been more or less perfect; hardware was existing, drawings and explosion charts were available, also a large body of experience and know-how by the design and test engineers; objectives were strength of certain components of the hydrostatic system during mulching and ploughing and vibration problems; models were comprehensive and complex; after successful verification with the company's measurements problems could be solved by parameter variations and corresponding design modifications.*

with similar features as our real world case, but only in a more qualitative sense. This might help sometimes to understand the physical background of a problem.

From all this we know, that modeling mechanical systems requires insight and intuition. The course concerned with such a topic aims at presenting some rules for mechanical models in a more general systematic way always in combination with small and large examples, many from industry, able to illustrate the most important features of modeling. It will be not a course presenting mathematical solution algorithms, but discussing the best and most efficient way to a good solution. The course has a strong focus on the art of modeling.

2 Mechanical Modeling

With respect to Technical Mechanics the aspect of *modeling* becomes one of the most important issues of mapping real world problems. Technical Mechanics is an engineering science, which considers motion or deformations of technical systems. They generate loads on machines, mechanisms and structures, which must be known for the design of such items. *Mechanical modeling* includes the replacement of a real machine, of real machine components or of real structures by certain basic elements. Considering mechanics this concerns for example masses, springs, dampers, frictional elements or finite elements with their shape functions which according to the topology of a structure must be interconnected in a physically correct way, usually leading to certain types of constraints. This process requires a deep insight into the operational problems of a machine and a sound knowledge of practice on the one and of mechanical theories on the other side.

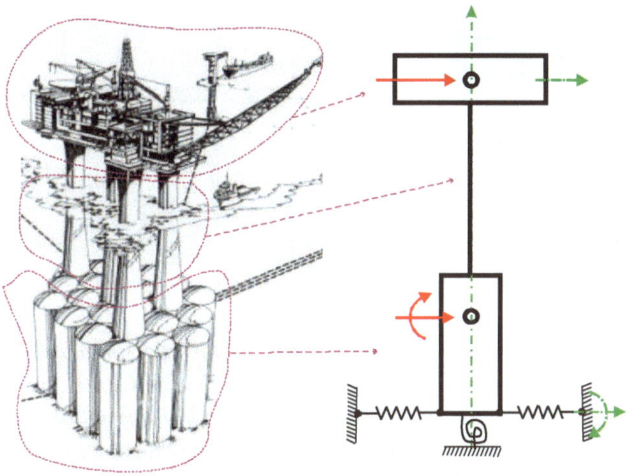

Figure 2: *A convincing example of "simple" modeling, but parameter estimates from a large FE-model: capsizing of an oil platform in the seventies in spite of large efforts during the design phase; for finding the reason with respect to this catastrophe Duncan [7] developed a simple model of the platform with 2-3 degrees of freedom and 3 different soil models for the sea ground; an additional FE-model has been used for finding the global parameters for the simple model; simple model results were convincing and explained the capsizing by some resonance behavior with the north sea wave spectrum, occurring very seldom.*

The quality of modeling decides on time and costs during a product development process, at least to a certain amount. Good models not only lead to quicker solutions, but also to better transparency of the problem under consideration and with it to accelerated achievements for a technical problem.

What is a *good model*, or better, what is a *good mechanical model*? A mechanical model will be a good one, if the mathematical model based on it gives us informations close to reality or for many cases close to a restricted reality, which might be of special interest to us. We have to anticipate, that reality is known, that it might be measurable or at least that it might be precisely describable. Therefore a good model should help us to come to a deeper understanding of the technical problems involved and of the design ideas behind them. To produce only numbers and charts will be not enough, we want to produce insight. Performing models has to keep that in mind.

How can we achieve a mechanical model? Usually we can assume, that every machine, machine component or structure offers some important operational functions, which are easy to describe and to model. With regard to our mechanical

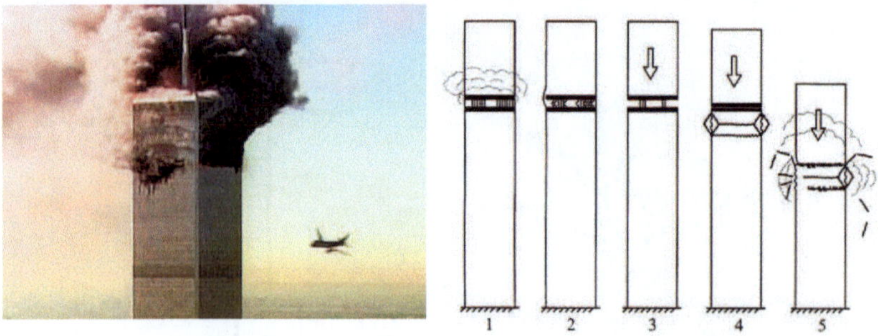

Figure 3: *An intelligent model for understanding the basic principles of a complete destruction: "The 110-story towers of the World Trade Center were designed to withstand as a whole the forces caused by a horizontal impact of a large commercial aircraft. So why did a total collapse occur? The cause was the dynamic consequence of the prolonged heating of the steel columns to very high temperature. The heating lowered the yield strength and caused viscoplastic (creep) buckling of the columns of the framed tube along the perimeter of the tower and of the columns in the building core." (from Bazant [2])*

systems these are for example some idealized motion sequences or vibrations, some effects from kinematics and kinetics. We start with that. Looking a bit deeper into such a structure, we might realize, that machines cannot be built in an ideal way, that we are confronted with disturbances, with "dirty effects", which in many cases cannot be modeled straightforwardly and the mechanics of which are often not understood really. Exactly at that point the typical work of an engineer starts to go on, which as a matter of fact possesses more an intuitive-empirical character than a scientific one, for example the question, what can be neglected. A good mechanical model is always a minimalistic model, too, not smaller than necessary, but also not larger than adequate to the problem involved. Finding intuitively neglects we may consider the geometric and kinematic situation, the order of magnitude of forces and torques or of work done and energy. Establishing a good model always needs an iteration process, which leads us with every step to a better solution. In his famous lecture on "Clouds and Clocks" from 1965 Karl Popper [31] tells us, that iterations are not only characteristic features of every intellectual work, but that they lead also from step to step to a deeper insight of the problem and to new questions finally achieving a really innovative solution, which at the beginning of such a process could not be perceived:

- mechanical modeling (theoretically and/or experimentally),
- examination with respect to plausibility, comparisons with reality,
- adaptation and improvements of models.

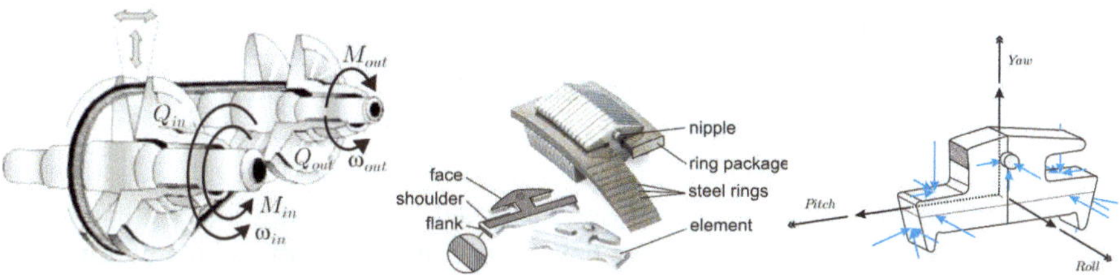

Figure 4: *A challenge of modeling, starting it even an efficient theory was missing: "Power transmission in automotive systems is classically carried out by gear trains, which transmit power by form-closure. In recent times an increasing number of continuous variable transmissions (CVT) are applied. They transmit power by friction, and at the time being they compete more and more with automated gears or with hand-shifted gears. The advantage of gear trains with gear wheels consists in a better component efficiency due to power transmission by form closure, the disadvantage in an only stepwise approximation of the drag-velocity hyperbola. This disadvantage is significantly reduced step by step by introducing automatic gear boxes with up to eight gear stages. The advantage of a CVT configuration consists in a perfect adaptation to the drag-velocity hyperbola, the disadvantage in a lower efficiency due to power transmission by friction and in a somehow limited torque transmission. An additional advantage of the CVT's is the possibility of very smoothly changing the transmission ratio without any danger of generating jerk. These features had to be evaluated by very detailed models, from belt approximations to every small component." (from Pfeiffer [24])*

Before doing this we have to define as precisely as possible, what should it be used for, what should be the outcome of our model, where should it be applied. With respect to these considerations we should regard the following aspects and problems:

- Motion (time response, frequency response): motion, motion patterns, frequency, damping, stability, amplitude and phase response functions.
- Control: If the system will be controlled, questions of observability, controllability, control quality, control stability and control optimization have to be answered.
- Perturbation: perturbation of the system, sensitivity of parameters, deterministic and stochastic perturbation.
- Optimization: Optimization of the dynamic system as a whole (process + controller) with respect to certain performance criteria, design strategies for the optimization of parameters and structures regarding sensitivities or other criteria.

The necessary methods involve a wide range of mathematics, system and control theory. They have multidisciplinary character.

3 Mathematical Modeling

Establishing a mechanical model we additionally have to watch some important aspects concerning mathematical and numerical modeling , that means the whole sequence of steps to a final solution of our mechanical model ideas:

- *discretization*: Can we compose our model only by rigid body elements or even by point masses or do we have to use also elements with a continuum-mechanical character? How shall we model such non-rigid bodies?
- The character of expected motion: Does some basic motion exist or do we have a type of *reference motion*? Is there some state of rest? Is it possible to describe the motion as one with a (usually nonlinear) reference motion and small deviations from it? Can we linearize, completely or in parts?
- *coordinates*: How many degrees of freedom includes our model? Can we find a set of coordinates, which corresponds directly to these degrees of freedom? If not, what sets of coordinates offer a formulation of constraints in a most simple way?
- *numeric*: What solution methods fit best to our problem, analytically (if any) or numerically? Can we put the mathematical formulation in a form, which corresponds in an optimal way to our solution possibilities? Is it possible to discover within our mathematical model and formulation already some qualitative or even quantitative results?

A perfect mechanical model, even a complicated one, will always be the simplest one possible, according to the well-known statement, that technology will be perfect if you cannot let out anything more. Especially for very complex systems we recommend to always start with a drastically simplified model for a better overview of the problems involved, if it is feasible. Then, in a second step, establishing a large model will be easier. In some cases it might be reasonable to go the other way round, namely to start with a large model and after some insights to break it down into smaller models. This depends on the particular problem, but as always also on the efforts which can be spent for a solution. Dresig [6] calls the first way inductive modeling and the second way deductive modeling. Anyway, a good comprehension of the problems will always come out with a better and faster development process.

Whatever method we use in mechanics, we always come out with a set of nonlinear differential equations of motion of first or second order, which are linear with respect to the accelerations but nonlinear with respect to velocities, positions and orientations. Very frequently we have to consider additional constraints and connections, which in many cases lead to differential-algebraic equations, the solution of which are nowadays also standard. All these equations represent the *mathematical model* of our problem and are an intermediate activity of our sequence (real world problem - mechanical modeling - mathematical modeling - numerical modeling - simulation). It should be kept in mind, that mathematics can give only results and information on a basis, which we have defined beforehand for example as assumptions and constraints of the mechanical model. Therefore, establishing the mechanical model requires extreme care, empirical knowledge and instinctive feeling combined with a good understanding of the mechanical problems, at least from a qualitative point of view. With respect to this step large expenditure can be generated but also omitted, if done intelligently.

The degrees of freedom (DOF) as expressed by minimal coordinates determine the size of the mathematical, of course before also of the mechanical model. Additional simplifications might be feasible by linearization, by using invariants of motion like energy integrals of conservative systems or by modifying the equations of motion, for example by transforming the differential equations. Anyway, we should try to find a set of minimal coordinates and if this is not possible, we have to add the relevant constraints or the relevant connections with force laws, but again trying to find a minimalistic formulation.

One could say, that this represents an old-fashioned procedure in the face of modern commercial computer codes, but it does not for two reasons: Firstly, technological progress is not possible without understanding the underlying problems, and the process described above helps significantly to increase our understanding. Secondly, also for computer codes the users have to establish a mechanical model, and it is advisable that this model will be carefully established on the basis of the same thoughts as discussed above. The quality of the results depend on such considerations. Commercial codes usually do not use a minimal formulation but a structure, which leads to fast and efficient numerical algorithms [41]. But interpretation of results depends as a matter of fact on the thorough understanding of the mechanical model and of the real world situation.

What we have mentioned does not depend on the choice of the mechanical laws we have applied, Newton-Euler, Lagrange, Hamilton, whatsoever. But on the other side, the choice of the mechanical foundation for the derivation of the equations of motion considerably influences the expenditure in establishing these equations. This has to be regarded very carefully. We shall discuss it in the following.

With respect to all possibilities for the derivation of the equations of motion we must perform kinematic groundwork, by the way one of the most frequent sources for errors and mistakes. The first step will choose coordinate systems, not to be underestimated, because a good choice helps to reduce effort, a bad choice produces effort. The second step must determine positions, orientations, velocities and accelerations, on the basis of these coordinate systems. In a third step we try to find minimal coordinates, and, if necessary, we establish the constraints. Velocities and accelerations usually are needed in an absolute form and with respect to inertial and body-fixed coordinate frames. This depends on the problem under consideration. Systems with small degrees of freedom might conveniently be represented with respect to inertial coordinates, which simplifies all derivations with respect to time. Performing kinematics does not depend on the methods of kinetics, which themselves offer a broad variety of possibilities.

Moment and momentum equations according to NEWTON-EULER and not applying any other mechanical principles can be used preferably for small systems and in combination with Euler's cut principle. We get a set of relations for each free body diagram including all reaction forces and torques. The elimination of these reaction forces without the D'ALEMBERT principle might be cumbersome for large systems. Therefore such a direct method makes sense only for smaller systems of clear kinematic structure.

Combining the moment and momentum equations with the D'ALEMBERT principle results in a very efficient method, which came out after very long discussions within the multibody system community. The constraint forces can be eliminated by the D'ALEMBERT principle, and the equations of motion include finally only the applied forces. The Jacobians of translation and rotation project the motion into the free directions (not into the constrained ones). After solving these equations we additionally can go back to certain free body diagrams for an evaluation of reaction forces (inverse kinetics). More flexibility with respect to constraint forces or to contact forces, for example, offer the LAGRANGE equations of first kind. As a summary we state, that NEWTON-EULER together with D'ALEMBERT on the one side, and LAGRANGE on the other side are, at the time being, the best procedures for treating large dynamical systems [24], where the projection method following NEWTON-EULER - D'ALEMBERT is without any doubt the number one.

The application of LAGRANGE's equations of second kind or of HAMILTON's principle requires the evaluation of the kinetic and potential energies expressed by generalized coordinates. Knowing the energies the equations of motion follow by corresponding differentiations. This process is the most important argument against an automized application of the analytical methods as given by LAGRANGE and HAMILTON, because computing time for differentiation is significantly larger than for a determination of the Jacobians of the NEWTON-EULER method. Nevertheless analytical methods are very convenient for mechanical systems with a small number of degrees of freedom, especially for those cases, where a treatment "by hand" will be achievable.

Another class of problems have been dealt with since the eighties, namely dynamical systems with contacts including friction. Non-smooth mechanics was established by MOREAU and PANAGIOTOPOULOS [22] and has been worked out up to now to a nearly comprehensive building of theories. For systems of that kind we have to consider Moreau's measure differential equations in combination with

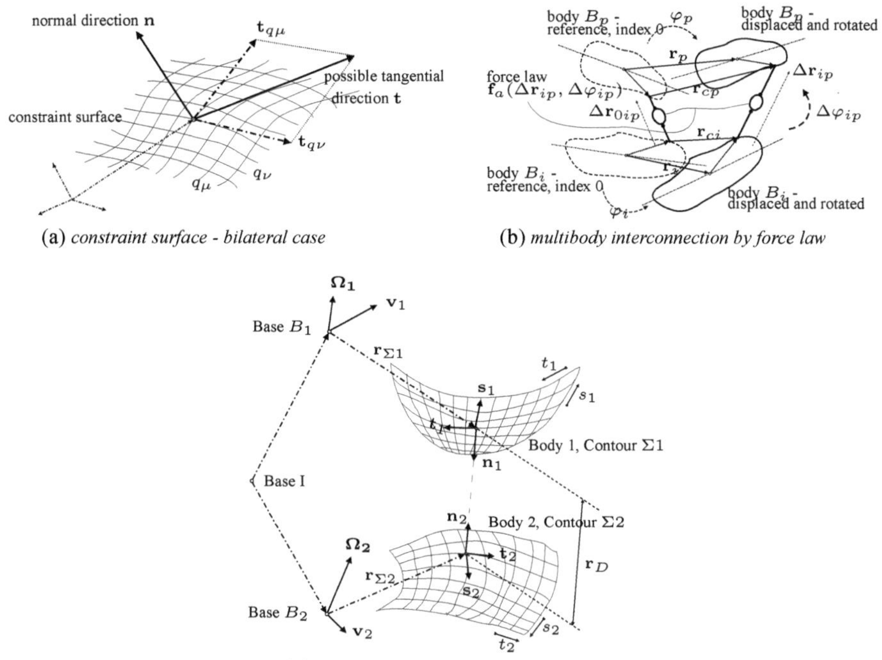

(c) *contact features - unilateral case*

Figure 5: *Constraints and interconnections form the structure of mechanical systems: "An important aspect of motion are constraints, which very often are treated as an unpopular appendage to the momentum and moment of momentum equations, but which in reality represent the most important interfaces gluing together masses of bodies and forcing them to a required performance of motion. If momentum and moment of momentum represent the motors of motion, then constraints are the controllers, which tell the dynamical system where to go. This is in perfect correspondence to the geometric and kinematic character of such constraints, which in fact realize the functioning of a machine or any kind of technical or physical system." [24]*

set-oriented constraints in the form of prox-functions.

The following table gives a survey of some methods quite frequently used in mechanical modeling.

- momentum and moment of momentum (NEWTON, EULER):

$$\dot{\boldsymbol{p}}+\tilde{\omega}_R\boldsymbol{p}=\boldsymbol{F}, \qquad \dot{\boldsymbol{L}}+\tilde{\omega}_R\boldsymbol{L}=\boldsymbol{M}$$
$$(\boldsymbol{p}=m\boldsymbol{v}_s, \quad \boldsymbol{L}=\boldsymbol{I}^s\boldsymbol{\omega}_s, \quad \text{s: mass center})$$

  representation in an arbitrarily moving reference frame R
- principle of d'(ALEMBERT), LAGRANGE, JOURDAIN:

$$\int_S (\ddot{\boldsymbol{r}}dm - d\boldsymbol{F}^e)^T \delta\boldsymbol{r} = 0\,.$$

  inertial representation; $\delta\boldsymbol{r}$: virtual (displacement) velocity (Lagrange), S: system under consideration
- projection equations:

$$\sum_{i=1}^{N}[(\frac{\partial \boldsymbol{v}_{si}}{\partial \dot{\boldsymbol{s}}})^T \quad (\frac{\partial \boldsymbol{\omega}_{si}}{\partial \dot{\boldsymbol{s}}})^T]\begin{pmatrix} \dot{\boldsymbol{p}}+\tilde{\boldsymbol{\omega}}_R\boldsymbol{p}=\boldsymbol{F}^e \\ \dot{\boldsymbol{L}}+\tilde{\boldsymbol{\omega}}_R\boldsymbol{L}=\boldsymbol{M}^e \end{pmatrix}_i, \qquad \dot{\boldsymbol{s}}=\boldsymbol{H}(\boldsymbol{q})\dot{\boldsymbol{q}}$$

  N: number of bodies, arbitrary reference frame R, (nonholonomic) minimal velocities $\dot{\boldsymbol{s}}=\boldsymbol{H}(\boldsymbol{q})\dot{\boldsymbol{q}}$, minimal coordinates $\boldsymbol{q}$, special solution (holonomic system): $\dot{\boldsymbol{s}}=\dot{\boldsymbol{q}}$
- LAGRANGE's equations of first kind:

$$\begin{pmatrix} \boldsymbol{M} & \boldsymbol{W} \\ \boldsymbol{W}^T & \boldsymbol{0} \end{pmatrix}\begin{pmatrix} \ddot{\boldsymbol{z}} \\ \boldsymbol{\lambda} \end{pmatrix}+\begin{pmatrix} \boldsymbol{Q}^g-\boldsymbol{Q}^e \\ \bar{\boldsymbol{w}} \end{pmatrix}=\begin{pmatrix} \boldsymbol{0} \\ \boldsymbol{0} \end{pmatrix}.$$

  $\boldsymbol{\Phi}(\boldsymbol{z})=\boldsymbol{0}$: constraint $\Rightarrow \boldsymbol{W}=-(\frac{\partial\boldsymbol{\Phi}}{\partial\boldsymbol{z}})^T, \bar{\boldsymbol{w}}=\dot{\boldsymbol{W}}\dot{\boldsymbol{z}}$, $\boldsymbol{Q}^g$: Coriolis- and centrifugal forces, $\boldsymbol{Q}^e$: (generalized) impressed forces
- LAGRANGE's equations of second kind:

$$\left[\frac{d}{dt}\left(\frac{\partial T}{\partial\dot{\boldsymbol{q}}}\right)-\frac{\partial T}{\partial\boldsymbol{q}}+\frac{\partial V}{\partial\boldsymbol{q}}\right]^T=\boldsymbol{Q}_{NK}\,.$$

  T: kinetic energy, V: potential, $\boldsymbol{Q}_{NK}$ (generalized) non-conservative forces
- HAMILTON's canonical equations:

$$\dot{\boldsymbol{q}}^T=\frac{\partial H}{\partial\boldsymbol{p}}, \qquad \dot{\boldsymbol{p}}^T=-\frac{\partial H}{\partial\boldsymbol{q}}\,.$$

  holonomic-conservative: $\boldsymbol{p}^T=(\frac{\partial T}{\partial\dot{\boldsymbol{q}}})$, $H=\boldsymbol{p}^T\dot{\boldsymbol{q}}-T+V$

- MOREAU's measure differential equations:

$$\begin{aligned}\mathbf{M}d\mathbf{u} =& \mathbf{h}dt + d\mathbf{r},\\ d\mathbf{u} =& \dot{u}dt + (\mathbf{u}^{+} - \mathbf{u}^{-})d\eta, \quad d\mathbf{r} = \mathbf{W}(\boldsymbol{\lambda}dt + \boldsymbol{\Lambda}d\eta), \quad d\eta = \sum d\delta_i.\end{aligned}$$

dt: Lebesgue measure, $d\eta$: atomic measure, $\boldsymbol{W}$: constraint matrix, $\boldsymbol{\lambda}$: constraint forces, $\boldsymbol{\Lambda}$: constraint impulses

4 Building Models

4.1 Foreword

Karl Popper, the great philosopher of the 20th century, established a picture of science and of scientific processes, which possesses substantial validity for all areas of science and even for many human activities in general, and therefore also for our concern here, technology and especially mechanics [31]: "The method of science is the method of bold conjectures and ingenious and severe attempts to refute them. A bold conjecture is a theory with a great content - greater at any rate than the theory which, we are hoping, will be superseded by it." And in addition: "Theories are nets cast to catch what we call "the world": to rationalize, to explain and to master it. We endeavor to make the mesh ever finer and finer." The same statements are true for models as one very important part of any scientific doing. Even the process of thinking about any problem applies models of thought. And the cognition processes of human beings and of animals of any kind uses "world models" in their brains to finally evaluate an "action selection" as basis of the subsequent motion.

He also tells us, that problem solving will be always an iteration process, going from one step to another one with improved solutions, but as a part of each step, also with an outcome of new problems. His arguments are the following [31]:

"Using **P** for problem, **TS** for tentative solution, **EE** for error elimination we can describe the fundamental evolutionary sequence of events as follows: But

$$\mathrm{P} \rightarrow \mathrm{TS} \rightarrow \mathrm{EE} \rightarrow \mathrm{P}$$

this sequence is not a cycle: the second problem is, in general, different from the first: it is the result of a new situation which has arisen, in part, because of the tentative solutions which have been tried out, and the error elimination which controls them. In order to indicate this, the above schema should be rewritten: But even in this form an important element is still missing: the multiplicity of the

$$\mathrm{P}_1 \rightarrow \mathrm{TS} \rightarrow \mathrm{EE} \rightarrow \mathrm{P}_2$$

tentative solutions, the multiplicity of the trials. Thus our final schema becomes something like this: (end citation Popper)". Everybody dealing with science as

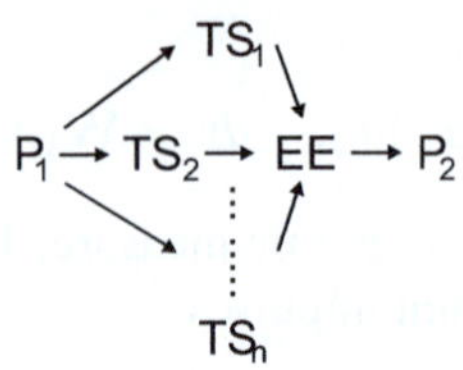

such and specifically with models and computer codes will confirm this process. It is a fact and a fundamental truth.

Generally spoken, models are always models of something; they are mappings and representations of natural or artificial objects, which themselves might be models. Models do not include all attributes of the objects they consider, but only those features, which are important for their creators and/or users. Models offer their typical replacement character of objects for recognizing, acting, model accepting and applying subjects, for certain times and situations and in the presence of certain restrictions concerning intellectual and/or real situations (H. Stachowiak, Allgemeine Modelltheorie, Wien 1973, in [32]).

We shall consider in the following three examples of modeling including three different types of problems and indicating some possibilities of simple and of sophisticated concepts. These are: the oil-rig Stratfjord B with an explanation of its capsizing by a simple model, the World Trade Center with an explanation of its collapse by an intelligent model and a V-12 Diesel Engine including an iterative evaluation process to improve an insufficient physical assumption at the beginning (see Figure 6). A really good model, whatever it is and any way it might be realized, must improve insight and understanding according to [23]: "You understand a subject when you have grasped its structure, not when you are merely informed of specific numerical results". To grasp the structure is the main problem, at the beginning and at every time of model evaluation. If this is not substantially elaborated at the starting point, then all the following work and diligence might be for

Figure 6: Model examples: Oil Rig, World Trade Center, Diesel-Engine

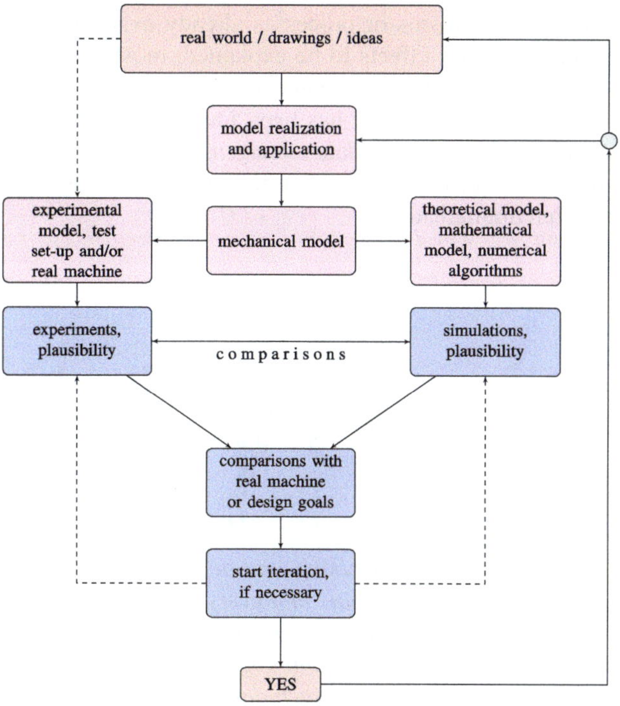

Figure 7: Modeling Process

nothing. And these considerations do not depend on the decision, which way to "model-realization" we choose, either self-made algorithms or commercial codes. In any case we have to deal with the structure of a problem.

4.2 Real World Situation

Whatever we model and wherever we start, we should not forget the basic aim and objective of a machine, namely to realize a process or processes. The word process should be taken in a broad way, for example, milling is of course a process, but moving a transportation system along a road or a track is also a process. That means, that modeling machines or mechanisms always includes modeling the process, for which the machine has been built. Considering real world technology or biology we have to start with some kind of hardware, even for that case, where we solve problems in a virtual environment, because it finally has also its origin in real world problems. Typical starting points are the following (see Figure 7):

Real machines, mechanisms or processes already exist, which include problems to be solved or cloudy effects to be explained, mostly with the goal of design modifications and design improvements. One of frequently underestimated problems is data acquisition, which has been improved considerably by the application of computer systems and data management systems, but without that it may become difficult and cumbersome. For biological modeling we step into these problems very often dramatically.

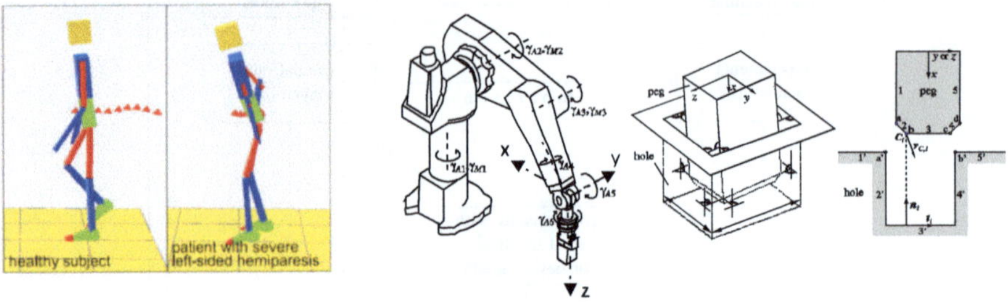

Figure 8: *Healthy / hemiparetic walking and peg-in-hole-insertion as examples for process realization: Every machine and every other artefacts are built or the biological evolution has realized them for performing some kind of process, for example drilling, milling or grinding in technology or walking, swimming or flying in biology. Therefore modeling machines includes always modeling of the processes they are built for. This might be cumbersome, but the only way to find out the real loads and operational issues.*

No real machines exist, and we are amidst the product development process with the request of avoiding potential problems with the help of models already during that phase. Usually data problems do not exist, but models have to be very flexible with respect to parameter variations, and with respect to physical and technological understanding and insight. Modern simulation techniques are an important pillar for product development. This pillar is based on models.

One of the most exciting application of models consists in testing ideas and concepts by models. This might go to a point, where new findings can be achieved by a heuristic use of models. These types of models must be very versatile with quick modification capabilities. Typically, their basic character will be heuristics.

We should not forget process modeling, which is necessary to generate realistic data for a machine, data including topological and structural aspects and including motion and loads. The topological structure of a machine is always designed for a limited variety of processes, which not only have to be known but also which have to be described precisely. Peg-in-hole problems are standard for many production lines in mechanical and automotive engineering industries. In many cases it is a

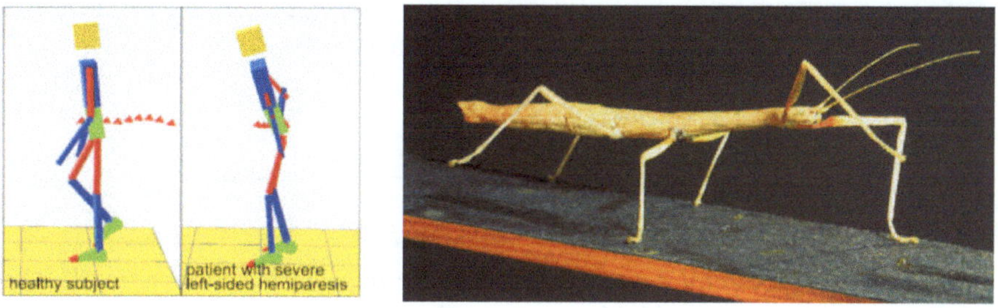

Figure 9: *Real world examples requiring models for the search of construction principles: It is an exciting objective to use models mainly for the detection of biological construction principles or biological load behaviour and only incidentally for direct simulation purposes. For hemiparatic walking the load anomalies in the knee or hip joints can only be evaluated by models, but they give a clear indication of the degree of hemiparesis. The load distribution principles in the legs of insects can be determined by a large optimization problem with the insect model as one small component but with unknown optimization criteria to be evaluated by comparison of simulations and measurements.*

cumbersome task to model such problems. An example is given in Figure 8 depicting a hole insertion for a rectangular and a circular peg. For the rectangular peg alone we get more than 40 contact situations, which must be modeled properly applying for example non-smooth mechanical theories and which must be combined with the dynamics of a robot to achieve efficient control for that case. It has been a matter of a whole dissertation.

One important issue of model application consists in a reduction of experimental testing. Many questions with respect to performance, efficiency, safety, noise and wear can be answered by simulations using intelligently selected parameter variations. Additionally we may optimize the tests by choosing the best parameter combination with regard to representing the system performance and giving comprehensive informations. This is standard in chemistry, for example. On the other hand we should keep in mind, that model-based simulations represent the real world only to a certain extent, because there always remain uncertainties and imponderables with respect to the model itself and the necessary data.

4.3 Mechanical Models

Before starting with building a model we must establish a thorough and a precise description of our object to be modeled. From my own experience it always turned out to be very helpful. This should be as far-reaching and profound as possible,

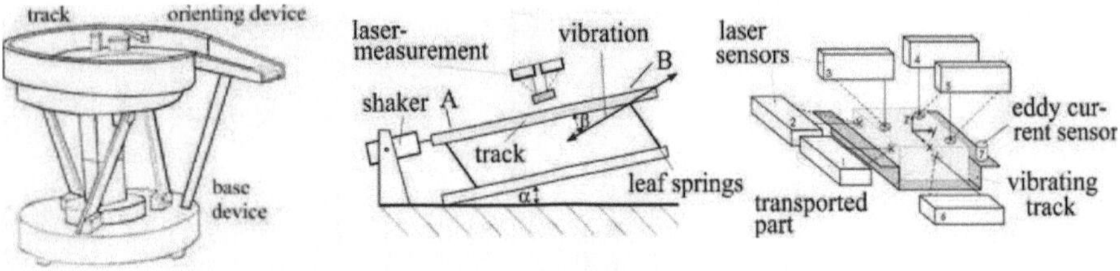

Figure 10: *Experimental set-up for a conveyor transportation system: The operation of conveyors is a highly non-smooth process, because transportation is realized by impacts with friction. To measure the dynamic behaviour of every individual part is difficult requiring a sophisticated arrangement of various sensors. It also requires an exact speed and vibration control to be able to compare simulations and measurements.*

as far as possible at the beginning of such an activity. We establish something like a model of thoughts including problems like constraints, the best mechanical elements for the model, including also first ideas for the mathematical model to be build up at a final stage and a first estimate of problems to be expected and hopefully to be avoided.

Such thoughts may result also in first indications of the type of model we begin with. Do we want a theoretical or an experimental model or both. Do we start with a large comprehensive model or with a small and simplified one, giving us some first important insights. This all takes time, but a time we get back in the course of our research by a minimum of detours and time-consuming surprises. Let us start with model elements.

Every machine, mechanism, transportation system and also biological systems are constrained with respect to their motion possibilities. We may state without restrictions, that the functionality and the operational features are exclusively realized by a collection of intelligent constraints or by interconnections with given force laws. They prescribe motion, but they also prescribe the local positions and orientations with no motion and with some load carrying capabilities. Kinematic constraints determine together with force-based interconnections the performance and the efficiency of machines, therefore they are a central and fundamental component of every mechanical model, at least with respect to dynamics. Extreme care has to be taken to establish the constraints with convenient coordinates and taking into account the logical structure of the system to be modeled. All interconnections by some linear or nonlinear force laws, as already mentioned above, need to be defined geometrically but also by their interconnecting forces and torques being given either by linear or nonlinear physical laws. This always includes for many cases complicated evaluations with respect to relative kinematics and kinetics.

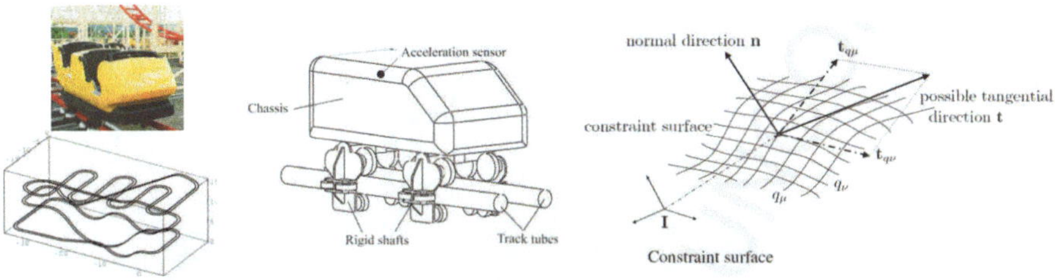

Figure 11: *A safety-oriented constrained system: Safety is an issue with highest priority when designing fair equipment. Models must take that into consideration. Roller coasters for example are constrained not to leave the track by wheel packages, each with five wheels. They form a system of bilateral and unilateral constraints assuring safe riding. Additionally the track design must watch certain limits with respect to acceleration and force loads on the passengers. Such systems are also non-smooth.*

One important decision concerns the properties of masses, rigid or elastic, linear elastic or nonlinear elastic. This depends very much on various excitation sources in the system, for example on parametric and self-exciting mechanisms or on external excitations due to operational requirements. Any machine component with elastic eigenfrequencies within the frequency regime of the just mentioned mechanisms must be modeled as elastic parts with "sufficient" eigenmodes to cover the given excitations. Fortunately, many systems possess only linear elasticity, but considering for example crash behavior requires highly nonlinear elasticity.

Whatever we model we need a complete set of data describing our machine, for example. This is by no means a simple task and can afford weeks or months. It includes data of geometry and topology, design data, masses and mass moments of inertia, data of the materials, operational data like frequencies and speeds, control data and data for describing requirements. The list is long and can contain thousands of magnitudes. Data require continuous care during the realization of models.

5 Modeling Examples

5.1 Example: Power train of a tractor

5.1.1 Objectives

As most of the modern tractors also the Agco/Fendt concept possesses a very complex power transmission system including many different gear components and a hydrostatic system with hydraulic motor and pump. The largest loads arise by ploughing and mulching, which caused at the beginning of the product development some failures of the hydrostatic components.

From the very beginning it was clear to establish a large comprehensive model of the power transmission comprising every detail of its structure, especially with respect to the hydrostatic system. The results of the corresponding simulations should either confirm the existing design and/or additionally give some recommendations for the improvement of components. A further objective consisted in better insight into the powertrain operation, again with a focus on the hydrostatic system (see [4], [24]).

5.1.2 Introduction

Power transmissions of tractors represent the most complicated drive train units of nearly all transportation systems. They must meet a large variety of very hard requirements, and in the last decades the demands for more versatility, flexibility and comfort for the driver increase continuously. Tractors are supposed to do all kind of agricultural work, but also to fulfill many requirements of forestry, and additionally tractors should be able to move along normal roads with reasonable speed and safety. The above Figure pictures a typical modern tractor concept with a very sophisticated drive train system.

This altogether leads to very complex drive trains, for which modern CVT-concepts offer significant advantages. They may operate with some given speed independently from the engine speed thus running the motor at the point of best fuel efficiency. It offers advantages for all implements by generating an optimal power distribution between driving and working.

The German company Agco/Fendt as one of the leading enterprises of modern tractor technologies developed for that purpose a very efficient power transmission including for low speeds a hydrostatic drive and for larger speeds a gear system with the possibility of mixing the power transmission according to external requirements. The hydrostat system possesses the great advantage to develop very large torques especially at low speeds and at standstill.

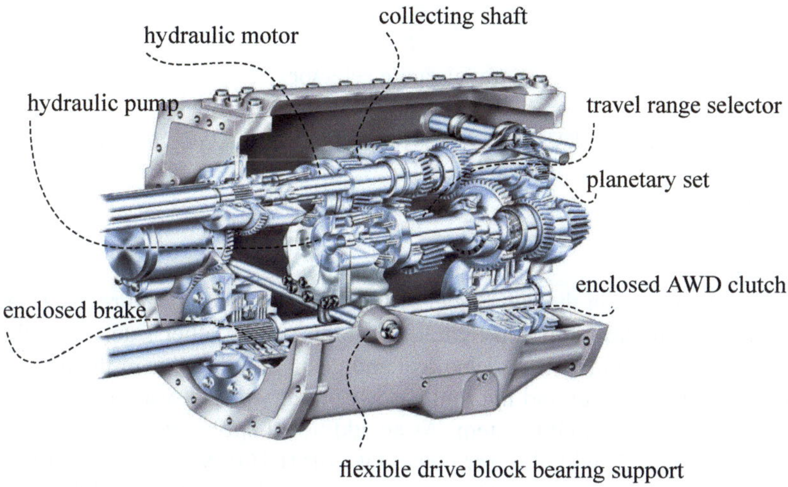

Figure 12: ML Power Transmission of the VARIO series (courtesy AGCO/Fendt)

Figure 12 ilustrates the kernel of the power transmission, and Figure 13 presents a sketch of the overall system. Following this sketch we recognize that the torque of the Diesel engine is transmitted via a torsional vibration damper (1) to the planet carrier (5) of the planetary gear set (2). The planet gear distributes the power to the sunwheel (4) and the ring gear (3). This ring gear drives via a cylindrical gear pair the hydraulic pump (6), which itself powers the two hydraulic motors (7), where the oil flow from pump to motor depends on the pump displacement angle α. The hydraulic motors (7) generate a torque according to the oil flow and the motor displacement angle β. Furtheron, the sun wheel transmits its own torque via a gear pair to a collecting shaft (8), which adds the torques coming directly

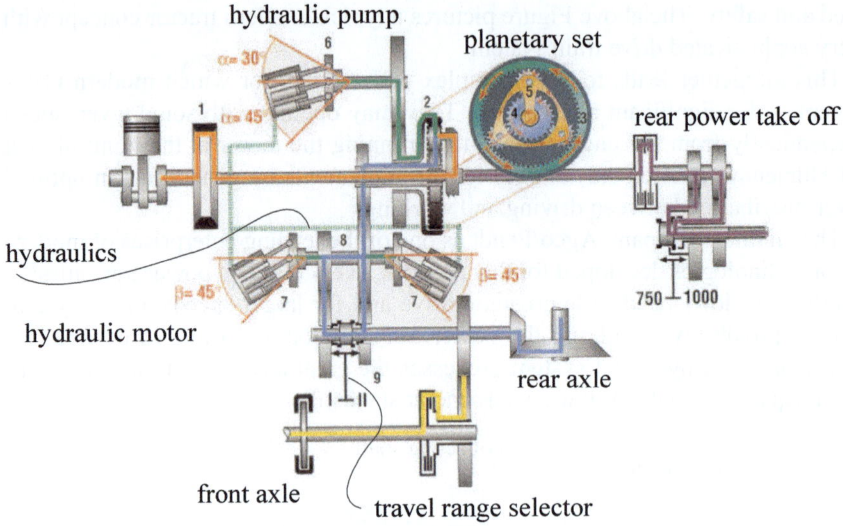

Figure 13: Sketch of VARIO-ML System (courtesy AGCO/Fendt)

from the Diesel engine via the sun wheel on the one and coming indirectly from the chain ring gear-hydraulic pump-hydraulic motors on the other side. Thus the collecting shaft (8) combines the mechanical and the hydraulic parts of the torque. This splitting of mechanical and hydraulic power in an optimal way is the basic principle of the stepless VARIO system. As an additional option we may also connect the planet carrier (5) with the rear power take off (PTO) system, which drives all possible rear implements for performing processing of agriculture and forestry. The travel range selector (9) allows switching between slow and fast operation. The outgoing torque will be transmitted to the rear axle by a pinion and to the front axle by a gear pair.

The system allows an adaptation of the tractor speed only by an appropriate combination of the displacement angles α and β without changing the engine's speed, which means, that the engine's speed may be kept constant at an optimal fuel efficiency point in spite of varying tractor speeds, of course within a limited speed range. This stepless VARIO-concept represents a type of CVT-system (**C**ontinuous **V**ariable **T**ransmission), which results in significant improvements with respect to handling and working performance.

5.1.3 Modeling

For the investigation of dynamic loads of certain critical components and additionally for the assessment of the vibrational behaviour an appropriate model has to be established, which includes as most important elements the mechanical parts, the Cardan shafts and the hydrostat components, and also the hydraulics. Many mechanical parts like shafts, elastic connection and the like can be assumed to behave linearly, but the Cardan shafts, the hydraulic pumps and motors as well as the hydraulics itself will generate non-linearities. Therefore we shall focus on those components. We start with a consideration of the complete system representing afterwards the equations of the components, as far as necessary.

Figure 14 depicts the model of the ***complete power transmission*** of the tractor under consideration. Partly and as far as possible we model the mesh of two gear wheels by two rigid bodies with a fixed transmission ratio resulting in a reduction to one degree of freedom (DOF). The crankshaft of the Diesel engine is described by a rigid body (2) with one DOF, which is loaded by the torque from the combustion pressure. They must be determined from measurements of the combustion process and correctly projected to the crankshaft model. The front PTO (power take off, 20, 21) is connected to the engine by a vibration damper. Also the planetary set (7) and all auxiliary equipment (5) are driven more or less directly by the engine. The model follows the diagrammatic sketch of Figure 13 with some extensions. They concern mainly the two power take off systems, front PTO and rear PTO, and also the front and rear axles with the tires. We shall come back to all components.

Figure 14 pictures a very classical mechanic-hydraulic system, where the physical relations are obvious. Therefore it represents a good example how to establish a mechanical (or physical) model, which is equivalent to the real world problem. As already mentioned, engineering mechanics nor engineering physics are not deductive sciences thus requiring as a rule bundles of assumptions and neglections without destroying the principal information base of a system. From this we shall try also in this case to come to simple equations.

Dealing with ***components*** we start with rigid and elastic shafts. For the problem of loads and vibrations we assume for the rigid shaft a single rotating mass with one degree od freedom, and for an elastic shaft simply two rotating masses connected by a spring with two degrees of freedom. We get

- for the rigid shaft

$$J\ddot{\varphi} = \sum_i M_i \qquad \text{with} \qquad M_i = M_i(t, \varphi_k, \dot{\varphi}_k). \tag{1}$$

J is the mass moment of inertia, φ the absolute rotational angle and M_i an external torque depending on time, the angles and the angular velocities of some neighboring bodies or the environment.

1 vibration damper
2 engine model
3 mass damper
4 elastic shaft
5 auxiliary equipment
6 planetary gear set
7 planetary gear set
8 hydraulic pump
9 hydraulic pump
10 hydraulic motor
11 collecting shaft
12 distributor shaft
13 rear drive
14 front drive
15 front drive
16 front drive
17 tractor mass
18 rear PTO
19 rear PTO
20 front PTO
21 front PTO

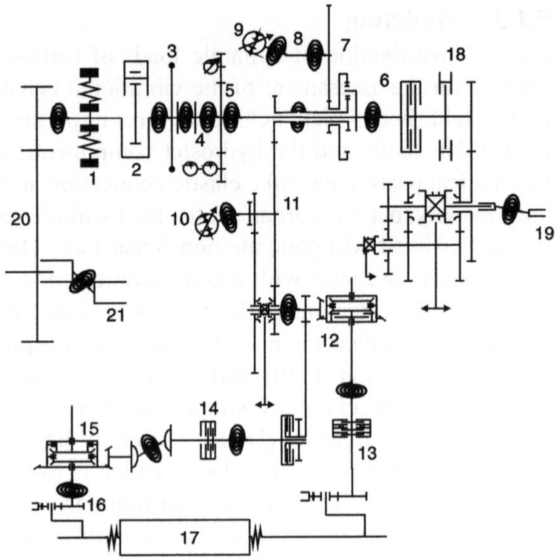

Figure 14: Mechanical Model of the VARIO Power Transmission

- for the elastic shaft

$$\begin{pmatrix} J_1 & J_{12} \\ J_{12} & J_2 \end{pmatrix} \begin{pmatrix} \ddot{\varphi}_1 \\ \ddot{\varphi}_2 \end{pmatrix} + \begin{pmatrix} d_T & -d_T \\ -d_T & d_T \end{pmatrix} \begin{pmatrix} \dot{\varphi}_1 \\ \dot{\varphi}_2 \end{pmatrix} + \\ + \begin{pmatrix} c_T & -c_T \\ -c_T & c_T \end{pmatrix} \begin{pmatrix} \varphi_1 \\ \varphi_2 \end{pmatrix} = \begin{pmatrix} \sum_k M_{1,k} \\ \sum_k M_{2,k} \end{pmatrix}. \tag{2}$$

The quantities in this equation follow from

$$\begin{aligned} c_T &= \left(\frac{1}{c_1} + \frac{1}{c_2} \right)^{-1}, \qquad \text{with} \qquad c_i = \frac{GJ_p}{l_i} \qquad , \; l_2 = l - l_1, \\ &\begin{pmatrix} J_1 & J_{12} \\ J_{12} & J_2 \end{pmatrix} = \frac{1}{3} \rho J_p l \begin{pmatrix} 1 & \frac{1}{2} \\ \frac{1}{2} & 1 \end{pmatrix}, \\ d_T &= 2D\sqrt{c_T J}, \end{aligned} \tag{3}$$

where c and d are spring and damper coefficients, respectively, J are mass moments of inertia, D is the Lehr attenuation constant (practically $D \approx 0.02$-0.05 for our case), G the modulus of shear and ρ the material density. The

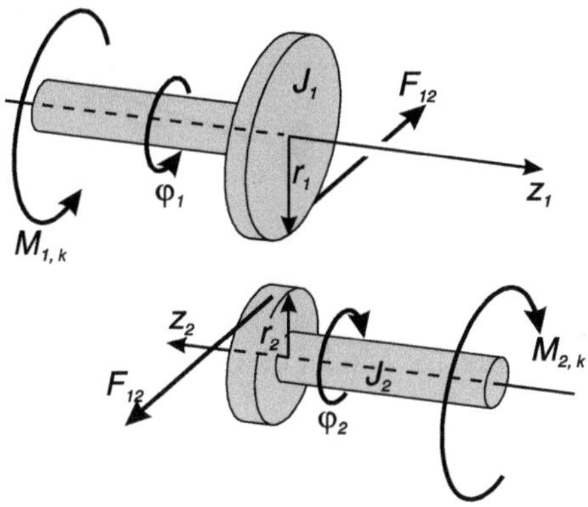

Figure 15: Kinematic Connection, for example Gear Mesh

torqes $M_{1,k}, M_{2,k}$ come again from couplings to the neighboring bodies or environment.

All interconnections with elasticity must be examined with respect to the magnitude of these elasticities. In our case the stiffness of all tooth meshings is so large, that the coressponding frequencies exceed the frequency range of interest by far. Therefore we are able to model such gear meshes as a purely kinematical connection. Figure 15 illustrates such a connection. The two parts are under the load of the contraint force F_{12} and the torques $M_{1,k}, M_{2,k}$. If we have a kinematical connection with a ratio $(i_{1,2} = \frac{\varphi_1}{\varphi_2} = \frac{r_2}{r_1})$, for example, we are able to define a Jacobian by

$$\dot{\boldsymbol{q}} = \begin{pmatrix} \dot{\varphi}_1 \\ \dot{\varphi}_2 \end{pmatrix} = \begin{pmatrix} 1 \\ \frac{1}{i_{1,2}} \end{pmatrix} \dot{\varphi}_1 = \boldsymbol{Q}\,\dot{\varphi}_1 \tag{4}$$

The equations of motion of the configuration of Figure 15 write

$$\left.\begin{array}{lcl} J_1\ddot{\varphi}_1 & = & M_{1,k} + F_{12}r_1 \\ J_2\ddot{\varphi}_2 & = & M_{2,k} - F_{12}r_2 \end{array}\right\} \hat{=} \boldsymbol{M}\ddot{\boldsymbol{q}} = \boldsymbol{h}, \tag{5}$$

which can be transformed with the help of the Jacobian (eq. (4) and with the reduced mass moment of inertia $J = J_1 + \frac{J_2}{i_{1,2}^2}$ to yield

$$\boldsymbol{Q}^T\boldsymbol{M}\boldsymbol{Q}\,\ddot{q} = \boldsymbol{Q}^T\boldsymbol{h} \implies J\ddot{\varphi}_1 = M_{1,k} + \frac{M_{2,k}}{i_{1,2}}, \tag{6}$$

The constraint force F_{12} eliminates by the multiplication with the Jacobian $\mathbf{Q}$.

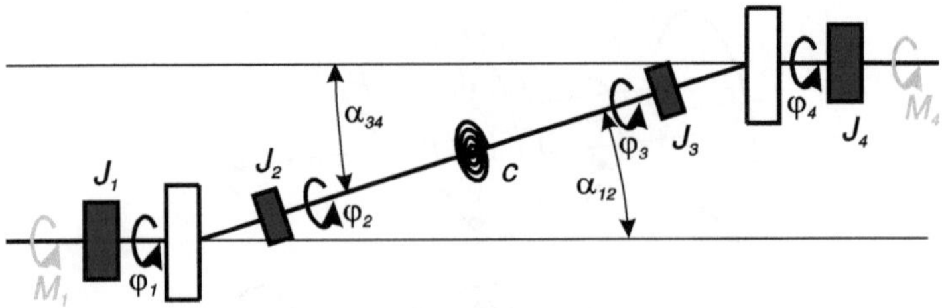

Figure 16: Model of a Cardan Shaft

The tractor power transmission system includes several ***Cardan shafts***, especially for all PTO systems. Cardan shafts are sources of parameter excited vibrations with their sub- and super-harmonic resonances, which can become dangerous. Therefore good models are obligatory. An excellent survey is given in [40]. We choose a model with four bodies interconnected by springs and dampers (see Figure 16). The four equations of motion write

$$\left.\begin{array}{rcl} J_1\ddot{\varphi}_1 & = & +M_{G12,1} + M_1 \\ J_2\ddot{\varphi}_2 & = & -M_{G12,2} + c(\varphi_3 - \varphi_2) \\ J_3\ddot{\varphi}_3 & = & +M_{G34,3} + c(\varphi_2 - \varphi_3) \\ J_4\ddot{\varphi}_4 & = & -M_{G34,4} + M_4 \end{array}\right\} \hat{=} \boldsymbol{M}\ddot{\boldsymbol{q}} = \boldsymbol{h} \tag{7}$$

$$\text{with } M_{G12,2} = M_{G12,1}\frac{\sin^2\varphi_1 + \cos^2\varphi_1\cos^2\alpha_{12}}{\cos\alpha_{12}}$$

$$\text{and } M_{G34,3} = M_{G34,4}\frac{\sin^2\varphi_4 + \cos^2\varphi_4\cos^2\alpha_{34}}{\cos\alpha_{34}}$$

To not loose a good overview we have left out the damping terms. The complete force element of the shaft between "1" and "2" comprises a linear damper d and linear spring c. Between the shaft parts (1 - 2) and (3 - 4) we have a kinematical relation of the well-known form

$$\varphi_{i+1} = \varphi_{i+1}(\varphi_i) = \arctan(\frac{\tan\varphi_i}{\cos\alpha}), \tag{8}$$

where α is the Cardan angle ($\alpha = \varphi_{12} = \varphi_{34}$). The above relation reduces the four degrees of freedom to two, for example $\mathbf{q} = (\varphi_1, \varphi_2, \varphi_3, \varphi_4) \Longrightarrow (\varphi_1, \varphi_4) = \mathbf{q}_{red}$. The corresponding Jacobian follows from equation (8):

$$\boldsymbol{Q}^T = \begin{pmatrix} 1 & \frac{\cos\alpha_{12}}{\sin^2\varphi_1+\cos^2\varphi_1\cos^2\alpha_{12}} & 0 & 0 \\ 0 & 0 & \frac{\cos\alpha_{34}}{\sin^2\varphi_4+\cos^2\varphi_4\cos^2\alpha_{34}} & 1 \end{pmatrix}$$

$$\dot{\boldsymbol{q}} = \begin{pmatrix} \dot{\varphi}_1 \\ \dot{\varphi}_2 \\ \dot{\varphi}_3 \\ \dot{\varphi}_4 \end{pmatrix} = \boldsymbol{Q}\dot{\boldsymbol{q}}_{red} = \boldsymbol{Q}\begin{pmatrix} \dot{\varphi}_1 \\ \dot{\varphi}_4 \end{pmatrix} \tag{9}$$

With these relations we receive the final equations of motion of the Cardan shaft

$$\boldsymbol{M}_{red}\ddot{\boldsymbol{q}}_{red} + \boldsymbol{b}_{red} = \boldsymbol{h}_{red}$$

$$\boldsymbol{M}_{red} = \begin{pmatrix} J_1 + J_2(\frac{\cos\alpha_{12}}{\sin^2\varphi_1+\cos^2\varphi_1\cos^2\alpha_{12}})^2 & 0 \\ 0 & J_3(\frac{\cos\alpha_{34}}{\sin^2\varphi_4+\cos^2\varphi_4\cos^2\alpha_{34}})^2 + J_4 \end{pmatrix}$$

$$\boldsymbol{b}_{red} = \begin{pmatrix} J_2\frac{\cos^2\alpha_{12}\sin 2\varphi_1(\cos^2\alpha_{12}-1)}{(\sin^2\varphi_1+\cos^2\varphi_1\cos^2\alpha_{12})^3}\dot{\varphi}_1^2 \\ J_3\frac{\cos^2\alpha_{34}\sin 2\varphi_4(\cos^2\alpha_{34}-1)}{(\sin^2\varphi_4+\cos^2\varphi_4\cos^2\alpha_{34})^3}\dot{\varphi}_4^2 \end{pmatrix}$$

$$\boldsymbol{h}_{red} = \begin{pmatrix} \frac{\cos\alpha_{12}}{\sin^2\varphi_1+\cos^2\varphi_1\cos^2\alpha_{12}}\, c\,(\arctan(\frac{\tan\varphi_4}{\cos\alpha_{34}}) - \arctan(\frac{\tan\varphi_1}{\cos\alpha_{12}})) + M_1 \\ \frac{\cos\alpha_{34}}{\sin^2\varphi_4+\cos^2\varphi_4\cos^2\alpha_{34}}\, c\,(\arctan(\frac{\tan\varphi_1}{\cos\alpha_{12}}) - \arctan(\frac{\tan\varphi_4}{\cos\alpha_{34}})) + M_4 \end{pmatrix} \tag{10}$$

The dependency on the rotational angles enters these equations by the Jacobian $\mathbf{Q}$ (equation (9). All quantities of the equations of motion change with the rotational angles and speeds. Furtheron they depend strongly on the kink angles α_{12} and α_{34}. For these angles being zero, the dependency disappears, and with increasing angles their influence increases nonlinearly.

The homokinematical configuration gives a balancing effect for $\alpha_{12} = \alpha_{34}$. The parameter influences disappear for equal rotational angles $\varphi_2 = \varphi_3$, which is only possible for a completely rigid shaft ($c \Rightarrow \infty$). But Cardan shafts are usually very elastic leading to oscillations between the two shaft masses and thus to parameter excitation. The largest influence comes from the nonuniformity of the stiffness terms in $\mathbf{h}_{red}$ of equation (10). Nevertheless we can expect small angular displacements, which allows us to develop a linearized shaft stiffness. From equation (10, last term) and a linearization with respect to the difference of the angles φ_1 and φ_4 and finally with the assumption $\alpha_{12} = \alpha_{34} = \alpha$ we come out with:

$$c_{lin} = \left(\frac{\cos\alpha}{\sin^2\varphi_1 + \cos^2\varphi_1\cos^2\alpha}\right)^2 c, \qquad M_c = c_{lin}(\varphi_4 - \varphi_1). \tag{11}$$

Figure 17 confirms the model and allows to use for the simulation of the PTO Cardan shafts the linearized approach. The graph on the right side depicts the dependency of the Cardan shaft stiffness on the angular orientation φ and on the

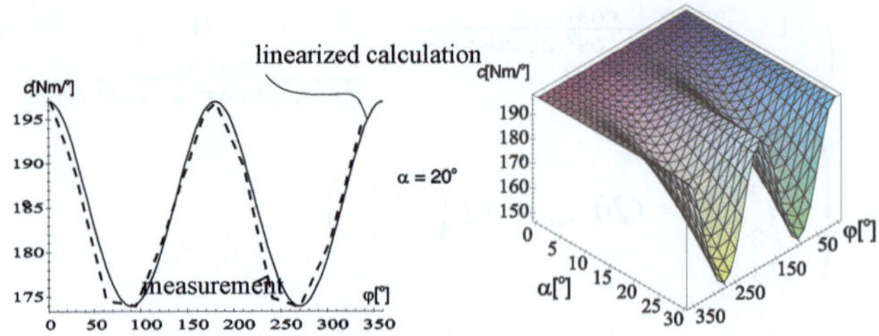

Figure 17: Linearized Stiffness of the Cardan Shaft

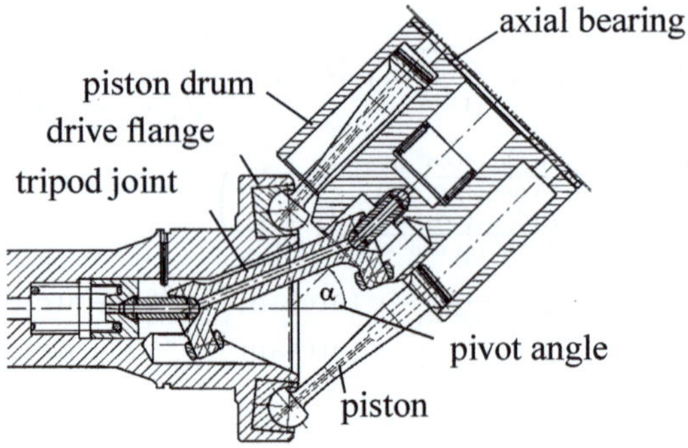

Figure 18: Pump/Motor Configuration of the Hydrostat

kink angle α. For an angle $\alpha = 10°$ the difference of minmal and maximum stiffness is 6%, for $\alpha = 20°$ about 10% and for $\alpha = 30°$ already 30%. The parameter excitation influences as a matter of fact the whole power transmission system of the tractor.

The heart of the VARIO power drive is the ***hydrostatic system*** with its hydraulic pumps and motors operating on the basis of a piston type machine. They are heavily loaded, especially for processes like ploughing and mulching. Therefore it makes sense to know these loads already during the design phase to find the correct lay-out. Figure 18 shows a drawing of the pump/motor configuration as used in the VARIO system. The piston drum is rotationally displaced with respect to the shaft axis by an angle α, the magnitude of which determines the oil fluid flow. The translational motion of the nine pistons within the corresponding cylinder liners increase with increasing pivot angle. Drive flange, pistons and piston drum can be modeled as rigid bodies. The tripod joint is an elastic part, it transmits the rotational motion of the drive flange to the piston drum. It is coupled to the drive flange and to the piston drum by a kind of three "feet", which allow relative motion in axial and radial but not in circumferential tangential direction. These three feet

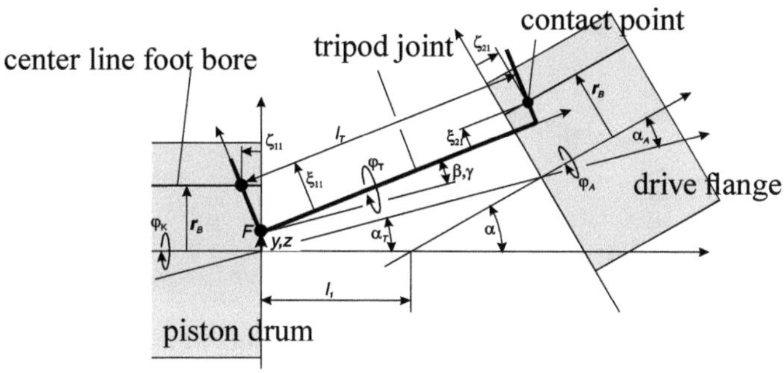

Figure 19: Pump/Motor Model

consist of three pins perpendicular to the tripod axis and three rings with a spherical outer surface moving in corresponding bushings in the flange or the drum, which allow axial and radial motion, but which can transmit the full torsional torques. Therefore the tripod joint is a really critical part, which has to be designed properly.

Figure 19 depicts a tripod model with the relative kinematics of the tripod axis, the position and orientation of which we need to know. As a first step we state, that the contact points of the six tripod feet in the drum and in the flange can be described twice, using the body coordinates of the flange and the drum on the one and using the body coordinates of the tripod itself on the other side. Putting $\mathbf{r}_{drum,flange} = \mathbf{r}_{tripod}$ results altogether in (6x3 = 18) algebraic equations, because everyone of the six $\mathbf{r}$ possesses of course three components.

These 18 equations correspond to the unknown displacements ξ_{ij} and ζ_{ij} of the contact points to the tripod axis and to the front sides of drum and flange, respectively (i=1 for drum, i=2 for driving flange, j=1,2,3 for the three contact points of the three feet on each side of the tripod). They correspond furtheron to the three unknown positions (x, y, z) of the tripod joint, where the axial displacement x can be neglected, it corresponds to its tilt angles γ and β, its rotational orientation φ_T and finally to the rotational orientation ϕ_K of the drum. These are 18 unknowns covered by 18 algebraic equations of the form

$$
\begin{aligned}
& y \sin\alpha_T + l_T(i-1) - \xi_{ij}\gamma\cos[\varphi_T + \frac{2\pi}{3}(j-1)] + \xi_{ij}\beta\sin[\varphi_T + \frac{2\pi}{3}(j-1)] = \\
& \quad = l_1(i-1)\cos\alpha_T + \zeta_{ij}\cos(\alpha_{i-1}-\alpha_T) - r_B\sin(\alpha_{i-1}-\alpha_T)\cos[\varphi_i + \\
& \qquad + \frac{2\pi}{3}(j-1)]
\end{aligned}
$$

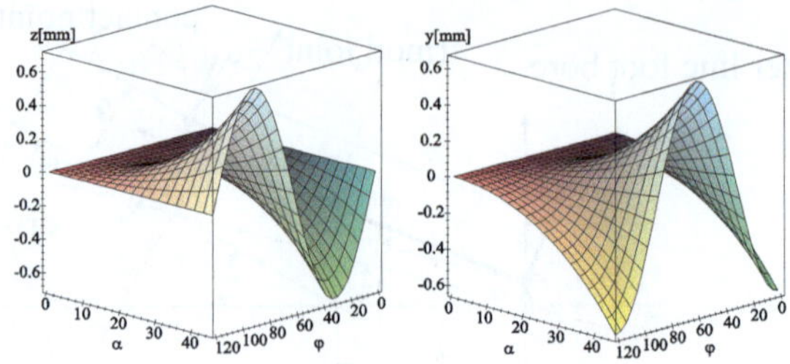

Figure 20: Translational Displacements of a Tripod Foot

$$
\begin{aligned}
&y \cos\alpha_T + l_T(i-1) - \xi_{ij}\gamma\cos[\varphi_T + \frac{2\pi}{3}(j-1)] = \\
&= -l_1(i-1)\sin\alpha_T + \zeta_{ij}\sin(\alpha_{i-1}-\alpha_T) - r_B\cos(\alpha_{i-1}-\alpha_T)\cos[\varphi_i + \\
&\qquad + \frac{2\pi}{3}(j-1)]
\end{aligned}
$$

$$
z - \beta l_T(i-1) - \xi_{ij}\gamma\sin[\varphi_T + \frac{2\pi}{3}(j-1)] = r_B\cos[\varphi_i + \frac{2\pi}{3}(j-1)] \tag{12}
$$

Again we have i=1 for drum, i=2 for driving flange, j=1,2,3 for the three contact points of the three feet on each side of the tripod.

The unknown quantities ξ_{ij} and ζ_{ij} are contained in the equations (12) in a linear form. Therefore we are able to reduce the set to one with six degrees of freedom only, namely to the six quantities y, z, β, γ, φ_T and $\varphi_K = \varphi_1$. The evaluation is performed numerically. The considerations and relations are equally valid for the hydromotors and the hydropump, where in both cases the kinematics of the tripod joint is of special interest.

The tripod kinematics depends mainly on the displacement angle α of the piston drum and on the rotational orientation φ_T of the driving flange. The three tripod feet bring in a periodicity of 120° for all tripod results. The translational quantities (y, z) and the tilt angles (β, γ) are 90° dephased. The Figures 20 and 21 illustrate the kinematical properties. The influence of these characteristics on the dynamics of the tripod itself is not very large, which corresponds to experience. Therefore a detailed analysis will be only necessary for special evaluations required for the tripod joint.

The ***hydrostat piston kinematics*** produces oil flow and oil pressure. The pistons perform a stroke motion within the drum cylinders, while rotating around the tripod joint. This motion depends on the displacement angle α, which at the same time represents a control quantity for flow and pressure. According to Figure 22

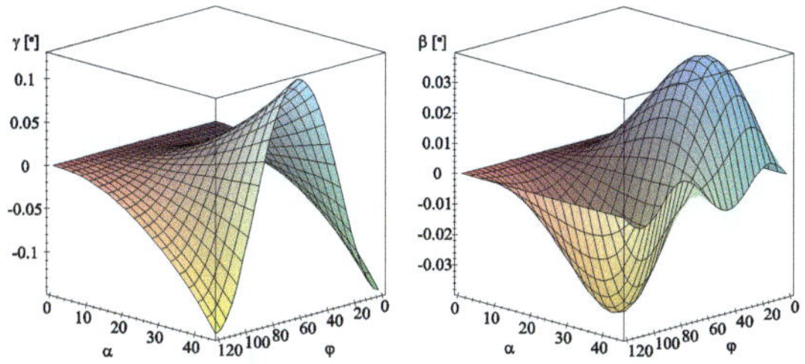

Figure 21: Rotational Displacements of a Tripod Foot

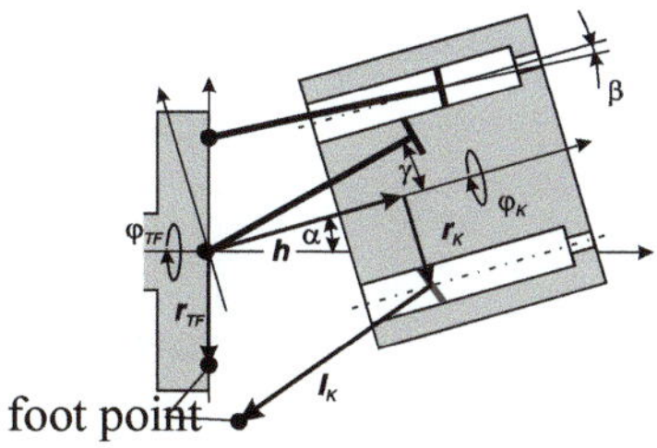

Figure 22: Piston Kinematics

we get the following vector chain

$$\boldsymbol{r}_{TF} = \boldsymbol{h} + \boldsymbol{r}_K + \boldsymbol{l}_K, \tag{13}$$

which can be decomposed to the three equations for the unknowns (h, β, γ)

$$\begin{aligned}
\gamma &= \arcsin\left(\frac{r_{TF}}{l_K}(\sin\varphi_{TF}\cos\varphi_K - \cos\varphi_{TF}\sin\varphi_K\cos\alpha)\right)\\
\beta &= \arcsin\left(\frac{r_K - r_{TF}(\sin\varphi_{TF}\sin\varphi_K + \cos\varphi_{TF}\cos\varphi_K\cos\alpha)}{l_K\cos\gamma}\right)\\
h &= r_{TF}\cos\varphi_{TF}\sin\alpha + l_K\cos\beta\cos\gamma
\end{aligned} \tag{14}$$

The results of these equations are illustrated in Figure 23. To get an idea of the forces resulting from the above kinematical consideration we investigate a simplified case with no rotational speed difference, $\varphi_T - \varphi_K = 0$, and a piston pressure according to Figure 24. The pistons possess at one end spherical bearings, which allow a force tranfer only in the direction of the piston axis to the driving flange.

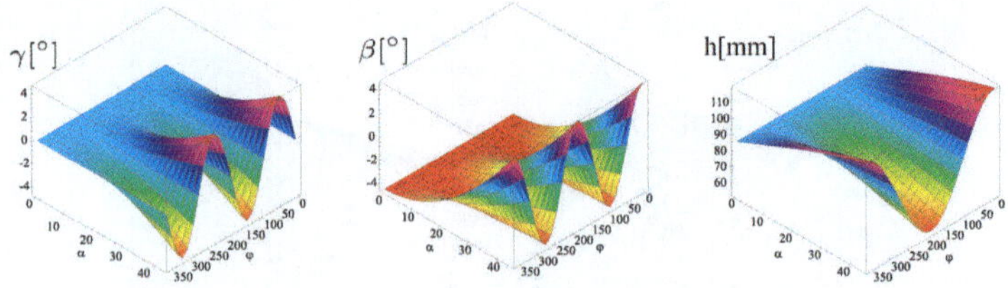

Figure 23: Results of Piston Kinematics

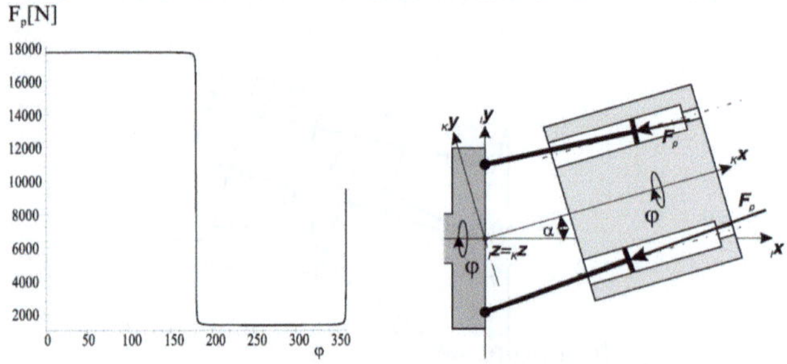

Figure 24: Load Example

This direction for one piston is defined by

$$ {}_D\boldsymbol{e}_{K,k} = \begin{pmatrix} -\cos\beta\cos\gamma \\ -\sin\beta\cos\gamma \\ \sin\gamma \end{pmatrix}_k \quad \Longrightarrow \quad {}_D\boldsymbol{F}_{K,k} = F_{p,k} \begin{pmatrix} -\cos\beta\cos\gamma \\ -\sin\beta\cos\gamma \\ \sin\gamma \end{pmatrix}_k , \quad (k = 1, 2, \cdots 9) \tag{15} $$

The reference coordinates are piston drum fixed. The forces can be decomposed into three components and then summarized for all nine pistons, regarding their individual position and orientation. Before showing the results we transform the forces from the piston drum "D" to the driving flange "F" by the transformation (Figure 24)

$$ {}_F\boldsymbol{F}_{K,k} = F_{p,k} \begin{pmatrix} -\cos\alpha\cos\beta\cos\gamma \\ -\sin\alpha\cos\beta\cos\gamma\cos\varphi \\ +\sin\alpha\cos\beta\cos\gamma\sin\varphi \end{pmatrix}_k , \qquad (k = \quad 1, 2, \cdots 9) \tag{16} $$

A numerical evaluation indicates, that axial forces of the pistons are large on both, drum and flange; whereas tangential and radial forces are only large on the driving flange, see Figures 25 and 26. To establish the equations of motion of ***the complete system*** we have to combine the above models and to supplement them

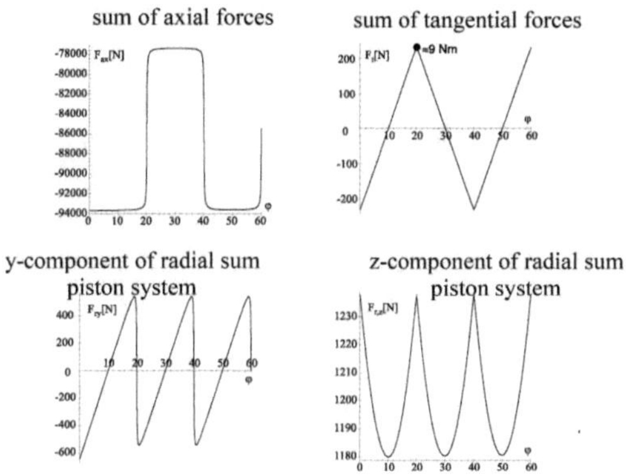

Figure 25: Summarized Piston Forces on the Drum

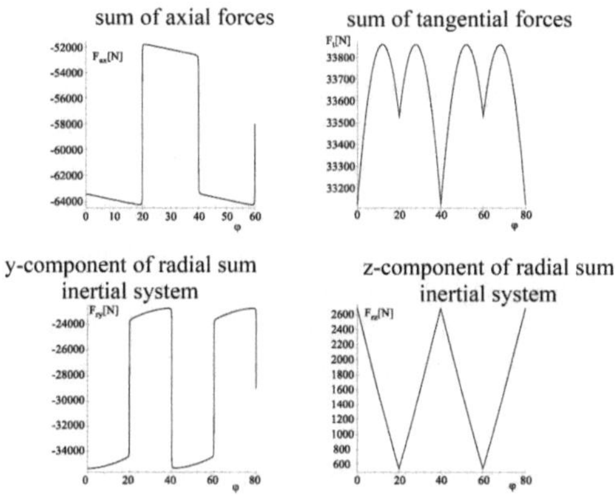

Figure 26: Summarized Piston Forces on the Driving Flange

by the appropriate models for the oil hydraulics and the axial drum bearing. Such models follow the methods as discussed in [24]. It will not be presented here. In spite of the fact that the system is complicated we can avoid both, bilateral and unilateral constraints, expressing instead all interconnections by linear or nonlinear force laws. In a first step this comes out with

$$\mathbf{M}\ddot{\mathbf{q}} - \mathbf{h}(\mathbf{q}, \dot{\mathbf{q}}, t) = \mathbf{0}, \tag{17}$$

which can be approximated by linear and nonlinear parts yielding

$$\mathbf{M}\ddot{\mathbf{q}} + \mathbf{D}\dot{\mathbf{q}} + \mathbf{K}\mathbf{q} = \mathbf{h}(\mathbf{q}, \dot{\mathbf{q}}, t), \qquad \dot{\mathbf{q}}_{hydraulic} = \mathbf{h}_{hydraulic}. \tag{18}$$

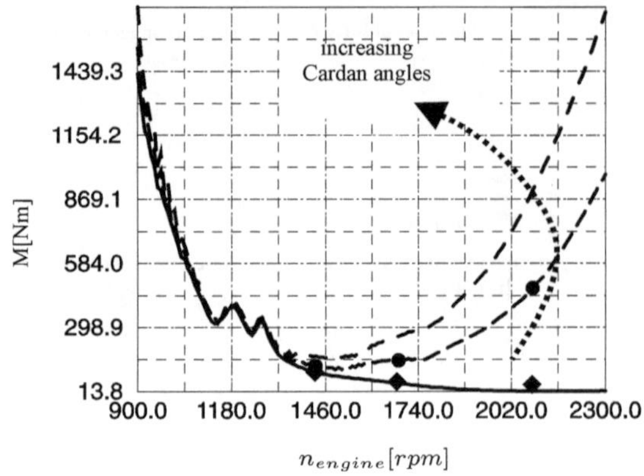

Figure 27: Torque Amplitudes at the Rear PTO, Comparison Simulation/Measurements (•, ♦), (courtesy AGCO/Fendt)

Nonlinear terms come mainly from the hydrostat system and from the hydraulics. Many elastic parts are linear. The above separation makes sense, because for stationary operation the nonlinear terms do not have much influence, and the linear part of the equations of motion can be used simply for evaluating the eigenbehaviour, which gives important informations of what components oscillate with respect to other ones and where potential resonances could be expected. This presupposes a linearization around the operation point under consideration. The eigenvalues and the eigenforms result from $\mathbf{M}\ddot{\mathbf{q}} + \mathbf{K}\mathbf{q} = \mathbf{0}$. We shall give examples.

5.1.4 Numerical and Experimental Results

Experiments were performed by the company AGCO/Fendt at all PTO systems. Figure 27 depicts a comparison of simulation and measurements for the rear PTO. These measurements were performed with a very stiff Cardan shaft and Cardan angles of $\alpha \leq 14°$. The process considered is mulching. The stiff Cardan shaft leads to large shaft eigenfrequencies of about 40 Hz, which corresponds to an engine speed of about 2400 rpm. This value exceeds the operational speed range and is therefore not dangerous. Simulations and experiments agree very well.

The equations of motion (18) were linearized around a stationary operation point, which gives for our case a reasonable approximation of the real dynamic behaviour including nonlinearities. With $\mathbf{M}\ddot{\mathbf{q}} + \mathbf{K}\mathbf{q} = \mathbf{0}$ we are able to evaluate eigenforms and eigenfrequencies, which has been done for 20 eigenfrequencies up to ≈ 2000 Hz. We shall consider only a few of them.

The matrix $\mathbf{K}$ contains all linearized stiffnesses of the system, which are of particular interest for the hydrostat, because its stiffness depends on the inclination

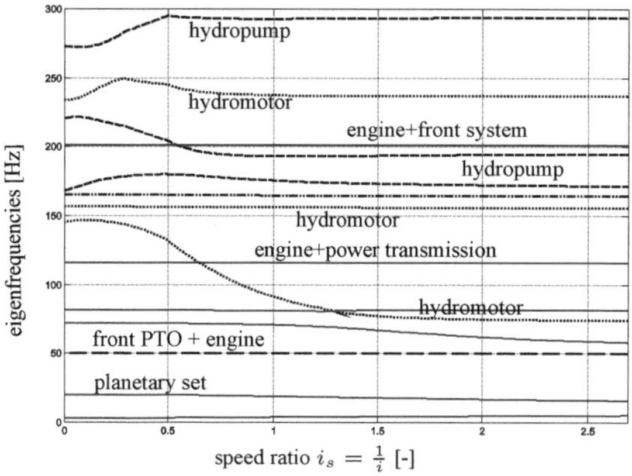

Figure 28: Influence of the Speed Ratio on some Eigenfrequencies

angles α and β (Figure 22) and thus on the speed ratio $i_s = n_{collectingshaft}/n_{engine} = 1/i$. As a consequence all eigenforms and eigenfrequencies resulting from an analysis of $\mathbf{M}\ddot{\mathbf{q}} + \mathbf{K}\mathbf{q} = \mathbf{0}$ depend also on the ratio i_s. Figure 28 depicts the dependencies of some typical eigenfrequencies on the speed ratio. With the exception of hydropump and hydromotor themselves the influence on all other components is very small.

For system design one need to know the topological vibration behaviour, that means what components vibrate with respect to other ones. We shall focus our consideration on front and rear PTO loads. They possess significant influence on the second, the third, the 17th and the 18th eigenforms and -frequencies. While these loads influence both, the second and the third eigenmodes, they have an effect on the 17th and 18th eigenmode only locally, front load on front PTO system and rear load on rear PTO system. Therefore we show only one case for the second eigenmode (19.7 Hz without load, $\leq$10 Hz with load) in Figure 29, the numbering of which follows Figure 14 on page 22. The masses of Figure 14 are represented by small rhombi, and the difference between solid and dotted lines indicate the vibration amplitudes. The loads reduce of course the eigenfrequencies, and the complete transmission system vibrates down to the collecting shaft (11) (Figure 14).

With respect to the higher eigenmodes with about 1600 Hz without load we have similar tendencies, drastic reduction of frequencies to values of 30-50 Hz and local vibrations at front and rear PTO's. Figure 30 illustrates the situation of the eigenforms. Loads at the front or rear PTO's excite only vibrations of the front or rear regime, which do not influence each other.

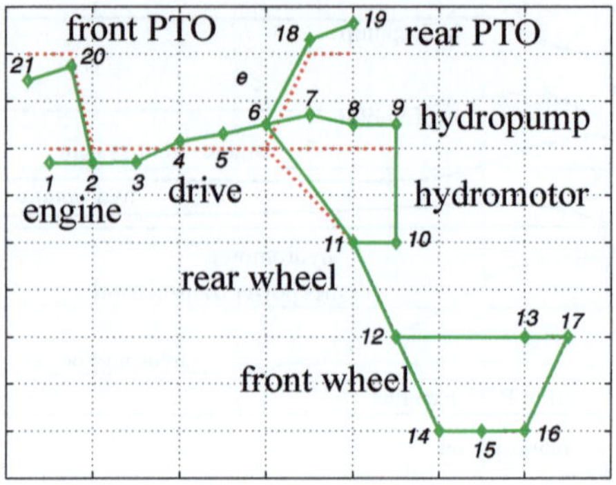

Figure 29: Second Eigenform (both PTO moment of inertia $J \geq 2\text{kgm}^2$)

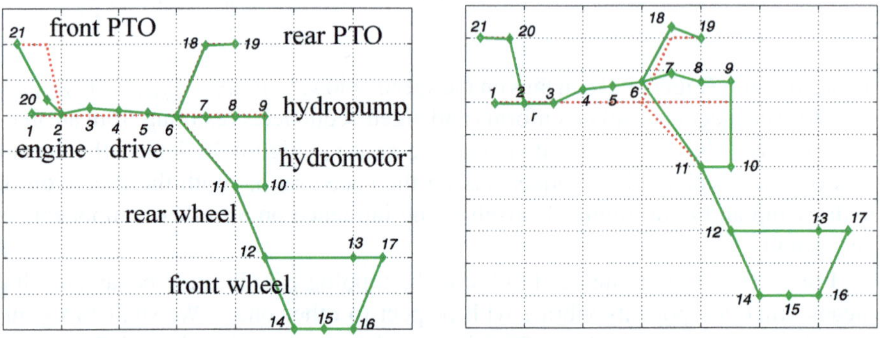

Figure 30: PTO Eigenforms at 30-50 Hz (both PTO moment of inertia $J \geq 2\text{kgm}^2$), left figure for front load, right figure for rear load

At this point a general remark for practical applications: The Figures 29 and 30 represent eigenvectors of a linearized system, which describe parts of the approximated eigenbehaviour of the tractor drive. These curves are characterized by the fact, that the eigenforms are not those of a continuum but of a discrete system with discrete masses. It is sometimes very convenient to use such eigenvectors, also in the case of a discrete mass system, as shape functions for a Ritz-approach, because we get then a very simple first-step approach to very complicated system structures. Usually eigenbehaviour analysis is performed only for continuum systems.

As a last result we present some Campbell diagrams illustrating the excitation structure in certain areas of the system. We consider again the case of a rear load at a Cardan-angle of 22° and 25°. The engine is a four-stroke configuration. Figure 31 illustrates the torque at the rear PTO shaft. We recognize the second

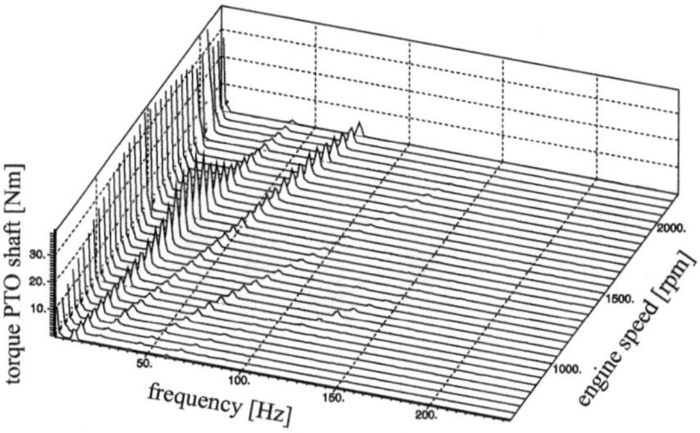

Figure 31: Campbell Diagram of the Rear PTO Torque (with rear implement and a PTO shaft with 22° and 25°)

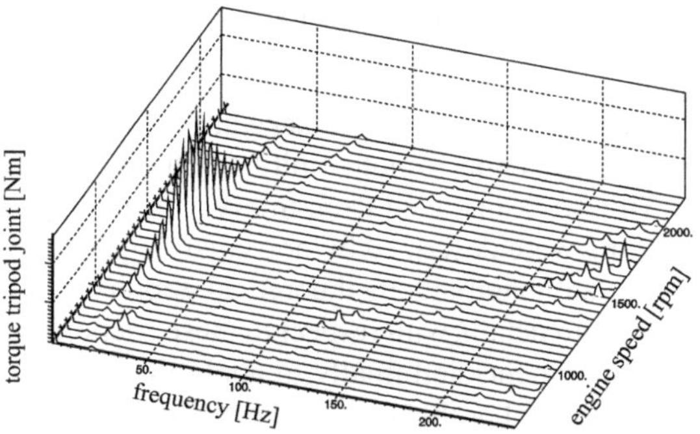

Figure 32: Campbell Diagram of the Tripod Joint Torque of the Hydraulic Motor (with rear implement and a PTO shaft with 22° and 25°)

order influence of the rear PTO and the second and fourth engine order. Higher engine orders do not posses an influence. Due to the relatively small stiffness of the Cardan shaft for the case considered the second PTO shaft order is dominant. The corresponding load on the tripod is shown by Figure 32. Also here and to a larger extent the second PTO system order dominates the vibrations. From these results we conclude that the internal excitation by the PTO systems at the front and rear side of a tractor lead to significant vibrations. This is confirmed also by additional simulations and by practical experience.

5.1.5 Realization of Objectives

The power transmission system has been simulated in all details and with respect to performance, loads on components, vibration behaviour as well as noise and wear. A couple of recommendations concerning design improvements and future developments were evaluated.

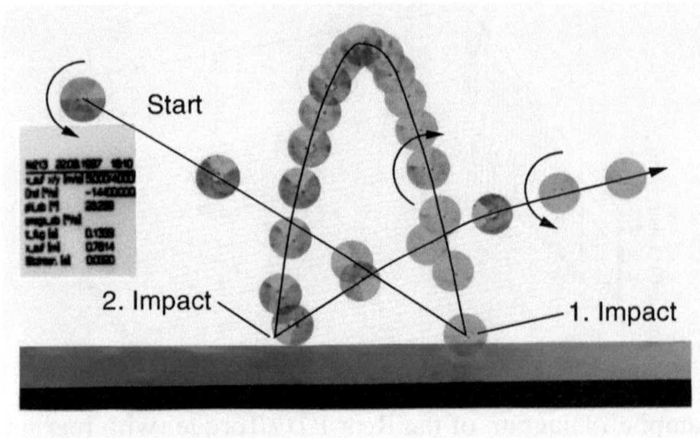

5.2 Example: Impacts with Friction

5.2.1 Objectives

The research on impact with friction took place at the author's Institute since the eighties, when we developed non-smooth multibody theory for large systems with contact problems. This work included from the very beginning also impact problems, which at that time were described by not satisfying theories. The objectives consisted in achieving an unambiguous theoretical approach to impacts with and without friction, to go into details with respect to these frictional impact processes, to verify theory by experiments and finally to apply that to large problems of industrial reality. Such practical problems were known in many industries, but could not really be solved in a satisfying way.

5.2.2 Introduction

Many machine and mechanism processes are accompanied by impacts with friction. They arise by short-time contacts between two or more bodies, and they generate energy losses mainly due to friction in tangential contact directions. During the last two decades a couple of impact models based on the theory of rigid body contact were established connected with the names of Moreau, Fremond and Glocker, which all work quite satifactorily with respect to practical applications, although some examples indicate deviations requiring more investigations with respect to the impact models and the type of examples considered. We shall focus on Glocker's model, for which some experimental verifications are available by Beitelschmidt. A missing link are energy considerations, which are available, but nevertheless do not provide us with a complete information for all possible cases. We shall fill a bit this gap by founding the investigations on a combined phenomenological and theoretical basis.

We consider rigid bodies as part of a multibody system. which come into contact including normal and tangential features, and we focus especially on short-time contacts being interested for the energy losses accompanying such processes. The principal situation is illustrated in Figure 33. Starting with the models as developed in [15], [3] and [25] we use the following classical assumption for impacts and also for impacts with friction:

- The duration of the impact is so short, that the mathematical description may assume a zero impact time.
- As a consequence we neglect wave processes, which would take place in a finite time interval. Important indicators are the impact time in comparison to the wave progressing time, precisely defined wave refelection patterns and no dispersion. In very many technical problems such wave phenomena do not play a role.
- Following these assumptions the mass distribution of the body does not change during the impact, the bodies remain rigid.
- All positions and orientations of the impacting bodies remain constant. The translational and rotational velocities of the bodies are finite and may change jerkyly during the impact.
- Accordingly the position of the impact point and that of the normal and tangential vectors remain constant.
- All forces and torques, which are not impulsive forces and torques, remain also constant during the impact.
- All during the impact evolving impulses act during the impact in a constant direction. Their lines of action do not change and correspond to the normal and tangential vectors in the impact point.
- The impact can be divided into two phases: the compression phase and the expansion phase.
- The compression phase starts at time t_A and ends at time t_C. The end of the compression equals the start of the expansion phase. Expansion is finished at time t_E, which is also the end of the impact.

During compression impulses in normal and tangential directions of the contact are stored, and during expansion these stored impulses are released, both processes accompanied by losses. The losses result from an application of Poisson's friction law. A detailed description of these processes may be found in the literature [24], [15], [16], [20], [17] and [18] with increasing depth of mathematical representation.

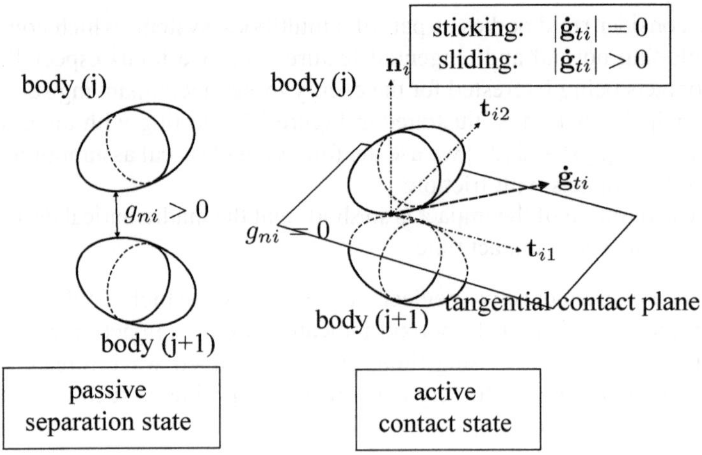

Figure 33: Principal Situation in a Multibody Contact (i)

5.2.3 Impact Characteristics

According to Moreau ([22], [20]) we may express the dynamics with and without impacts by one measure differential equation in the form

$$\mathbf{M}d\mathbf{u}+\mathbf{h}dt - \mathbf{W}d\mathbf{\Lambda} = \mathbf{0} \quad \Longleftrightarrow$$
$$\begin{cases} \mathbf{M}\dot{\mathbf{u}} + \mathbf{h} - \mathbf{W}\lambda = \mathbf{0} & (t \neq t_i) \\ \mathbf{M}(\mathbf{u}^+ - \mathbf{u}^-) - \mathbf{W}\mathbf{\Lambda} = \mathbf{0} & (t = t_i) \end{cases} \tag{19}$$

The part $\mathbf{W}\lambda$ contains all contact reactions due to non-impulsive contacts and the part $\mathbf{W}\Lambda$ all impulsive contact reactions. The time $t_i \in I_{kl}$ represents one of the instants (i), where an impact takes place. The vector $\mathbf{h}$ includes all non-impulsive and applied forces, whatsoever, and for multibody systems without closed loops we also include in the generalized coordinates $(\mathbf{q}, \dot{\mathbf{q}})$ all bilateral constraints.

We start with the compression phase and the normal impact direction. At the end of compression the relative normal velocity is either zero or non-negative, $\dot{g}_{Ni} \geq 0$. The tangential compression phase is characterized mainly by friction. At the end of compression we may have three states: Firstly, sliding in a positive tangential direction ($\dot{g}_{NC} > 0$), where the tangential impulse acts during this phase in opposite direction with $\Lambda_{TC} = -\mu\Lambda_{NC}$, secondly, sticking at the end of compression ($\dot{g}_{NC} = 0$), where the tangential impulse is small enough to generate sticking during the whole compression phase, and thirdly, sliding in a negative tangential direction ($\dot{g}_{NC} < 0$), where the tangential impulse acts during this phase in opposite direction with $\Lambda_{TC} = +\mu\Lambda_{NC}$. The processes for these two directions are depicted by the well-known graphs of Figure 34.

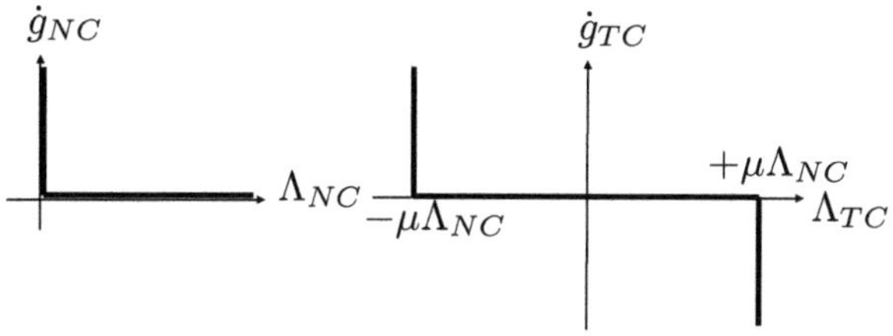

Figure 34: Contact Laws for Impacts

The impulse stored during compression is released with a loss governed by Poisson's law. Restoring the tangential impulse affords some additional considerations. According to Poisson's law we get back the stored tangential impulse Λ_{TCi} of the (i)th contact with a certain loss, that is $(\varepsilon_{Ti}\Lambda_{TCi})$, where Poisson's losses are quantified by $(0 \leq \varepsilon_{Ti} \leq 1)$. The tangential friction coefficient ε_{Ti} must be measured. But this contains not all losses during expansion. The restoration of the tangential impulse possesses another quality compared with the restoration of the normal impulse, because it cannot take place independently from the normal impulse, which as a matter of fact represents the driving constraint impulse for the generation of tangential friction forces. Therefore we shall assume, that the restoration of the tangential impulse is additionally accompanied by losses in "normal direction" expressed by ε_{Ni}. Figure 35 illustrates these processes, see also [15] and [25].

5.2.4 Energy Losses

The loss of energy is the difference of the total system energy after an impact and before an impact [24]. In terms of the generalized velocities $\dot{\mathbf{q}}$ we write

$$\begin{aligned} \Delta \mathrm{T} =& \mathrm{T}_E - \mathrm{T}_A \quad \leq \quad 0 \\ \Delta \mathrm{T} =& \frac{1}{2}\dot{\mathbf{q}}_E^T \mathbf{M} \dot{\mathbf{q}}_E - \frac{1}{2}\dot{\mathbf{q}}_A^T \mathbf{M} \dot{\mathbf{q}}_A = \frac{1}{2}(\dot{\mathbf{q}}_E + \dot{\mathbf{q}}_A)^T \mathbf{M} (\dot{\mathbf{q}}_E - \dot{\mathbf{q}}_A). \end{aligned} \tag{20}$$

These are expressions considering scleronomic systems without an excitation by external kinematical sources. Applying the relations as presented in [24], we get for the energy expression the form

$$\begin{aligned} 2\Delta \mathrm{T} =& 2\Delta \mathrm{T}_1 + \Delta \mathrm{T}_2 = +2\begin{pmatrix} \dot{\mathbf{g}}_{NE} \\ \dot{\mathbf{g}}_{TE} \end{pmatrix}^T \left[\begin{pmatrix} \mathbf{\Lambda}_{NC} \\ \mathbf{\Lambda}_{TC} \end{pmatrix} + \begin{pmatrix} \mathbf{\Lambda}_{NE} \\ \mathbf{\Lambda}_{TE} \end{pmatrix}\right] - \\ & - \left[\begin{pmatrix} \mathbf{\Lambda}_{NC} \\ \mathbf{\Lambda}_{TC} \end{pmatrix} + \begin{pmatrix} \mathbf{\Lambda}_{NE} \\ \mathbf{\Lambda}_{TE} \end{pmatrix}\right]^T \mathbf{G} \left[\begin{pmatrix} \mathbf{\Lambda}_{NC} \\ \mathbf{\Lambda}_{TC} \end{pmatrix} + \begin{pmatrix} \mathbf{\Lambda}_{NE} \\ \mathbf{\Lambda}_{TE} \end{pmatrix}\right] \end{aligned}$$

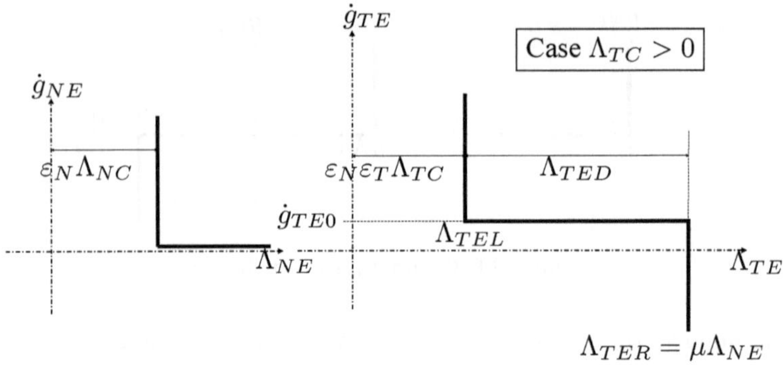

Figure 35: Shifted Normal and Tangential Characteristics for Impact Expansion

$$\text{with} \quad \boldsymbol{G} = \begin{pmatrix} \boldsymbol{G}_{NN} & \boldsymbol{G}_{NT} \\ \boldsymbol{G}_{TN} & \boldsymbol{G}_{TT} \end{pmatrix}$$
$$\text{where} \quad \boldsymbol{G}_{ij} = \boldsymbol{W}_i^T \boldsymbol{M}^{-1} \boldsymbol{W}_j, \quad i, j \in \{N, T\} \tag{21}$$

$\boldsymbol{G}$ is the mass projection matrix, which is quadratic and positive definite with the exception of dependent constraints, where it is semidefinite. The $\dot{\mathbf{g}}$ are relative contact velocities and the $\boldsymbol{\Lambda}$ impulses. The indices N,T stand for normal and tangential, respectively, the indices C,E for the end of compression and the end of expansion, respectively. The second term of the energy equation is a quadratic form and for itself always positive or zero, and from this we have $\Delta \mathrm{T}_2 \leq 0$, always. The energy loss has to be negative, which will be decided by the first term of the above relations. If this term is negative or at least zero, the condition $\Delta \mathrm{T} \leq 0$ will hold. Therefore we shall concentrate on the first term which writes in more detail

$$\begin{aligned} 2\Delta \mathrm{T}_1 &= +2 \begin{pmatrix} \dot{\mathbf{g}}_{NE} \\ \dot{\mathbf{g}}_{TE} \end{pmatrix}^T \left[\begin{pmatrix} \boldsymbol{\Lambda}_{NC} \\ \boldsymbol{\Lambda}_{TC} \end{pmatrix} + \begin{pmatrix} \boldsymbol{\Lambda}_{NE} \\ \boldsymbol{\Lambda}_{TE} \end{pmatrix} \right] = \\ &= 2[\dot{\mathbf{g}}_{NE}^T (\boldsymbol{\Lambda}_{NC} + \boldsymbol{\Lambda}_{NE}) + \dot{\mathbf{g}}_{TE}^T (\boldsymbol{\Lambda}_{TC} + \boldsymbol{\Lambda}_{TE})] \end{aligned} \tag{22}$$

For the evaluation of this equation we have to discuss the models. The compression/expansion model as considered here is a very powerful one and approved by many applications providing us with the necessary informations for impulsive processes in multibody systems, but it does not provide us with the details within the compression and expansion phases necessary for energy considerations. We get informations at three points A, C, E (A=beginning, C=end compression, E=end expansion), but not between these points.

We know for example, that for an impact with sliding or sticking the relative normal distance and velocity have to be zero. Otherwise we do not get tangential impact and contact motion. But on the other hand the results for the points A, C, E would give us for E only a non-zero normal velocity, which appears in physical reality only at the very end of the impact and not during expansion. To solve this problem without disturbing the model concept it is sufficient to define the transition locations at the very end of the compression phase (transition (C/E)) and of the expansion phase (point E). Such transitions from sticking to sliding or vice vera and from contact to detachment are assumed to take place in an infinitesimal short time with no energy effects.

So it can be shown, that the first term $\dot{\mathbf{g}}_{NE}^T(\mathbf{\Lambda}_{NC} + \mathbf{\Lambda}_{NE})$ of the energy equation (22), last line, is not zero due to positive normal impulses $(\mathbf{\Lambda}_{NC} + \mathbf{\Lambda}_{NE})$ and due to a non-zero end velocity $\dot{\mathbf{g}}_{NE}$ after the impact, which is physically reasonable for a separation of the two contacting bodies. But on the other hand sliding or sticking during expansion requires a zero normal relative velocity $\dot{\mathbf{g}}_{NE}$ in the contact, which makes the above mentioned term to zero. The $\mathbf{\Lambda}_{NE}$-value slips into the corner of Figure (35) allowing the system to build up the necessary separation velocity. From this we assume, that during the expansion phase the term $\dot{\mathbf{g}}_{NE}^T(\mathbf{\Lambda}_{NC} + \mathbf{\Lambda}_{NE}) = 0$ is zero.

As a result of these arguments and of the last condition of continual contact during the impact we get for compression and expansion $\mathbf{\Lambda}_N > 0$ and $\dot{\mathbf{g}}_N = 0$, which is also part of the complementarity, and therefore simply

$$2\Delta \mathrm{T}_1 = 2\dot{\mathbf{g}}_{TE}^T(\mathbf{\Lambda}_{TC} + \mathbf{\Lambda}_{TE}), \tag{23}$$

the sign of which we have to investigate. For this purpose we consider this equation with respect to the following four cases, see for the arguments always the Figures 34 and 35:

- **sticking during compression, sticking during expansion**

  The tangential impulses have to be within the appropriate friction cones. The tangential velocities are zero, therefore we need not to consider the magnitudes of the impulses.

  $$\begin{aligned} &-\mathrm{diag}(\mu_0)\mathbf{\Lambda}_{NC} \leq \mathbf{\Lambda}_{TC} \leq +\mathrm{diag}(\mu_0)\mathbf{\Lambda}_{NC}, \\ &\mathbf{\Lambda}_{TEL} \leq \mathbf{\Lambda}_{TE} \leq \mathbf{\Lambda}_{TER} \\ &\Longrightarrow \quad \dot{\mathbf{g}}_{TE}^T(\mathbf{\Lambda}_{TC} + \mathbf{\Lambda}_{TE}) = 0 \end{aligned}$$

- **sliding during compression, sliding during expansion**

  Sliding means single-valued impulse laws according to Coulomb's law. Some difficulties will appear for the cases with reversed sliding, that means, with a tangential relative velocity the sign of which is different during compression and during expansion. Therefore we have to consider the two cases without and with tangential reversibility. For the first case we do not have a change of sign of the relative tangential velocity, which gives $\text{sign}(\dot{\mathbf{g}}_{TC}) = \text{sign}(\dot{\mathbf{g}}_{TE})$. This comes out with the relations:

$\dot{\mathbf{g}}_{TE}^T \mathbf{\Lambda}_{TC} = -\dot{\mathbf{g}}_{TE}^T[\text{diag}(\mu)\text{sign}(\dot{\mathbf{g}}_{TE})\mathbf{\Lambda}_{NC}] = -\text{diag}(\mu)|\dot{\mathbf{g}}_{TE}|\mathbf{\Lambda}_{NC} \leq \mathbf{0}$,

$$\Longrightarrow \quad \dot{\mathbf{g}}_{TE}^T(\mathbf{\Lambda}_{TC} + \mathbf{\Lambda}_{TE}) < 0$$

The case with tangential reversibility is more complicated, because it includes a change of sign of the tangential relative velocity and thus at least an extremely short stiction phase, which we put exactly at the point (end of compression)/(beginning of expansion). The sliding velocity during compression decreases until it arrives at one of the corners of Figure 34, then we get an extremely short shift from this corner to the other one, which allows the contact to build up a tangential velocity with an opposite sign, then valid for the expansion phase. Only by such a short stiction phase a reversal of tangential velocity is possible. On the other hand such a transition from stick to slip, as short as it might be, follows the same process as for the next case sticking/sliding. Therefore it is dissipative:

$$\Longrightarrow \quad \dot{\mathbf{g}}_{TE}^T(\mathbf{\Lambda}_{TC} + \mathbf{\Lambda}_{TE}) < 0$$

- **sticking during compression, sliding during expansion**

  The transition from sticking in compression and sliding in expansion follows the mechanism (Figure 34): If $\mathbf{\Lambda}_{TC} \gtrless \mathbf{0}$, then sliding is only possible for being at the very end of compression on the friction cone boundary with $\mathbf{\Lambda}_{TC} = \pm\text{diag}(\mu)\mathbf{\Lambda}_{NC}$ and $\dot{\mathbf{g}}_{TC-at} \lesseqgtr \mathbf{0}$ (at = after transition stick-slip). This results always in a negative sign of the expression $(\dot{\mathbf{g}}_{TE}^T \mathbf{\Lambda}_{TC})$. For the rest we assume a continuation of the signs after going from stick to slip $[\text{sign}(\dot{\mathbf{g}}_{TE}) = \text{sign}(\dot{\mathbf{g}}_{TC-at})]$. Then we arrive at:

$$\Longrightarrow \quad \dot{\mathbf{g}}_{TE}^T(\mathbf{\Lambda}_{TC} + \mathbf{\Lambda}_{TE}) < 0$$

- **sliding during compression, sticking during expansion**

  This case is again simpler, because we get sticking at the end with a zero relative tangential velocity. Therefore we need not to consider the impulses.

$$\Longrightarrow \quad \dot{\mathbf{g}}_{TE}^T(\mathbf{\Lambda}_{TC} + \mathbf{\Lambda}_{TE}) = 0$$

- **summarized result for all cases**

$$\Longrightarrow \quad \dot{\mathbf{g}}_{TE}^T(\mathbf{\Lambda}_{TC} + \mathbf{\Lambda}_{TE}) \leq 0 \quad \Longrightarrow \quad \Delta \mathrm{T}_1 \leq 0 \quad \Longrightarrow \quad \Delta \mathrm{T} \leq 0$$

One may object that the above considerations assume in the case of multiple impacts the same impact structure for all simultaneously appearing impacts, which is usually not true. But even any combination of the above four cases for simultaneous impacts gives a loss of energy. Practical experience indicates in addition that the simultaneous appearance of impacts is extremely scarce, it is an event, which nearly does not happen.

As a final result we may state that the above evaluation confirms the physical argument, that any impact processes are accompanied by energy losses. This confirms also the well-known statement of Carnot, that *"in the absence of impressed impulses, the sudden introduction of stationary and persistent constraints that change some velocity reduces the kinetic energy. Hence, by the collision of inelastic bodies, some kinetic energy is always lost"*.

The above considerations and the underlying theory have been confirmed not only by the experimental work of Beitelschmidt [3], but also by many industrial projects where the non-smooth methods were applied [39], [14]. A detailed and very advanced mathematical proof of these energy losses may be found in the recent work of Glocker [17], [18].

5.2.5 Verification of Impacts with Friction

With respect to large technical applications impacts with friction play such an important role, that a really applicable theory came amazingly late. The first ideas regarding large dynamical systems with frictional impacts are contained in the famous contribution of Moreau in the year 1988 [22], but then more or less applied to static or quasi-static (smaller) problems. During the nineties these ideas were included into multibody theory by ([15], [3], [25]) and applied to large industrial problems, which without exception confirmed the relevant theories. In addition a thourough and systematic experimental proof of the theory has been performed by Beitelschmidt [3].

In the following we shall focus on these experiments. In designing a test set-up for measuring impacts with friction a first principal decision had to be made with respect to the experiments and to the geometrical type of impact, plane or spatial. Colliding bodies moving in a plane include linear complementarity problems, spatial contacts generate nonlinear complementarities. Therefore motion in a plane was considered where one body is a disc and the other one the ground. On this basis some further requirements had to be defined:

• maximum translational velocity	10 m/s
• maximum rotational velocity	40 rps
• throw direction	$0^o - 90^o$
• release time	< 12 ms
• encoder main axis	1600 points
• encoder momentum axis	400 points
• throwing disc	diameter 50 mm
.	thickness 20 mm
.	weight 300 g
• continuous variable velocity control	

- translation and rotation decoupled
- disturbance-free support and release of disc
- mass balance, statically and dynamically
- electric drives (pulse width modulation with 250 steps)
- automatic control for the throwing process, the release of stroboscope and camera

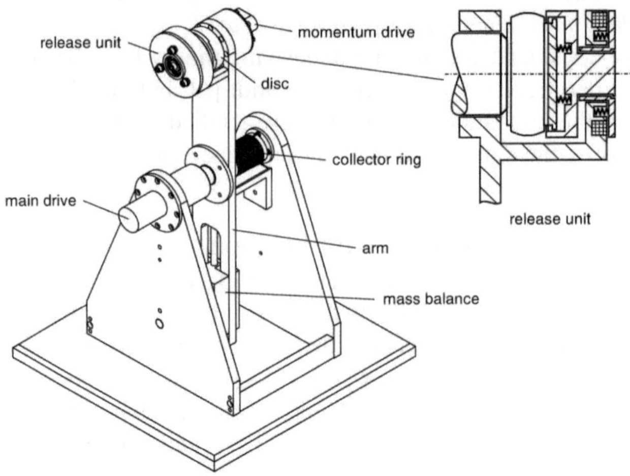

Figure 36: Principle of the Throwing Machine

As a result, the machine of Figure (36) was designed and realized, which meets all requirements. A release unit containing the disc is mounted at the end of a rotating arm with mass balance. The unit itself drives the disc giving it a prescribed rotational speed. The arm drive and momentum drive are decoupled allowing to control the two speeds independently. The rotation of the arm can be used to generate a translation, the rotation of the release unit realizes a rotation of the disc. Both mechanisms require an extremely precise time management of the release process. The flight of the body is photographed under stroboscopic exposure in a dark room before and after hitting his target. From the evaluation of the photographs one can calculate the velocities and the position of the body immediately before and after the impact.

Figure (37) depicts the structure of the test set-up. A computer performs all control calculations, processes sensor data, evaluates control torques, releases stroboscope and camera and records all measured data. Within this overall structure we find for each drive an individual control concept, which has thoroughly been optimized with regard to the above requirements [3]. Also, a typical sequence of events for the test procedure can be seen from Figure (38). All computer codes

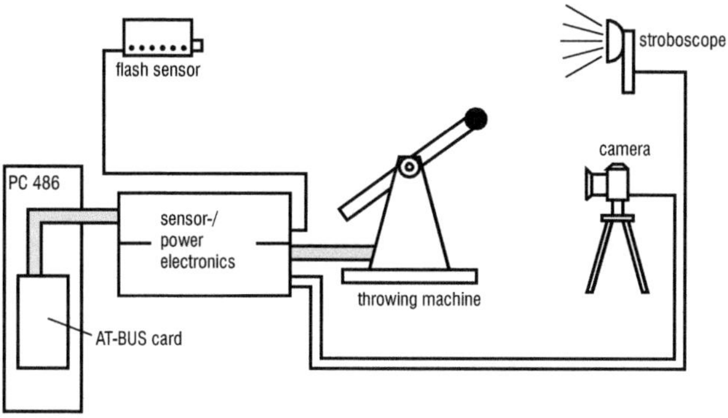

Figure 37: Structure of the Complete Test Set-Up

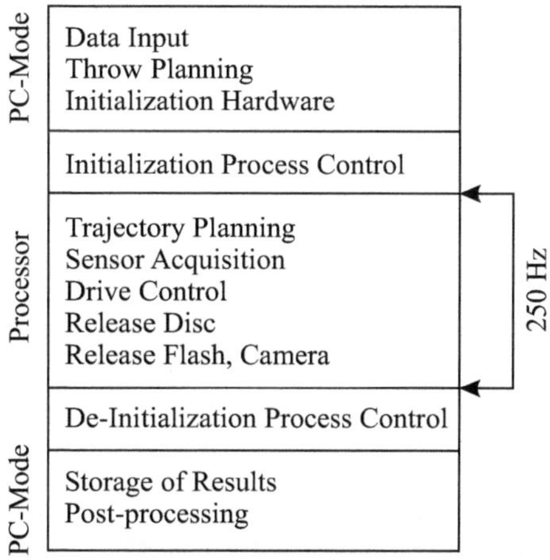

Figure 38: Sequence of Events of the Throwing Machine Control

have been realized in C++, which was feasible due to the fact that the PC-Mode activities are not critical with respect to time.

The evaluation of the measurements as recorded by the camera and the processor was straightforward. Figure (39) illustrates the method and depicts additionally two photographs of experiments. Especially the rubber disc experiment shows nicely a reversal of the trajectory due to the disc's rotation. The experimental process provided thus a very precise and well reproducible basis for determining the properties of impacts with friction.

a)

b)

c)

Figure 39: Disc Trajectory During an Experiment, a) method of evaluation, b) photograph steel, c) photograph rubber

In the following we shall give only a few examples out of more than 600 experiments performed with axisymmetric and with eccentric discs. In all cases the comparisons with theory come out with a good to excellent correspondence [3]. In the subsequent diagrams we shall use dimensionless velocities and impulses defined by

$$\gamma = \frac{\dot{g}_{TA}}{-\dot{g}_{NA}}, \qquad \gamma_{NC} = \frac{\dot{g}_{NC}}{-\dot{g}_{NA}}, \qquad \gamma_{TC} = \frac{\dot{g}_{TC}}{-\dot{g}_{NA}},$$
$$\gamma_{NE} = \frac{\dot{g}_{NE}}{-\dot{g}_{NA}}, \qquad \gamma_{TE} = \frac{\dot{g}_{TE}}{-\dot{g}_{NA}}, \qquad \gamma_{TE0} = \frac{\dot{g}_{TE0}}{-\dot{g}_{NA}}, \tag{24}$$

where the indices N, T refer to normal and tangential directions. The indices A, C, E are the beginning and the end of the compression phase, and the end of the expansion phase, respectively. The kinematical magnitude $\dot{g}$ is a relative velocity in the contact zone. Experiments usually generate a negative normal velocity $(-\dot{g}_{NA})$ at the beginning.

Figure (39) also indicates the evaluation process for all experimental results. For every small part of the trajectory we perform three stroboscope flashes thus achieving a certain redundancy for the measurements. The trajectory is approximately a parabola, and the velocity possesses a positive component in x- and a negative component in y-direction. The stroboscopic measurements in connetion with the marked sectors of the discs allow a safe evaluation of the translational and rotational velocities of the discs. To find the time and the point of impact, the measurements before and after such an impact are represented by a statistical interpolation scheme, which allows to determine the impact together with the dispersion of the results.

Figure (39) gives two examples. The material pairing steel on steel behaves conventional and in such a way, which one could expect. The picture (b) shows the steel disc approaching the ground with a translational and rotational velocity and leaving the ground with a mirrored but more or less a similar trajectory.

The part (c) of Figure (39) represents a spectacular case. The rubber disc appears from the left side with a horizontal velocity of 5 m/s and a vertical velocity of 4 m/s in negative y-direction. The rotational velocity in a counterclockwise direction amounts to 40 rps (2400 rpm). This results in a tangential relative velocity of 12.5 m/s at the point of impact. The impact process is depicted by the Figure part (c). After the first contact the velocities reverse by the impact jump, and the disc flies backwards with a clockwise rotation. At the second impact the velocities change again, and the disc flies forward with the original direction of rotation.

As a result we may state, that firstly for the rubber case the impact coefficient of restitution in normal direction depends much more on the velocities at collision than for stiff materials, that secondly we get a typical characteristic behavior in the sense of tangential reversibility, and that thirdly for soft materials like rubber we may have friction coefficients larger than one ($\mu > 1$). The theory describes this behavior very well, where especially for soft materials a correction is advantageous. The Figure (40) shows results of experiments with a PVC test body. The

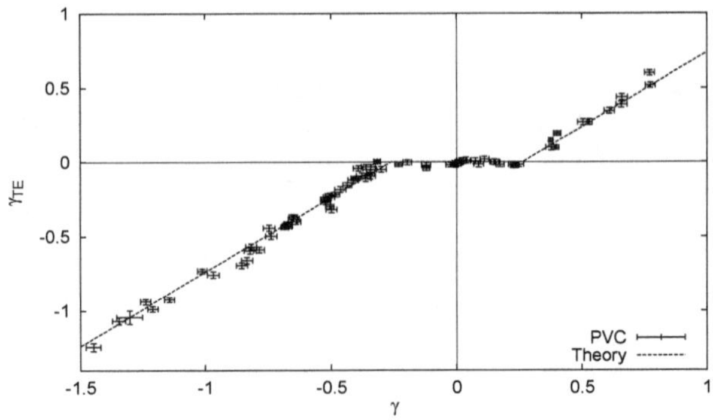

Figure 40: Dimensionless Tangential Relative Velocity, after vs. before the impact, PVC-body

experiments are marked by crosses, the dotted line shows the theoretical result. For small tangential relative velocities before the impact, sticking occurs, and the rolling constraint between disc and ground is fulfilled after the impact. If the relative velocity is big enough, the body slides throughout the impact and has a redcuced tangential relative velocity at the end of the impact. No tangential reversion occurs. In the area around zero tangential speed we get sticking.

A similar diagram for a rubber-body is shown in Figure (41). For most of the impacts the tangential relative velocity has changed during the impact: the bodies collide with a negative relative velocity and separate with a positive velocity. The inclination of the line through the origin is $-\varepsilon_N \varepsilon_T$. If ε_N is known from another simple experiment one can evaluate the coefficient of tangential reversibility from this plot. For this series of experiments the parameters $\varepsilon_N = 0.75$ and $\varepsilon_T = 0.9$ were identified. If the tangential relative velocity increases further, sliding occurs in the contact point during the impact. Then it is not possible to restore the elastic potential energy during the phase of expansion. For very high velocities the rubber body slides during the whole impact and the effect of tangential reversibility is not further visible.

In Figure (41) two lines are plotted for comparing theory with experiment. What is called "Theory old" corresponds to the original theory of impacts with friction as presented in Glocker's dissertation [15]. What is called "Theory new" includes the extension as given by Beitelschmidt [3], which applies mainly for very soft material pairings. If we consider the contact point of two bodies, where Coulomb's friction applies, and that point of the contact zone, where the spring

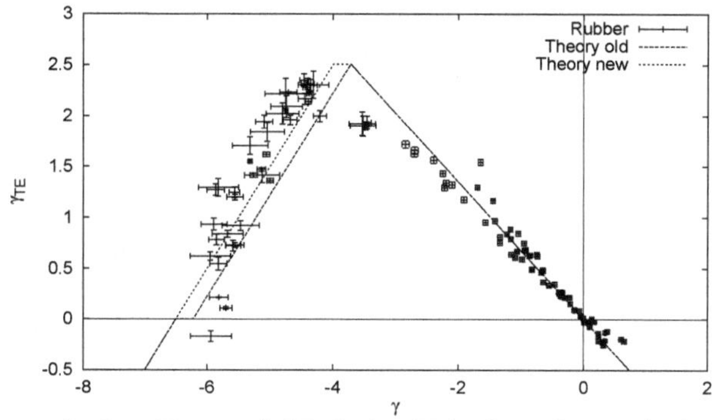

Figure 41: Dimensionless Tangential Relative Velocity, after vs. before the impact, rubber-body

force resulting from the storage of impulse applies, we come out with two force laws in series. This gives a modification of the complementarities with respect to the friction cone, and thus a modification of the final results.

Impacts with friction play an essential role in machines, mechanisms and also in biology. Therefore we need good models being verified by sound experiments. Such experiments have been performed by an especially designed throwing machine. The experimental results compare excellently with calculated values from existing theories. A slight improvement of the theory could be deduced from the experimental findings. It concerns the frictional complementarities during expansion, which have been slightly modified and are mainly applicable for soft material pairings.

5.2.6 Applications

A large body of industrial applications from power transmission to walking machines are presented in all details in [24]. In the following we shall focus on two typical examples, the vibration conveyor and the chimney damper.

Vibration Conveyor Vibratory feeders are used in automatic assembly to feed small parts. They are capable to store, transport, orient and isolate the parts. An oscillating track with frequencies up to 100 Hz excites the transportation process, which is mainly based on impact and friction phenomena between the parts and the track. Vibratory feeders are applied for a wide variety of parts and for lots of different tasks. In the majority of cases, the parts are available as a sort of bulk material that is stored in a container. The transportation process, starting in this reservoir, is often combined with orienting devices that orient parts, or select only these parts having already the right orientation (Figure 42 shows an example of a vibratory bowl feeder with an orienting device). Each kind of parts, with its special geometry and mechanical properties, requires an individual adaption of the feeder. This individual tuning comprises the development of suitable track and orienting device geometries and the adjustment of the excitation parameters frequency and amplitude. Due to the complex mechanics of the feeding process this design is usually done by trial and error without any theoretical background. A complete dynamical model of the transportation process allows a theoretical investigation and consequently an improvement of the the feeder properties [43].

Friction and impact phenomena between the parts and the track are the most important mechanical properties of transportation processes. Consequently, the required dynamical model has to deal with unilateral constraints, dry friction and multiple impacts. The mechanical model of the vibratory feeder can be split in two parts: the transportation process and the base device. The dissertation [43] focuses on modeling and simulation of the transportation process. The modeling of the base device can be done with well known standard technics for multibody systems.

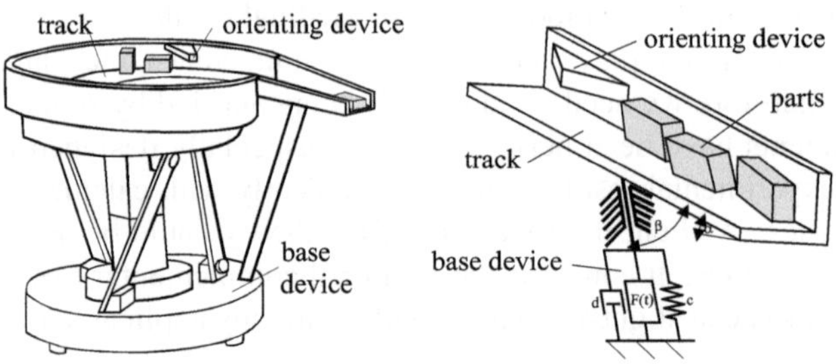

Figure 42: Vibratory bowl feeder and mechanical model

For the verification of the developed model of the transportation process an experimental vibratory feeder was built, allowing different measurements concerning the impact model and the average transportation rates. Figure 43 shows the principle of the device. The track, fixed on leaf springs is excited with an electro magnetic shaker with a frequency about 50 Hz. The eigenfrequency of the system is at 52 Hz. The resulting vibration amplitude reaches a maximum value of about 2 mm. The track has an inclination angle $\alpha = 3^o$, the angle between the track and the direction of the vibration is $\beta = 15^o$. For the accurate contactless measurement of the motion of the transported part six laser distance sensors were applied. For the vibration measurement of the track an eddy current sensor is used.

For a comparison of the theory and the measurements the averaged transportation rate was used. Figure 44 gives a result, which before the background of the complexity of the problem looks good. An interesting finding is the fact that the

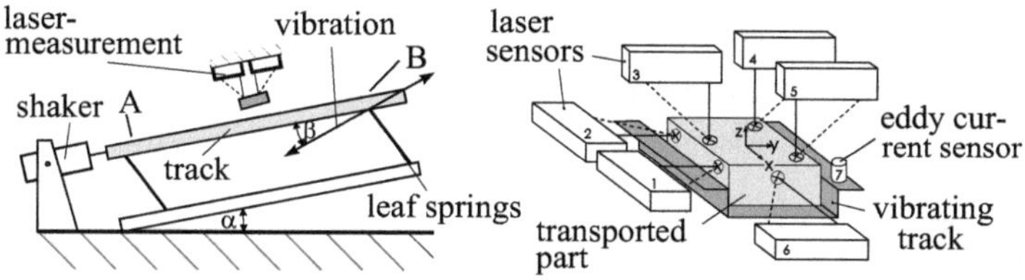

Figure 43: Test setup and part measurement

averaged transportation velocity does not depend very much on the number of parts and also not on the type of modeling, plane or spatial [43]. Therefore the design of vibration conveyors can be carried through considering one part only. For the layout of orienting devices we need of course a spatial theory.

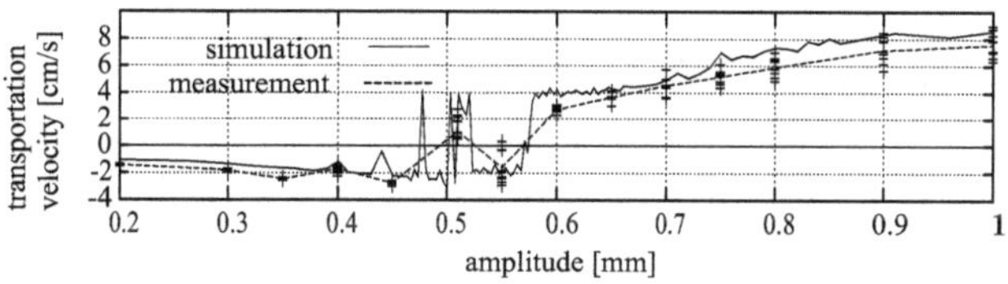

Figure 44: Simulation and measurement of the average transportation rate

Chimney Dampers Towerlike structures like steel chimneys may be excited to vibrations by vortex streets. Such a mechanism becomes dangerous if the first eigenfrequency of the structure is small enough to be exited by not too large wind velocities which appear frequently. Oscillations of that type are damped significantly by a simple idea. A pendulum is attached to the chimney top. Its mass and length is tuned to the chimneys first eigenfrequency by classical analysis, and additionally optimal viscous damping is also evaluated by classical formulas, given for example by Den Hartog's method. In a second step the thus determined ideal damping behavior is not realized by viscous means but by a package of circular plates which are moved by the end of the pendulum rod through internal circular holes of the plates. The approximate realization of viscous damping by dry friction is accompanied by impacts of the pendulum rod in the plate holes and by stick-slip transitions between the plates. To achieve a best fit to the ideal viscous case an optimization of the complete pendulum-plate-system is carried through applying multibody theory with unilateral contacts [27].

Figure 45 illustrates the basic principle. At the top of the chimney a pendulum is arranged which damps the oscillations. To achieve best damping efficiency the damper is optimized in two steps. In a first step a classical damper is assumed

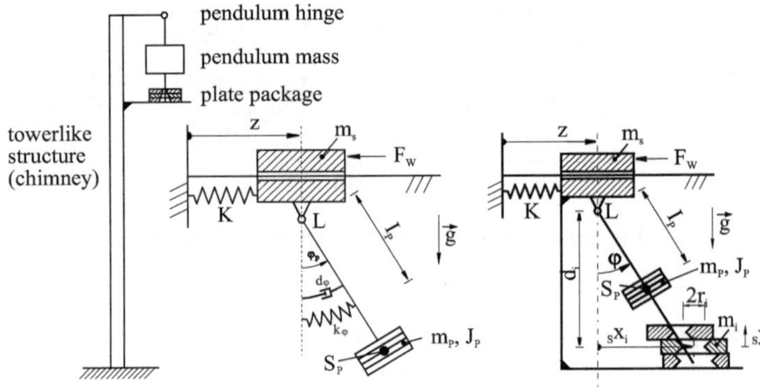

Figure 45: Principal configuration, viscous and plate model

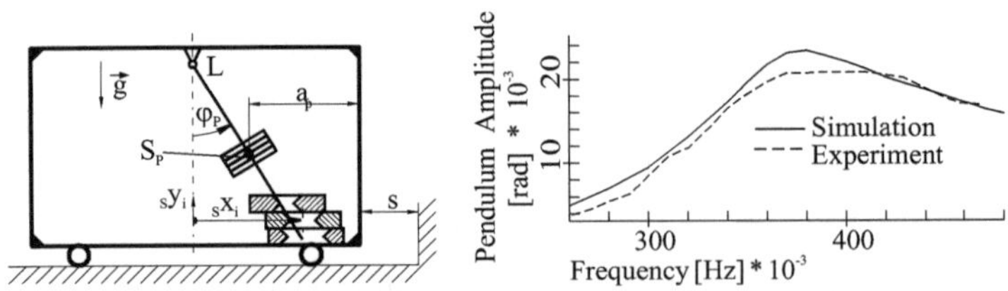

Figure 46: Vibrating test-setup and a comparison theory/ measurements

working with viscous damping. For the resonance frequency area such a damper can be optimized with respect to length, mass and viscous damping of the pendulum. As damping will be realized by dry friction within a plate package, the plates must be selected in such a way that they perform a damping effect coming as nearest as possible to the optimal viscous damping. Therefore, in a second step the plate package will be optimized with respect to damping efficiency.

To verify the theory two types of experiments have been performed. In a first test a pendulum-plate-configuration has been arranged in a car-like frame with wheels, which could be excited periodically. Figure 46 depicts the results which compare well with theory.

A second test has been performed with a real steel chimney, which was bended by a steel rope. By releasing suddenly this rope the chimney starts to vibrate. This process has been measured and compared with the corresponding theory based on the equations of chapter 2. Figure 47 illustrates a comparison between experiment and simulation, which also confirms the methods presented.

If we approximate the optimal viscous system of Figure 45 (middle picture) by a package of plates, we can do that in a best way by optimizing the number

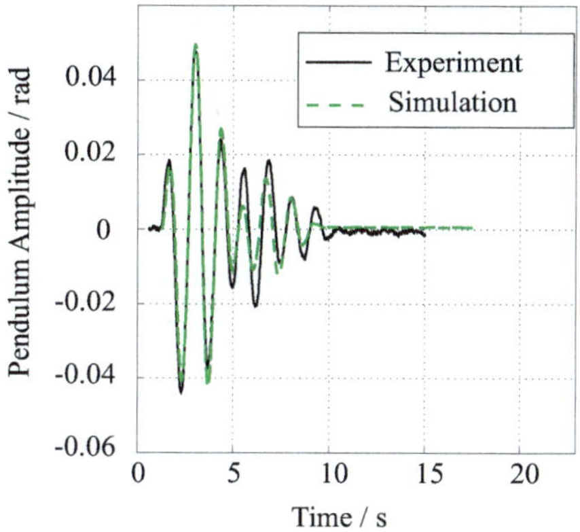

Figure 47: Pendulum vibrations in a real chimney

of plates, their radii and thicknesses. We find that the number of plates affect the damping efficiency only partly. On the left side of Figure 48 the total mass of the plates is put constant, also the radius of the top at the ground damper plate. The number of plates is varied from one to eight. Doing so the distribution of the plate holes is kept linear. In the case of one plate the diameter of the plate hole is half as big as the largest one in all other cases.

Figure 48 shows that the damping mechanism works best with one plate and worst with two plates. The reason lies in the fact, that in the case of two plates only the upper one is moved effectively. The ground plate is ineffective and hence its damping is too small. As an important result it should be noticed, that the application of a large number of plates does not make sense. Obviously the damping characteristic cannot be improved by using more than five or six plates.

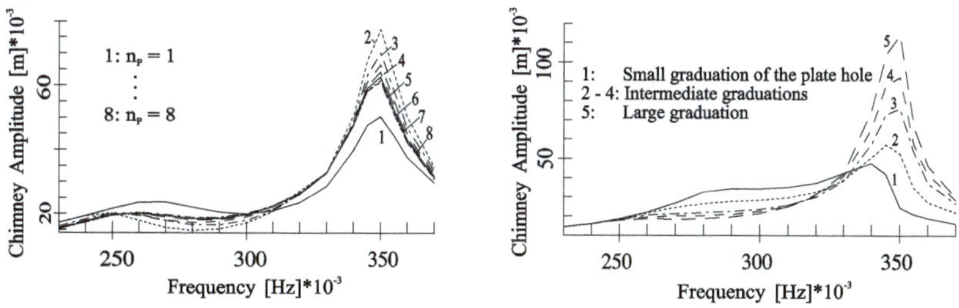

Figure 48: Variation of plate number (left) and of graduation (right)

A further sensible parameter is the distribution of hole diameters over all plates, called graduation. The right graph of Figure 48 shows with number 1 the behavior of a system with small and with graph number 5 with a big graduation. The other curves belong to some intermediate graduations. Keeping the total mass of the plates constant an increasing graduation (from 1 to 5) leads to a decrease of damping and for the considered configuration to rising chimney displacements.

5.2.7 Realization of Objectives

In the meantime a comprehensive theory concerning impacts with friction exists, but research is still going on, though there are not so many open questions left. This theory covers all details of impacts with or without friction, contact and detachment, sliding and sticking states and an unambiguous budgeting of energy. The approach has been experimentally verified in a very convincing way, and many large industrial applications have been successfully considered and evaluated. In this connection it should be noted, that numerical simulations for non-smooth systems are extremely time-consuming, and that this problem has been considerably reduced by the ideas of Alart/Curnier by introducing set-theoretical approaches for the non-smooth constraints [1]. The "non-smooth community" is still discussing the general friction laws, from Coulomb to more modern proposals, but the simple Coulomb friction law is still an extremely powerful tool to describe such phenomena. All efforts to go into more details using even molecular surface features were up to now not successful.

5.3 Example: CVT Gear Systems

5.3.1 Objectives

Every transportation system, whatever it is, needs a torque converter. To achieve a best fit to the power-resistance-behaviour of a car, for example, one has augmented the number of gear stages and at the same time introduced some control to get a smooth performance. Another approach consists in applying continuous gears like the CVT-gears. The problems in connection with the use of such transmissions consist in the limited torque which can be transmitted and in noise and wear features. Therefore it makes sense to investigate that in more details.

From the standpoint of modeling the CVT-chains are the most complicated components requiring very detailed approaches, which we learned in the course of more than 20 years applied research in this area. CVT-dynamics and especially chain dynamics represent a non-smooth mechanical system with thousands of contacts and with a complex process of torque transmission, which is performed by friction alone. Very detailed fundamental research has been necessary, also in close cooperation with industry, to be able to establish realistic models with convincing properties for practical applications. The success can be measured by an increasing good performance of such chains.

The main objectives were the same over the years: to understand the physical processes of a CVT, to model the transmission process in a realistic way by considering the plane and the spatial cases, to verify all details with appropriate test set-ups and to achieve results improving the hardware development from year to year. These requirements could not be fulfilled in a short time but only with research over decades. It is a typical modeling example in the sense of Karl Popper, that each finding with respect to a problem initiates new problems [31].

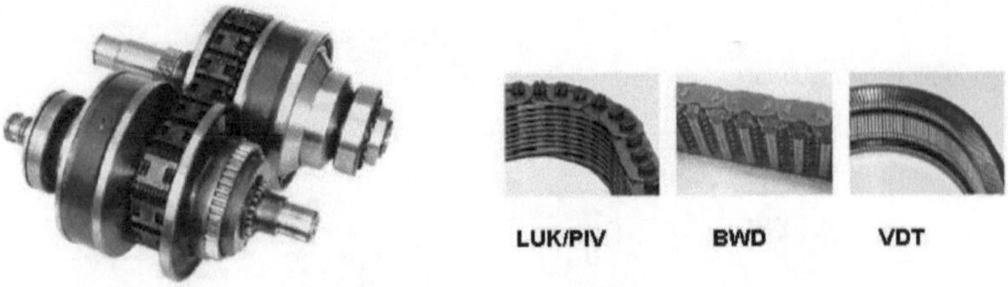

Figure 49: Typical CVT gear with the three chain/belt-types

5.3.2 Introduction

Before the background of increasing requirements for economic use of fuel in cars all ideas for realizing power transmissions with a nearly or completely continuous transmission ratio become very attractive. One of these ideas concern the transmissions operating with belts or chains, which are called **C**ontinuous **V**ariable **T**ransmissions (CVT) and which are used to a large extent in Japan, partly also in Europe.

These chains or belts operate within two disc pairs, where each pair consists of a fixed and a movable sheave (see Figure 49). The movable disc is displaced hydraulically according to some control laws, which must be given for simulating the system dynamics. Anyway, the sheaves must be modeled elastically, because on the one side already small elastic deformations influence significantly the contact performance between the chain elements and the discs, and on the other side these effects have considerable influence on the pressure ratio and the efficiency of the CVT. The sheave-shaft system as a whole is represented by a Ritz-approach with shape functions from a FEM-calculation. For all chain/belt types two- and three-dimensional models are available. An analysis in three dimension makes sense only, if details of irregularities must be known, for example with respect to the chain forces and from there with respect to wear and noise.

In the case of the VDT-push belt the situation is really complicated, therefore we shall focus here mostly on the simulation of this type of CVT. Each element possesses two contacts with neighbouring elements, two contacts with the ring system and two contacts with the elastic discs. Taken the contact dynamics in an exact way all contacts are unilateral with respect to contact/detachment and with respect to stick/slip. Even with the largest computers this is not feasible. Therefore we use as an approximation partly bilateral contacts between the components and partly unilateral contacts. All frictional processes are modeled by set-valued force laws. The results of that method are very nicely approved by experiments. Computing time is at least realizable.

	LUK	VDT
number of elements	63	382
element length (thickness)	9.85 mm	1.80 mm
element width	36 mm	29.6 mm
polygonal frequency	550 Hz	3000 Hz
elements/second (1000 rpm, i=1)	550	3000
lowest eigenfrequency	90 Hz	120 Hz
DOF, plane case	$\approx$200	$\approx$1300
contacts, plane case	$\approx$150	$>$1000

Table 1: Typical Data of CVT-Chain and -Belt

To give an impression of the size of such systems we recall some typical data (see also the Figures 50, 51, 52). The LUK-chain is a rocker pin chain possessing typically an element length of about 10 mm and an element width of 36 mm. The chain might have 60-70 elements and thus 180-210 DOF in the plane and 360-420 DOF in the spatial case. We have about 150 unilateral contacts between the rocker pins and the pulleys. At a speed of 1000 rpm with a transmission ratio of 1.0 about 550 elements enter and leave the pulleys in one second thus generating a polygonal frequency of 550 Hz. The lowest "eigenfrequency" of the overall system is about 90 Hz.

Due to very small elements the VDT-belt is an order of magnitude larger. The element length is about 1.8 mm, its width around 30 mm. The belt has typically 350-420 elements and thus 1050-1260 DOF in the plane and 2100-2520 DOF in the spatial modeling case. In addition the belt possesses 9-11 steel rings and about 3600 unilateral contacts, where the element-ring-contacts include the problem of a continuum with distributed unilateral contacts. At a speed of 1000 rpm with a transmission ratio of 1.0 about 3000 elements enter and leave the pulleys thus generating a polygonal frequency of 3 kHz. The lowest "eigenfrequency" is around 120 Hz for the overall system.

During the last 15 years a considerable number of publications on CVT problems were brought out. Sattler [33] develops a very useful theory for a reduced CVT-model, which is excellently suitable for a first design. We used his theory quite often for estimating the starting data for large model simulations. Srnik [42] has been the first one to develop a very detailed and sophisticated model of CVT rocker pin chains including all contact events as indicated above. His plane theory was extended to the spatial case by Sedlmayr [36]. Both theoretical models have very successfully been verified applying the measurements of Sauer [34] from the chair of agricultural technologies at the Technical University of Munich.

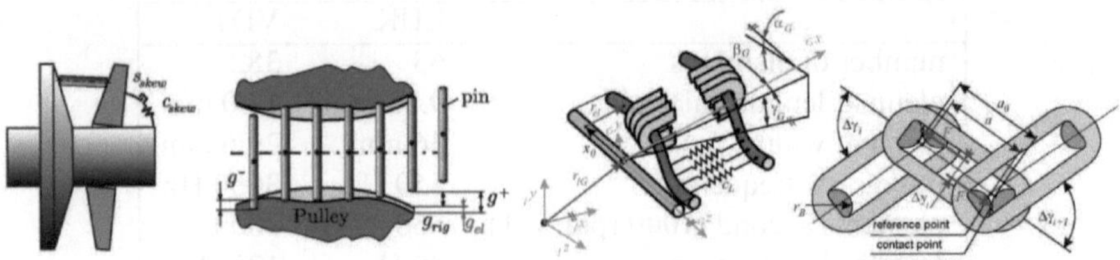

Figure 50: Contact Configurations of Rocker Pins

In the second half of the nineties we started at my former Institute research and development of CVT-push-belt systems, the dynamics of which is much more complicated than that of the rocker pin chains. A first contribution was elaborated by Bullinger [5] and then continued by Geier [13]. Pfeiffer [29] gives a first survey of the problems and the simulations.

One of the fundamental problems in modelling CVT's consists in the contact behavior of the elements. We have some hundreds up to some thousands of contacts, the behavior of which decide on noise and wear of the system. In the following we shall discuss the contact configurations of the rocker pin chain and the push-belt system considering their contribution to the equations of motion.

5.3.3 Models of the Rocker Pin Chain

The contact situation of the rocker pins is indicated in Figure 50. The deformed rocker pins enter the elastic sheaves and come into unilateral contacts with them. In addition we have contacts between pins and the elastic plates and partly some rolling without sliding of the two components of a rocker pin itself [36], see right pictures of Figure 50. The multibody system of the rocker pin chain CVT consists of two movable and two fixed sheaves and $n_L = 63$ chain link bodies. The sheaves as well as the links are elastic. The bodies are interconnected by force elements. See [29] for more details, especially more detailed formulas and Figures.

Pulley Set The contact forces between the chain and the pulleys cause a considerable deformation of the pulley's sheaves. The eigenfrequency of the sheaves is much higher than the operating frequencies. According to [42] the mass forces of the elastic deformations can be neglected in this case. With a FEM-analysis the sheave deformation can be calculated with the Reciprocal Theorem of Elasticity (Betti/Maxwell). Together with Hooke's law applied at the pin this results in a Linear Complementary Problem (LCP) in a standard form. The meaning is clear: If the pins get into contact, we have constraint forces, otherwise not. Both pulley sets have one rotational, two translational in-plane and one axial out-of-plane degree of freedom. The pulley misalignment mainly depends on the transmission ratio i and the length of the chain. In order to limit the contact pressure between pins and sheaves the disc curvature radius R must have a lower bound, which in our example is 1.7m.

Chain Links Each link represents an elastic body, having three translational rigid body and in addition the elastic degrees of freedom. The angles β_L and γ_L, shown in Figure 50, depend kinematically on the translational position of the successor link. The links are kinematically interconnected between pairs of rocker pins. The elasticity and the translational damping of the joint is taken into account by the link force element whereas the rotational damping and the axial friction between the pair of pins is concentrated in the joint force element. In the link force element each plate is taken into account as a spring. The effect of moving contact points relative to the plate spring reference points between a rocker pin and an adjoining plate are modelled as a contact torque.

Rigid and Elastic Bodies The cassical model for both, rigid and elastic bodies is the equation

$$\mathbf{M}(\mathbf{q})\ddot{\mathbf{q}} - \mathbf{h}(\mathbf{q}, \dot{\mathbf{q}}, t) = \mathbf{0} \tag{25}$$

with $\mathbf{M}$ the mass matrix, with $\mathbf{q} = (\mathbf{q}_{rigid}^T, \mathbf{q}_{elastic}^T)^T$ the generalized coordinates. The vector $\mathbf{h}$ contains all forces. The elastic generalized coordinates $\mathbf{q}_{elastic}$ originate from a Ritz-approach ($\mathbf{r}_{elastic} = \mathbf{w}(\mathbf{r})^T \mathbf{q}_{elastic}$) describing in our case the linearly assumed deformations of an elastic component. These are pulleys, plates and pins. However it turned out, that at least the elastic pulleys can be approximated very well in a quasistatic way using Maxwell influence numbers [42], [36]. Elasticity of plates and pins is necessary to get a good agreement with measurements, especially in the spatial case, where tilting will be possible.

Contact Forces and Torques We have unilateral contacts for the pairings (pin ends)/(pulley surfaces) and pins/plates. For these contacts a combination of regularized contact functions, instead of set-valued force laws, and of complementarities was chosen. The combination follows the requirement of minimum computing times. As a result we only used complementarities for the contacts within those zones, where the chain enters or leaves the pulley. All other contacts are regularized. From this forces are evaluated straightforward by the force laws and by Lemke's algorithm for the complementarities. Two types of torques are taken into consideration, the restoring torques from the non-symmetric local pressure distributions generated in the contacts pins/pulley and from the kinematical process of rolling without sliding of rocker pins on rocker pins [36].

5.3.4 Models of the Push Belt

Contacts of the push belt elements are more complicated (see Figure 51). We have a unilateral spatial contact between the element and the pulley, a plane unilateral

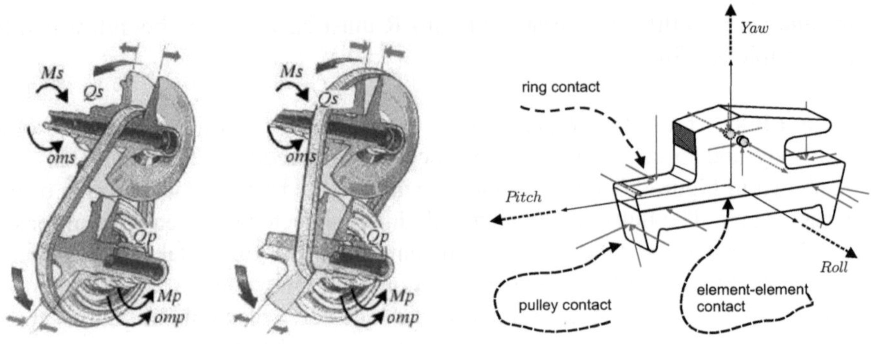

Figure 51: Contact Configurations of Push Belt Elements

contact between the element and the ring and five contacts between the elements themselves. The last ones are covered by an empirical nonlinear force law. All other contacts are unilateral contacts described by complementarities, then converted and solved by prox-functions [13].

Pulley Set The elastic pulley model follows more and less the same ideas as those used for the rocker pin chain. Pulley deformations and the contacts with the belt elements are similar. The pulley sheaves are modeled by rigid cones. For both cones we approximate the deformation by quasistatic force laws given by the Maxwell numbers. External excitations coming from the CVT environment and acting on the pulleys are taken into account. At the primary pulley a kinematic excitation is given by an angular velocity ω_{prim} (see Fig. 51). Accordingly, this pulley has no degree of freedom. At the secondary pulley a kinetic excitation is applied in form of an external torque M_{sec}. Accordingly, this pulley has one degree of freedom $\boldsymbol{q}_p = (\alpha_{sec})^T$, which is an angle of rotation. The pulley equations of motion write

$$\boldsymbol{M}_p \dot{\boldsymbol{u}}_p = \boldsymbol{h}_p + \boldsymbol{W}_p \boldsymbol{\lambda} \tag{26}$$

with the positive definite, constant and diagonal mass matrix $\boldsymbol{M}_p$. The vector $\boldsymbol{h}_p$ depends only on the time t: $\boldsymbol{h}_p = \boldsymbol{h}_p(t)$. Thus the matrices $\frac{\partial \boldsymbol{h}_p}{\partial \boldsymbol{q}_p}$ and $\frac{\partial \boldsymbol{h}_p}{\partial \boldsymbol{u}_p}$ used for the numerical integration are zero matrices.

Elements The elements are modeled by rigid bodies, describing each element m by three degrees of freedom $\boldsymbol{q}_m = (y_m, z_m, \alpha_m)^T$. The model of one single element is depicted in Fig. 52, where the center S of mass is determined by the translational positions y and z. The modeling of the whole CVT is performed in the plane containing the axes AA of all elements.

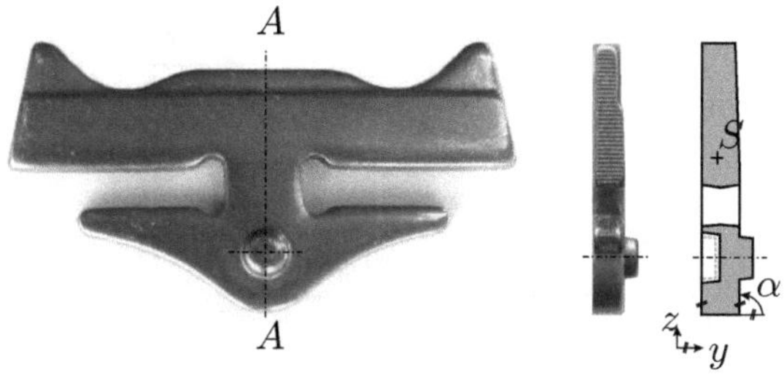

Figure 52: Model of an Element

The M elements are described by the generalized coordinates $\boldsymbol{q}_e = \left(\boldsymbol{q}_1^T, \ldots, \boldsymbol{q}_M^T\right)^T$, resulting in the equations of motion

$$\boldsymbol{M}_e \dot{\boldsymbol{u}}_e = \boldsymbol{h}_e + \boldsymbol{W}_e \boldsymbol{\lambda} \tag{27}$$

with the positive definite, constant and diagonal mass matrix $\boldsymbol{M}_e$. The vector $\boldsymbol{h}_e$ is constant and summarizes forces due to gravity. Thus the matrices $\frac{\partial \boldsymbol{h}_e}{\partial \boldsymbol{q}_e}$ and $\frac{\partial \boldsymbol{h}_e}{\partial \boldsymbol{u}_e}$ are zero matrices.

Ring Package For the plane model the two ring packages of the push belt are considered as one virtual ring package having double width. The 9 to 12 layers of each ring package are homogenized using (EA), (EI) and (Aρ) as parameters representing the longitudinal stiffness, the bending stiffness and the mass per length, respectively. The magnitudes E and ρ give the modulus of elasticity and the density, A and I denote the cross sectional area and the moment of inertia of the ring, which is treated as a one-dimensional continuum. Curling up for an unstressed configuration of the beam model is described by the curl-radius of the relaxed structure.

A finite element approach is used to describe the ring package elements with the coordinates $\mathbf{q}_i = (y_S, z_S, \varphi_S, \tilde{\varepsilon}, a_l, \beta_l, a_r, \beta_r)^T$. Several single finite elements are combined to construct a larger structural element of the ring. It has the global coordinate set $\mathbf{q}_g = (y_1, z_1, \varphi_1, a_1, a_2, y_2, z_2, \varphi_2)^T$. The first set y_S, z_S, φ_S describes the rigid body movements of the finite element, $\tilde{\varepsilon}$ gives an approximation for the longitudinal strain, and $a_l, \beta_l, a_r, \beta_r$ describe the bending deflections. The correlation of the two sets follows from the relation $\mathbf{q}_i = \boldsymbol{Q}(\mathbf{q}_g)$, the derivatives of which can be used to establish the equations of motion for the single element (i) and for the structural element (g). We come out with

$$\begin{aligned} \mathbf{M}_i(\mathbf{q}_i)\dot{\mathbf{u}}_i - \mathbf{h}_i(\mathbf{q}_i, \mathbf{u}_i, t) - \mathbf{W}_i(\mathbf{q}_i)\boldsymbol{\lambda} &= \mathbf{0}, \\ \mathbf{M}_g(\mathbf{q}_g)\dot{\mathbf{u}}_g - \mathbf{h}_g(\mathbf{q}_g, \mathbf{u}_g, t) - \mathbf{W}_g(\mathbf{q}_g)\boldsymbol{\lambda} &= \mathbf{0}. \end{aligned} \tag{28}$$

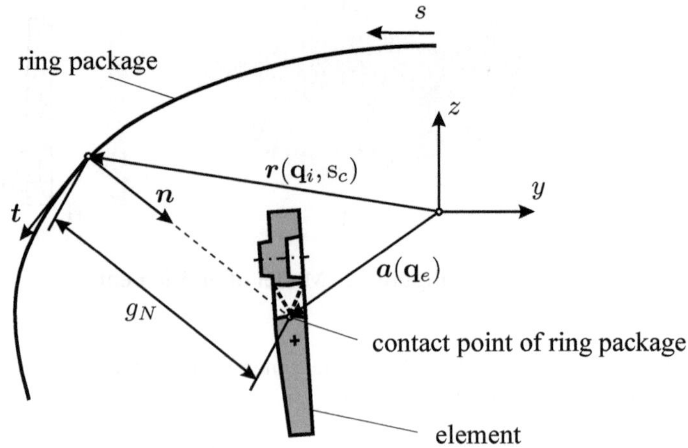

Figure 53: Contact between ring structure and element

The generalized velocity is $\mathbf{u} = \dot{\mathbf{q}}$. We may combine these two equations to give the complete set for the ring package ([13] and [14])

$$\mathbf{M}_r \dot{\mathbf{u}}_r = \mathbf{h}_r + \mathbf{W}_r \boldsymbol{\lambda}. \tag{29}$$

The constraints $\mathbf{W}_r \boldsymbol{\lambda}$ exist between the ring package and the belt elements.

Contacts Elements/Ring Package Constraints on elastic structures can be formulated using the schemes derived for rigid body constraints in classical multibody systems [44]. Figure 53 displays the contact points on the interacting bodies: the contact point is fixed on the element but sliding on the ring package, whereas the position is defined by the material coordinate s_c. It should be mentioned that uni- as well as bilateral constraints can be treated following the same geometrical concept. The constraint is specified by the corresponding proximal function [9],[20].

In case of the push belt, the contact between the ring package and the elements is assumed to be free of gaps in normal direction of the ring package, so that it can be modeled by a bilateral constraint. In tangential direction *Coulomb* friction is considered, which allows unilaterally defined stick-slip processes. Altogether we get the following equations:

$$\boldsymbol{\lambda}_{B,er} = \mathrm{prox}_{C_B}(\boldsymbol{\lambda}_{B,er} - r\boldsymbol{g}_{B,er}), \quad ; \quad \boldsymbol{\lambda}_{T,er} = \mathrm{prox}_{C_T(\boldsymbol{\lambda}_{B,er})}(\boldsymbol{\lambda}_{T,er} - r\dot{\boldsymbol{g}}_{T,er})$$

where the indices "B", "T" and "er" stand for bilateral, tangential and element/ring package, respectively.

Contacts Elements/Pulley The contacts between the elements and the pulley discs are unilateral in normal and in tangential directions. In spite of the fact that within the belt wrap no detachment of an element in normal direction takes place, these kind of events are very important for the belt/pulley zones, where the belt enters or leaves the pulley. We may have contact/detachment in normal and stick-slip by Coulomb friction in tangential directions. Therefore the contact laws write in these cases

$$\boldsymbol{\lambda}_{U,pe} = \text{prox}_{C_U}(\boldsymbol{\lambda}_{U,pe} - r\boldsymbol{g}_{U,pe}), \quad ; \quad \boldsymbol{\lambda}_{T,pe} = \text{prox}_{C_T(\boldsymbol{\lambda}_{U,pe})}(\boldsymbol{\lambda}_{T,pe} - r\dot{\boldsymbol{g}}_{T,pe})$$

Contacts Elements/Elements Figure 54 shows two elements being in contact. Only contact in normal direction is considered here, because tangential relative motion is circumvented by the guidance of the elements by the ring package and the nipple-hole connection between adjacent elements. The contact between adjacent elements is modeled by a rigid body contact

$$\boldsymbol{\lambda}_{U,ee} = \text{prox}_{C_U}(\boldsymbol{\lambda}_{U,ee} - r\boldsymbol{g}_{U,ee}) \tag{30}$$

with the gap distances $\boldsymbol{g}_{U,ee}$ between the considered contact points of the interacting elements.

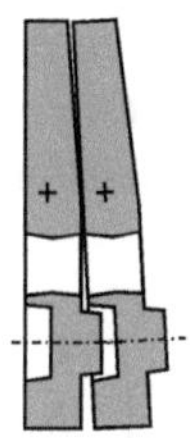

Figure 54: Contact between two elements

5.3.5 System Model

Within the scope of this article we shall not present the theory of unilateral multibody systems. We are dealing with the practical problems of CVT-chains and -belts and the corresponding models for representing the dynamics of such systems. The theory on unilateral systems may be found elsewhere, for example in [22], [25], [24], [16] or [10], and of course in many others.

The dynamics of the **rocker pin chain** will be represented by a set of the equations of motion (25) together with a second set of relative kinematics in normal and tangential directions and finally the set of all complementarity equations. This writes

$$
\begin{aligned}
&\boldsymbol{M}(\boldsymbol{q},t)\ddot{\boldsymbol{q}}(t) + \boldsymbol{h}(\boldsymbol{q},\dot{\boldsymbol{q}}t) - [(\boldsymbol{W}_N + \boldsymbol{W}_R) \quad \boldsymbol{W}_T] \begin{pmatrix} \boldsymbol{\lambda}_N(t) \\ \boldsymbol{\lambda}_T(t) \end{pmatrix} = \boldsymbol{0} \quad \in \mathbb{R}^f, \\
&\ddot{\boldsymbol{g}}_N = \boldsymbol{W}_N^T \ddot{\boldsymbol{q}} + \bar{\boldsymbol{w}}_N \quad \in \mathbb{R}^{n_N}, \\
&\ddot{\boldsymbol{g}}_T = \boldsymbol{W}_T^T \ddot{\boldsymbol{q}} + \bar{\boldsymbol{w}}_T \quad \in \mathbb{R}^{2n_T}, \\
&\ddot{\boldsymbol{g}}_N \geq \boldsymbol{0}, \qquad \boldsymbol{\lambda}_N \geq \boldsymbol{0}, \qquad \ddot{\boldsymbol{g}}_N^T \boldsymbol{\lambda}_N = 0, \\
&\boldsymbol{\lambda}_{T0} \geq \boldsymbol{0}, \qquad \ddot{\bar{\boldsymbol{g}}}_T \geq \boldsymbol{0}, \qquad \boldsymbol{\lambda}_{T0}^T \ddot{\bar{\boldsymbol{g}}}_T = 0.
\end{aligned} \tag{31}
$$

The abbreviations are clear, $\boldsymbol{q} \in \mathbb{R}^f$ are the generalized coordinates, $\boldsymbol{M} \in \mathbb{R}^{f,f}$ the symmetric and positive definite mass matrix, $\boldsymbol{W}$ the constraint matrices, $\boldsymbol{\lambda}$ the constraint forces, $\ddot{\boldsymbol{g}}$ the relative accelerations in the contacts, $\bar{\boldsymbol{w}}$ some external (excitation) terms and $\boldsymbol{\lambda}_{T0}$ the friction reserves with respect to the boundary of the friction cones [24].

The equations (31) for the dynamics of a rocker pin chain are solved numerically by evaluating the beginning or the end of a contact event, interpolating these points and solving the complementarity problems by Lemke's algorithm, rearranging the constraint matrices and treating the smooth equations between contact events by a Runge-Kutta 4/5 -scheme. In the meantime this method has been replaced by the prox-algorithms, which are more stable and safer and need considerably less computing time. The push-belt problem is evaluated that way.

The modeling of the **push belt CVT** by single components and their interconnections leads to the equations of motion for the plane overall system, consisting of the differential equations

$$
\begin{aligned}
\begin{pmatrix} \boldsymbol{M}_p & \boldsymbol{0} & \boldsymbol{0} \\ \boldsymbol{0} & \boldsymbol{M}_e & \boldsymbol{0} \\ \boldsymbol{0} & \boldsymbol{0} & \boldsymbol{M}_r \end{pmatrix} \begin{pmatrix} \dot{\boldsymbol{u}}_p \\ \dot{\boldsymbol{u}}_e \\ \dot{\boldsymbol{u}}_r \end{pmatrix} = \begin{pmatrix} \boldsymbol{h}_p \\ \boldsymbol{h}_e \\ \boldsymbol{h}_r \end{pmatrix} &+ \boldsymbol{W}_{B,er} \boldsymbol{\lambda}_{B,er} + \boldsymbol{W}_{T,er} \boldsymbol{\lambda}_{T,er} \\
&+ \boldsymbol{W}_{U,pe} \boldsymbol{\lambda}_{U,pe} + \boldsymbol{W}_{T,pe} \boldsymbol{\lambda}_{T,pe} \\
&+ \boldsymbol{W}_{U,ee} \boldsymbol{\lambda}_{U,ee}
\end{aligned} \tag{32}
$$

and the constraints:

$$
\begin{aligned}
&\boldsymbol{\lambda}_{B,er} = \mathrm{prox}_{C_B}(\boldsymbol{\lambda}_{B,er} - r\boldsymbol{g}_{B,er}) \quad ; \quad \boldsymbol{\lambda}_{T,er} = \mathrm{prox}_{C_T(\boldsymbol{\lambda}_{N,er})}(\boldsymbol{\lambda}_{T,er} - r\dot{\boldsymbol{g}}_{T,er}) \\
&\boldsymbol{\lambda}_{U,pe} = \mathrm{prox}_{C_U}(\boldsymbol{\lambda}_{U,pe} - r\boldsymbol{g}_{U,pe}) \quad ; \quad \boldsymbol{\lambda}_{T,pe} = \mathrm{prox}_{C_T(\boldsymbol{\lambda}_{N,pe})}(\boldsymbol{\lambda}_{T,pe} - r\dot{\boldsymbol{g}}_{T,pe}) \\
&\boldsymbol{\lambda}_{U,ee} = \mathrm{prox}_{C_U}(\boldsymbol{\lambda}_{U,ee} - r\boldsymbol{g}_{U,ee})
\end{aligned} \tag{33}
$$

The indices (p, e, r) stand for pulley, element and ring, respectively. The indices (U, B, T) indicate unilateral, bilateral and tangential, respectively. The $\boldsymbol{M}$ are the

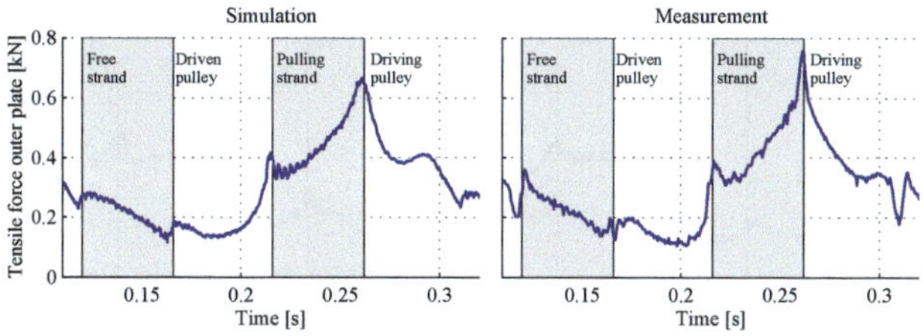

Figure 55: Tensile Force of an Outer Plate with Clasp Plate, misalignment Δz=+1.5mm

mass matrices, $\boldsymbol{W}$ the constraint matrices, $\boldsymbol{u} = \dot{\boldsymbol{q}}$ are velocities, and $\boldsymbol{\lambda}$ are constraint forces. The magnitudes $\boldsymbol{g}, \dot{\boldsymbol{g}}$ indicate relative displacements and velocities in the contacts, r is an auxiliary variable, a kind of slack variable (see [9]).

The equations (32) are solved numerically by a time stepping scheme including some prox-algorithms [13]. The modular configuration of the model comprising subsystems and constraints enables a refinement or even a substitution of models for single components and interactions in an easy manner. By this, both bodies and contacts can be modeled rigidly or flexibly in a hybrid way. The resulting differential equations have stiff character.

The research including Industry goes on. Extensions are more refinements with respect to spatial dynamics, first steps into design optimizations, better comparisons with the newest experiments, to give a few points. The dissertation [35] and some important publications of the same author characterize these progresses.

5.3.6 Results

Results for the Rocker Pin Chain The results of the dynamic model are computed for a uniform motion with constant driving speed n_1 and an external output torque T_2. Figures 55 and 56 show the tensile force of an outer plate of a chain with clasp plates for two different pulley misalignments (T=150Nm, n=600rpm, i=1). The comparison of simulation and measurement ([29], [37], [36], [38], [34]) confirms the mechanical model. Due to the bending forces the misalignment induces a large gradient of the tensile forces in the spans. Entering a pulley the shape of a pin changes abruptly because of the sudden growing contact forces. Thus especially at the beginnings and at the ends of contact arcs great force peaks appear.

In [29] many parameter variations are considered with the goal to show the possibility of influencing the chain performance significantly by even small changes of chain component geometry. For this purpose the paper concentrates only on the forces of the outer plates, which are the most stressed plates on one side, but represent also a good measure of the chain performance as a whole on the other side. With longer rocker pins and more plates the tensile forces of a plate can be

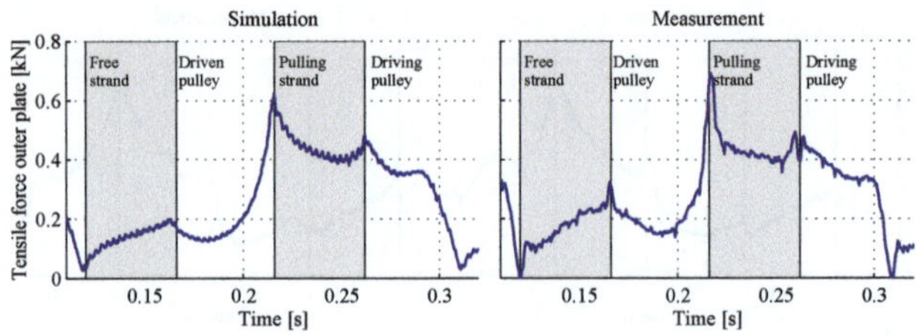

Figure 56: Tensile Force of an Outer Plate with Clasp Plate, misalignment Δz=-0.5mm

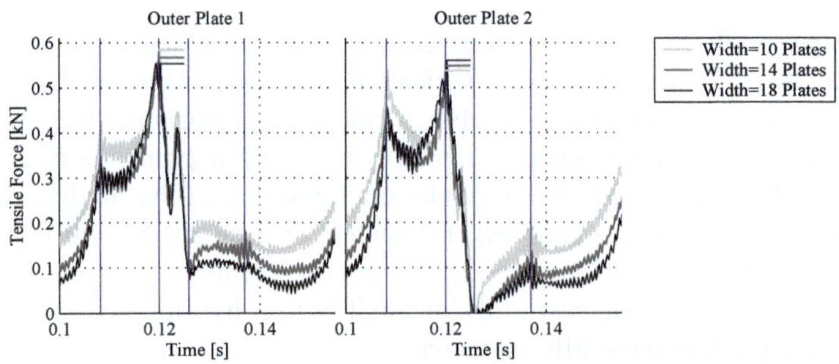

Figure 57: Variation of the Number of Plates of Each Link $(i_{red} = 1, i_{CVT} = 2.3, n_1 = 4000rpm)$

reduced because of a load distribution on more plates. But if we keep at the same time the cross section of the pins the same, parts of this reduction by load splitting are lost due to the larger pin bending. Thus the gain connected with more plates is not so large as often expected. Figure 57 gives an example.

Results for the Push Belt Rocker pin chains are manufactured by various companies with a certain focus at LUK, Germany. Push belt systems are mainly manufactured by VDT-Bosch in the Netherlands, which also operate production facilities in Japan. Measurements are up to now rare. Very good experiments have been performed by Honda in Japan (see [11], [12] and [19]), which are available by the cited papers, and also very good measurements have been performed by VDT-Bosch. Therefore we shall give some general results of simulations and a comparison of theory with the Honda measurements.

Geier [14] evaluated many cases by simulation and performed a rather comprehensive study of comparisons with measurements. We shall give only two examples. The first one depicts some simulations for the contacts between the

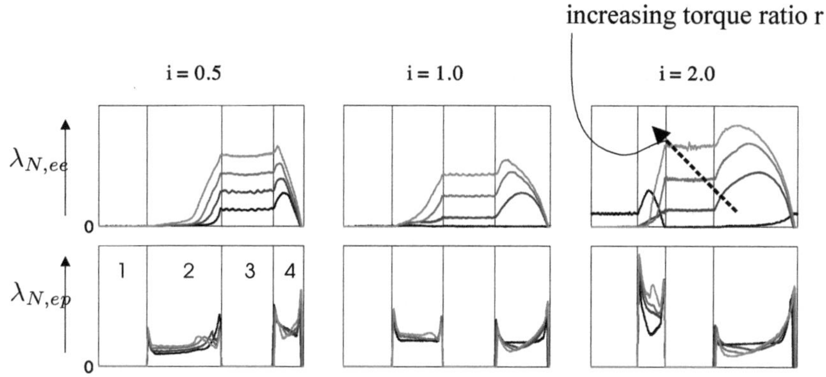

Figure 58: Simulation Results for two Contact Forces $\lambda_{N,ee}$, $\lambda_{N,ep}$, four Torque Ratios r and three speed ratios i (N=normal, ee=contact element/element, ep=contact element/pulley)

elements themselves and between pulley and element. Figure 58 illustrates the contact forces. Zone 1 corresponds to the free trum, zone 2 to the primary pulley, zone 3 to the pushing trum and zone 4 to the secondary pulley.

The force $\lambda_{N,ee}$ reflects the normal force in the element/element contact. This force is zero in the free trum and remains zero within the beginning of the primary pulley due to the fact, that the elements remain separated when entering the primary pulley. But then the contact force is built up and leaves the primary pulley with a value, which is maintained during the complete push trum phase making this trum to a kind of pressure bar. Entering the secondary pulley (zone 4) results in a maximum of the force $\lambda_{N,ee}$ firstly due to the influence of the pressure transmitted to the belt by the deformed pulley and secondly due to a prestressing effect by the ring package. This is all very plausible frome physical arguments and also confirmed by the simulations.

The contact force $\lambda_{N,ep}$ between pulley and element is of course zero for the two trums in the zones 1 and 3 and non-zero within the pulleys in the zones 2 and 4. The results illustrate the typical characteristics known also from rocker pin chains. In entering and leaving a pulley we always get a sharp rise of the contact forces between pulley and the elements, which are then a bit reduced within the pulley wrap arc. These forces do not depend very much on the torque ratio r, whereas the element/element contact force $\lambda_{N,ee}$ increases considerably with the torque ratio.

The second example taken from [14] concerns a comparison between simulations and the Honda measurements published in [12] and [19]. Figure 59 gives a comparison, which is very good. The measurements within the push trum are not constant but show a dip, which according to the papers ([12], [19]) does not

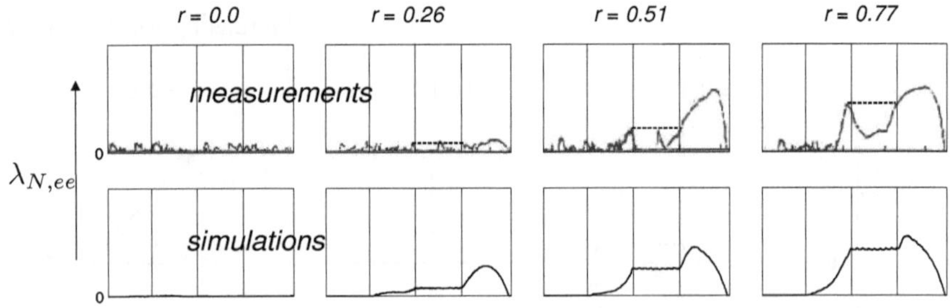

Figure 59: Verification of the Element/Element Contact Force $\lambda_{N,ee}$ for four Torque Ratios r (N=normal, ee=contact element/element)

correspond to reality. The measurements, which are really complicated to perform, could not give realistic values for that regime.

5.3.7 Realization of Objectives

As indicated at the beginning already the models, theories and algorithms accompanied during more than twenty years the product development of CVT-Gears at VDT, the Bosch-van Doorne company in Holland. Typical for such developments is the following procedure: a company starts with an innovation in form of an invention, pushing forward performance and price and thus reaching a certain market position. During that phase technology is frequently done by pure practicians mostly with an excellent feeling for the solution of problems. But such a process reaches limits making the use of more sophisticated and more scientific approaches necessary.

This has been the point, where we stepped into this CVT-development, at the very beginning also with not completely satisfying methods. From dissertation to dissertation the models were improved, mechanically and mathematically. A large step consisted in establishing the first models applying non-smooth dynamics to the system, which in one go brought more clearness and concision for all models to follow. An additional jump of numerics could be realized by set-oriented constraints, see [1] and [9].

With a growing degree of ripeness concerning the mechanical and mathematical models it was possible to look deeper into the CVT-processes: from plane models to spatial models, because such spatial models could explain much better disturbances of normal operation, for example out of plain dynamical effects; furthermore, increasing contact details of the elements with up to 17 local contacts per element, giving better information of noise and wear processes. Test results have been rare for a long time, because measurements are difficult and expensive. The situation has improved, and all comparisons theory/measurements are very convincing. In addition, verification efforts go on (see [35]).

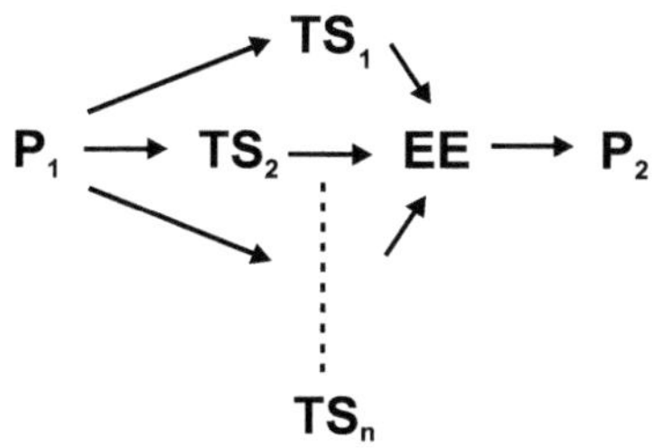

Figure 60: Popper's iteration process [31] (P for problem, TS for tentative solution, EE for error elimination)

The CVT-modeling is typical for the iterative character of large and long-lasting developments, where the objectives are not only developing a product and then producing it for a long time, but it is an incessant accompaniment of a continuously improved product: CVT becomes better with new problems, models become better with some new problems, CVT becomes better etc. Karl Popper [31] has expressed this process by the scheme of Figure 60.

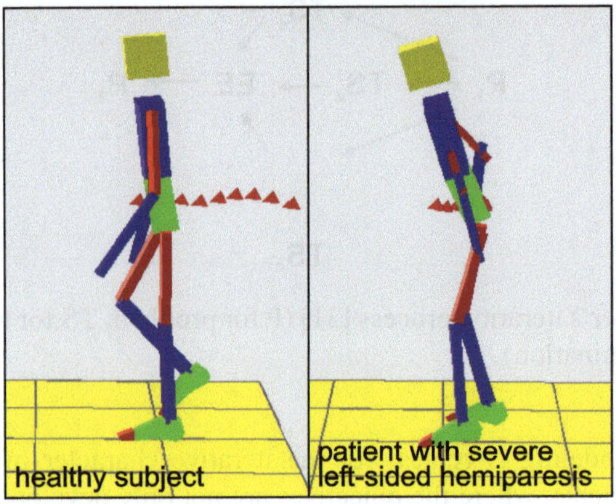

5.4 Example: Hemiparetic Walking

5.4.1 Objectives

Research on hemiparetic walking was part of a large research project of the German Research Foundation (DFG) including scientists from medicine, mechanical and electrical engineering. Hemiparese usualy is generated by a stroke and results in more or less severe problems of walking, amongst other possibilities. It is well known, that an improvement of the walking capabilities of a hemiparetic patient can be best realized during the first three months. Afterwards it becomes more and more difficult.

This problem has been the starting point for a research project with the goal to accelerate the treatment within these three months by establishing a control system for the doctor based on a data system. This data bank should collect all relevant data of the patient enabling the doctor to optimize his daily treatment. One important point of this overall system was a model of hemiparetic walking with an adaptation possibility for the individual data of the patient. This included also, as a new finding, to develop new measures for the degree of hemiparese. Measures could be joint torques or energies, which are magnitudes not being accessible for a measurement but only by modeling evaluations. Forces and torques at the feet can be measured, and some external kinematical body vectors can be measured, the rest must be done by a model.

5.4.2 Introduction

The idea of a model-based control of rehabilitation for hemiparetic patients requires efficient models of human walking, healthy walking as well as hemiparetic walking. Such models are presented including 42 degrees of freedom and allowing especially the evaluation of kinetic magnitudes with the goal to evaluate measures for the hardness of hemiparesis. As far as feasible the simulations have been compared successfully with measurements thus improving the confidence level for an application in clinical practice. The paper is mainly based on the dissertation [21], see also the paper [30]. Both contributions give a comprehensive collection of literature.

Hemiparesis with violent impairment of motricity of one side of the body occurs after stroke or skull-brain trauma. At the beginning and during rehabilitation hemiparetic patients undergo analysis of their gait. In clinical gait analysis the pathological state of the disturbed gait pattern is evaluated first qualitatively and quantitatively, in order to obtain a diagnosis and to dertermine a suitable therapy. Up to now mostly temporal variables such as stride length, gait speed, walking symmetry , kinematic quantities such as joint angles and electromyograms were considered. Despite a large amount of studies, the analysis of kinetic quantities like joint torques, power, work and ground reaction forces was considered only in few studies. The reason therefore resides in the great effort to determine kinetic quantities, because they cannot be measured directly but have to be calculated by inverse models using expensive force platforms. Kinetic quantities, however, enable a further refinement of diagnosis and a better understanding of pathological gait. Kinetic quantities significantly help to separate primary abnormalities (caused by the neurological deficit) from secondary abnormalities (adaptations to circumvent the primary).

The human locomotor apparatus is modeled as a three-dimensional system of multiple rigid bodies, connected by ideal ball-and-socket links. Segmentation of the human body respects the anatomy of the locomotor apparatus and the characteristics of human walking. Figure 61 shows the mechanical model, consisting of 13 segments and 12 links, each with 3 degrees of freedom (DOF). With 6 DOF for the trunk, defining its position and orientation in space, the model comprises 42 DOF. The joints are assumed frictionless. A coordinate frame is assigned to each segment, the x axis pointing in the anatomical (upright standing) position forward, the y axis upward and the z axis laterally. In the anatomical position, the coordinate axes of the segments are parallel. The sequence of rotation of each joint is internal/external rotation - adduction/abduction - flexion/extension. The origin of the coordinate system lies in the center of mass of each segment.

5.4.3 Mechanical and Mathematical Models

The segments are described by their mass, moment of inertia with respect to the body fixed coordinate axes, the vector to the center of mass and the vector to the distal link. Muscles are not taken into consideration, therefore the mechanical model of the human locomotor apparatus is just a skeletal one.

Several external forces and torques act on the human body when walking, see Figure 61. Ground reaction forces $\mathbf{F}_{gr}$ and torques $\mathbf{M}_{gr}$, which result from the interaction of the foot with the floor, act on the foot in contact. Joint torques $\mathbf{M}_j$ summarise all the torques which are generated by the joint spanning muscles.

The mass and inertia properties of the legs and arms are calculated by approximating their shapes with geometric bodies and using average body densities. By this means, the thigh, shank, upper and underarm are approximated by frustums with circular sections. For the trunk, pelvis, feet and head regression equations are used. By these means the general model of the locomotor apparatus is adapted to each patient. The use of this technique is obligatory because parameters of subjects vary within a wide range, e.g. hemiplegics after stroke who suffer from heavy reductions of the muscle mass in contrast to healthy well-trained persons.

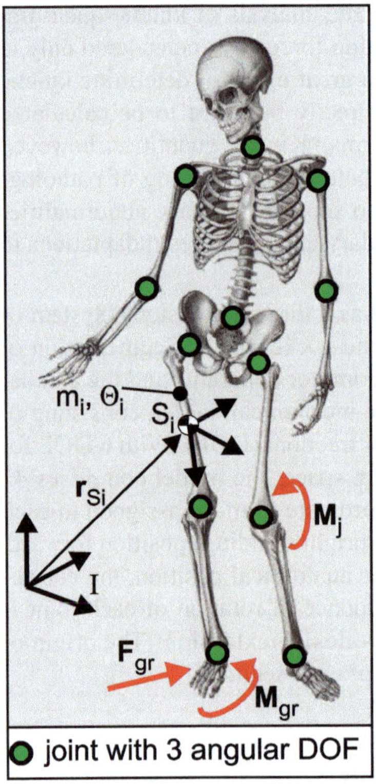

Figure 61: Mechanical model

The dynamics of the locomotor apparatus is described in configuration space by the equations of motion. The Newton-Euler method with the reference points in the centers of mass of the segments yields

$$\mathbf{M}(\mathbf{q},t)\ddot{\mathbf{q}}(t) - \mathbf{h}(\mathbf{q},\dot{\mathbf{q}},t) = \sum_{i=1}^{2}\left(\mathbf{J}_{T,i}^{T}\mathbf{F}_{gr,i} + \mathbf{J}_{R,i}^{T}\mathbf{M}_{gr,i}\right) + \sum_{k=1}^{12}\mathbf{J}_{R,k}^{T}\mathbf{M}_{j,k} \tag{34}$$

The vectors $\boldsymbol{q}, \dot{\boldsymbol{q}}, \ddot{\boldsymbol{q}} \in \mathbb{R}^{42}$ denote the generalized coordinates, velocities and accelerations, $\boldsymbol{M} \in \mathbb{R}^{42\times 42}$ is the mass matrix and accounts for the inertial properties of the system, $\boldsymbol{h} \in \mathbb{R}^{42}$ contains all gravitational and gyroscopic forces and torques. As above mentioned, muscle properties are not taken into consideration. Consequently the external forces acting on the multibody system, see Figure 61, are the ground reaction forces $\boldsymbol{F}_{gr,i} \in \mathbb{R}^3$ and torques $\boldsymbol{M}_{gr,i} \in \mathbb{R}^3$ and the 12 joint

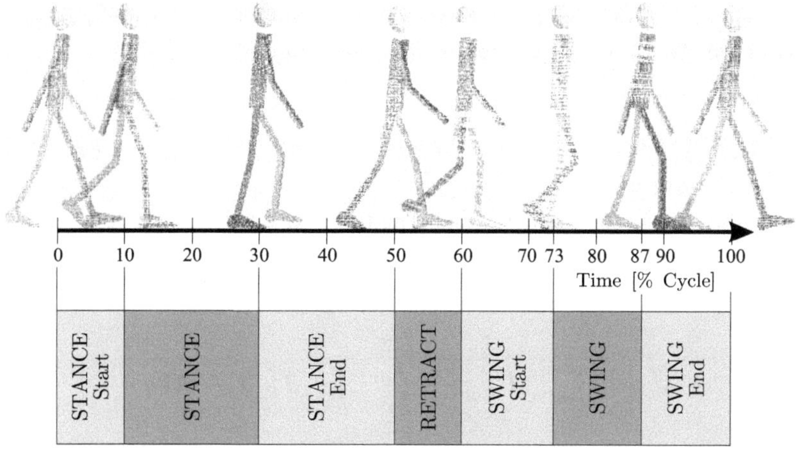

Figure 62: Human gait and its structure

torques $\boldsymbol{M}_{j,k} \in \mathbb{R}^3$, which are multiplied with the corresponding Jacobian matrices of translation $\boldsymbol{J}_T \in \mathbb{R}^{3\times 42}$ and rotation $\boldsymbol{J}_R \in \mathbb{R}^{3\times 42}$.

The equations of motion (eq. 34) can be solved in two different ways. First, assuming given forces and torques, the movement of the body is computed using numerical integration (direct dynamics). Unfortunately, control laws to compute forces and torques of normal and impaired human walking are unknown. Therefore only the second method, called inverse dynamics can be used. Inverse dynamics method supposes given movement and yields forces and torques.

Figure 62 depicts some typical details of the human gait pattern, which is characterized by single and double support phases. In single support phase, when only one foot is on the ground, we deal with one ground reaction force and one ground reaction moment. Assuming given generalized coordinates and their derivatives, equation 34 yields 42 known quantities on the left side. 3 components of $\boldsymbol{F}_{gr}$, 3 of $\boldsymbol{M}_{gr}$ and 3 of the 12 joint torques lead to 42 unknowns. We calculate the unknown quantities by matrix inversions and multiplications.

In double support phase, when both feet are on the ground, we deal with 2 unknown ground reaction forces and torques and 36 unknown joint torques, so that the number of 48 unknowns exceeds the number of available equations. Additional equations are delivered using optimization techniques, when we require that some objective functions have to be minimized. Since human walking is an optimized movement, not only the equations of motion have to be fulfilled, but also biological principles have to be taken into account. This method has already been applied to the walking cycle of a stick insect [8]. We assume the optimization criterion C linear in the squares of the unknown forces and torques:

$$C = f\left(\boldsymbol{F}_{gr}^2, \boldsymbol{M}_{gr}^2, \boldsymbol{M}_j^2\right) = \frac{1}{2}\boldsymbol{F}_{gr}^T \boldsymbol{C}_1 \boldsymbol{F}_{gr} + \frac{1}{2}\boldsymbol{M}_{gr}^T \boldsymbol{C}_2 \boldsymbol{M}_{gr} + \frac{1}{2}\boldsymbol{M}_j^T \boldsymbol{C}_3 \boldsymbol{M}_j \tag{35}$$

with the matrices $\boldsymbol{C}_i$ which weight the forces and torques.

The Lagrange multiplicators $\boldsymbol{\lambda}$ combine the criterion 35 with the equations of motion 34 and lead to the Lagrangian function L

$$L = C + \boldsymbol{\lambda}^T \left\{ \left[\boldsymbol{J}_{T,gr}^T \mid \boldsymbol{J}_{R,gr}^T\right] \begin{pmatrix} \boldsymbol{f}_{gr} \\ \boldsymbol{t}_{gr} \end{pmatrix} + \boldsymbol{J}_{R,j}^T \boldsymbol{t}_j - \boldsymbol{M}\ddot{\boldsymbol{q}} + \boldsymbol{h} \right\} . \tag{36}$$

According to Lagrange theory, a necessary condition for the minimisation of the criterion is that all partial derivatives of L with respect to the unknowns $\boldsymbol{f}_{gr}, \boldsymbol{t}_{gr}, \boldsymbol{t}_j$ and $\boldsymbol{\lambda}$ have to be zero. This leads to the following system of linear equations

$$\boldsymbol{J}\boldsymbol{f} = \boldsymbol{m} \tag{37}$$

$$\boldsymbol{f} = \left(\boldsymbol{f}_{gr}^T, \boldsymbol{t}_{gr}^T, \boldsymbol{t}_j^T\right)^T$$

which can easily be solved for the unknown ground reaction forces $\boldsymbol{f}_{gr}$, torques $\boldsymbol{t}_{gr}$ and joint torques $\boldsymbol{t}_j$ by inverting $\boldsymbol{J}$ [21], [30].

5.4.4 Measurements and Verification

The measurements performed with respect to healthy persons and hemiparetic patients have the goal firstly to evaluate all human parameters needed for the theoretical walking model and secondly to determine walking kinematics as a basis for inverse dynamics. All kinetic measures concerning hemiparesis can only be achieved that way. In the "Neurologische Klinik Bad Aibling", Germany, the disturbed gait patterns of hemiparetic patients are measured using the Optotrak System. This movement analysis system (see Figure 63) uses 58 active markers, which are applied on the skin of the patient. Two cameras measure the cartesian coordinates of the markers and a PC calculates relative angles between joints, using geometric relations. Joint velocities and accelerations are computed by differentiating twice joint angles and filtering with a 4th order Butterworth filter. The heel and toe markers are employed to determine stance and swing phase of the legs and thus single and double support phase.

Calculated ground reaction forces and joint torques show good agreement with measurements and calculations from literature. As an example Figure 64 shows

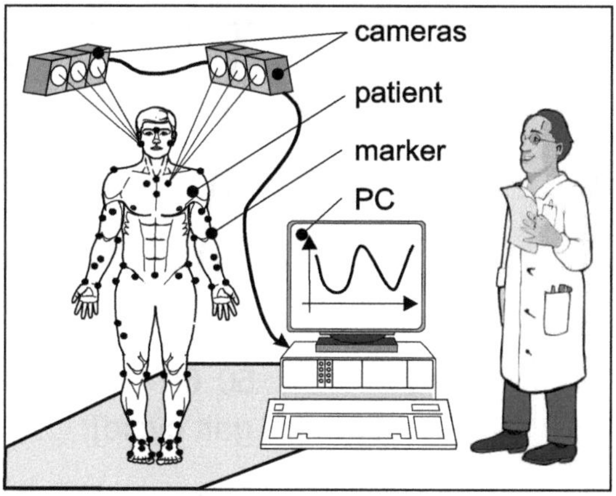

Figure 63: Measurement of human gait by a 3D motion analysis system

the calculated ground reaction force of a healthy subject during the single and double support phase of the right leg in comparison to data from other authors. The abscissa is time, normalized with the duration of the gait cycle and expressed in percent. A gait cycle starts with right heel contact (0%), the double stance phases are between 0% and 10% and between 50% and 60%. From 10% to 50% only the right foot and from 60% to 100% only the left foot is on the ground. The ordinate in Figure 64 is the ground reaction force normalized with body weight. The calculated ground reaction force in vertical direction (bold solid line in Figure 63) shows good agreement with measured data and calculations.

5.4.5 Results

The following sections show how the method of inverse dynamics is used in the above-mentioned neurological clinic for analysis of hemiparetic gait.

Animation Movement of the patient is automatically animated with “XAnimate”. The animation offers many advantages over common recording techniques like video tapes. The doctor or physical therapist can easily, quickly and at each time view the disturbed gait pattern under different angles and zoom on single segments. Furthermore the disturbed and a normal gait pattern can be opposed on the same screen and thus differences are quickly recognised. Animation of the body’s center of mass reveals excessive lateral movements of the patient. Two simple pictures of the animation of a healthy subject and a patient with severe left-sided hemiparesis are depicted in Figure 65.

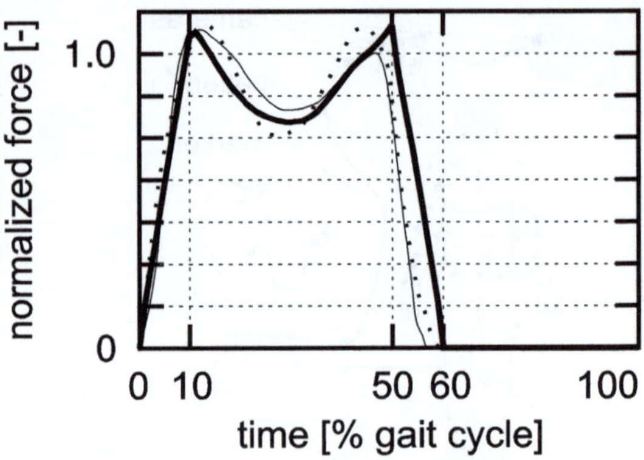

Figure 64: Vertical ground reaction forces (dotted line) and (solid line) in comparison with simulation results (bold solid line)

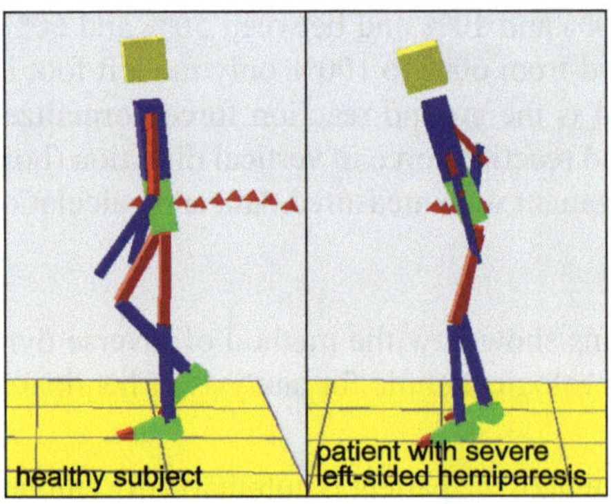

Figure 65: Animation of normal and hemiparetic gait

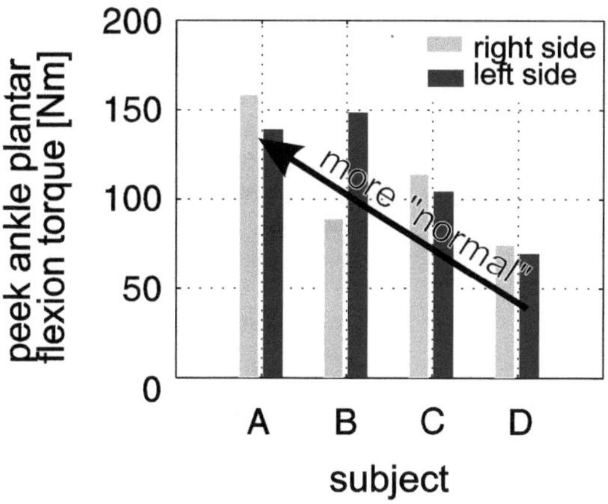

Figure 66: Peak ankle plantar flexion torque for 4 subjects

Evaluation of Degree of Hemiparesis and Outcome of Therapy Calculated kinetic quantities are used to quantify the degree of hemiparesis of the patients and to control outcome of therapy.

Figure 66 shows for example peek ankle plantar flexion torque for a healthy subject (A), patients with moderate (B) and severe (C and D) hemiparesis. One realises a close correlation between the degree of hemiparesis and this kinetic quantity. The more normal the gait pattern is, the higher and the more equal torques on the right and left side are.

Figure 67 depicts the calculated mechanical work in ankle for two steps for a hemiparetic patient at the beginning (D1), after 3 weeks (D2) and 7 weeks (D3) of physiotherapy in comparison to a healthy subject (A). One states an evident evolution of very low values of flexion work at the beginning to more normal values in course of rehabilitation. Here the outcome of therapy is evaluated by the method of inverse dynamics.

Detailed Analysis of One Patient's Gait Pattern Inverse dynamics method yields time course of all kinetic quantities for one gait cycle. These data are the base for a detailed analysis of the impaired gait. Time course of left knee flexion/extension torque for a healthy subject (solid line in Figure 68) and a patient with severe left-sided hemiparesis (dotted line in Figure 68) are for example studied. The patient exhibits in single support phase of the left leg (between 60 and 100% in gait cycle) a high and constant knee extension torque, whereas for the healthy subject the torque decreases very quickly to low values. Figure 69 shows

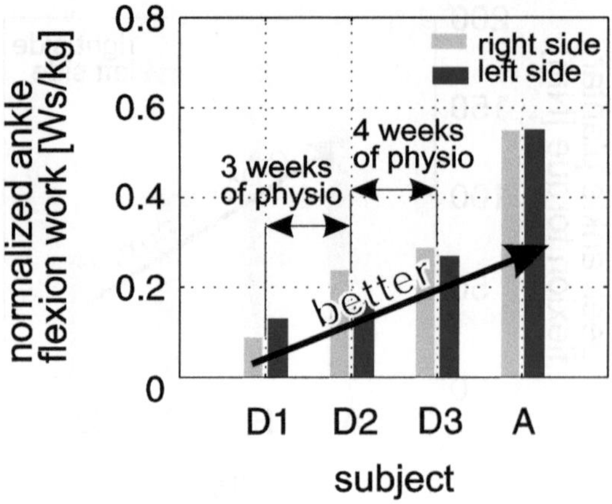

Figure 67: Normalized ankle flexion work in course of rehabilitation (physio=physiotherapy)

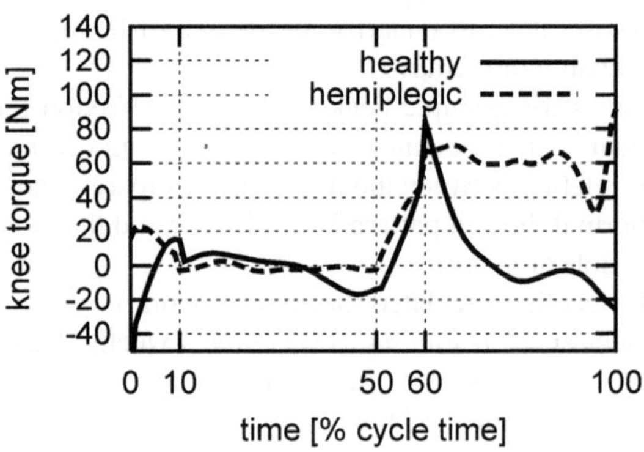

Figure 68: Left knee flexion/extension torque for a healthy subject (solid line) and a hemiplegic patient (dotted line)

the healthy subject and hemiparetic patient at 80% in gait cycle. In single support phase the left knee of the healthy subject is completely extended.

Due to the intact neural control of walking, the body is well controlled falling forward in order to reduce knee torques. In contrast to this perfect neural control of movement, the hip and knee of the patient, see Figure 69, are flexed, neural control

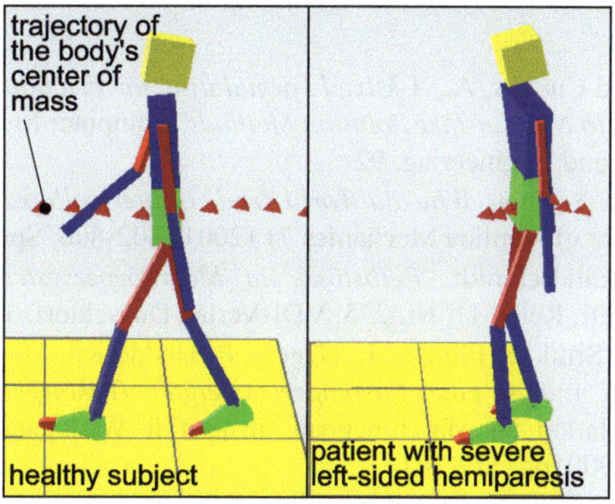

Figure 69: Animation of the two subjects of Figure 68

of movement is impaired and the patient has to generate a very high extension torque at the knee to maintain body upright and to progress. Here kinetic quantities give insight into impaired neural control of movement.

Kinematic Characteristics of Hemiparetic Gait Furthermore it was found that hemiparetic gait is characterised by excessive lateral movement of body's center of mass, high lateral ground reaction forces on affected side, low peak torques in all joints of the lower extremity in comparison to healthy subjects, lower peak torques on the affected than on the unaffected side and that energy for propulsion of the body results mainly from the unaffected side.

5.4.6 Realization of Objectives

The idea of better hemiparetic treatment and control was and still is a success. The application of electro- and magnetostimulation can be controled in a better way producing reduced stress to the patient. Establishing an overall control system is of course still a matter of ongoing research, but the basic concept turns out to be correct. The new measures for the degree of hemiparese by evaluating measurements in combination with a verified hemiparetic model are really an innovative tool to get a better assessment of a hemiparetic disease.

Bibliography

[1] Alart, P. and Curnier, A., *A Mixed Formulation for Frictional Contact Problems Prone to Newton-Like Solution Methods*. Computer Methods in Applied Mechanics and Engineering, 92

[2] Z.P. Bazant, Y. Zhou, *Why did World Trade Center collapse? - Simple analysis*. Archive of Applied Mechanics 71 (2001) 802-806, Springer 2001

[3] Michael Beitelschmidt, *Reibstösse in Mehrkörpersystemen.* Fortschritt-Berichte VDI, Reihe 11, Nr. 275, VDI-Verlag Düsseldorf, 1999

[4] H. Bork, J. Srnik, F. Pfeiffer, E. Negele, R. Hedderich, *Modellbildung, Simulation und Analyse eines leistungsverzweigten Traktorgetriebes*. Tagungsband Simulation im Maschinenbau, Institut fr Werkzeugmaschinen, TU-Dresden, 2000, pp. 329-347

[5] M. Bullinger, *Dynamik von Umschlingungsgetrieben mit Schubgliederband*, Fortschritt-Berichte VDI, Reihe 12, Nr. 593, VDI-Verlag Düsseldorf 2005

[6] Hans Dresig, *Schwingungen mechanischer Antriebssysteme.* Springer, Berlin, Heidelberg, 2001

[7] P.E. Duncan, *Simple models for the dynamics of deepwater gravity platforms.* Eng. Struct., 1979, Vol 1, 65-72

[8] Jürgen Eltze, *Biologisch orientierte Entwicklung einer sechsbeinigen Laufmaschine.* Fortschritt-Berichte VDI, Reihe 17, Nr. 110, VDI-Verlag Düsseldorf, 1994

[9] M. Foerg, L. Neumann, H. Ulbrich, *r-Factor Strategies for the Augmented Lagrangian Approach in Multi-Body Contact Mechanics*, III European Conference on Computational Mechanics, Lisbon, 2006

[10] M.Foerg, *Mehrkörpersysteme mit mengenwertigen Kraftgesetzen - Theorie und Numerik*, dissertation TU-Munich, Lehrstuhl für Angewandte Mechanik, 2007

[11] Fujii, T., Kurakawa, T., Kanehara, S., *A Study of Metal-Pushing V-Belt Type CVT - Part 1: Relation between Transmitted Torque and Pulley Thrust*, Int. Congress and Exposition Detroit, SAE Technical Paper Series, Nr. 930666, pp. 1-11, 1993

[12] Fujii, T.,Takemasa, K., Kanehara, S., *A Study of Metal-Pushing V-Belt Type CVT - Part 2: Compression Force between Metal Blocks and Ring Tension*, Int. Congress and Exposition Detroit, SAE Technical Paper Series, Nr. 930667, pp. 13-22, 1993

[13] Th. Geier, M. Foerg, R. Zander, H. Ulbrich, F. Pfeiffer, A. Brandsma, A. van der Velde, *Simulation of a Push Belt CVT considering uni- and bilateral Constraints*. Second Int. Conf. on Nonsmooth/Nonconvex Mechanics with Applications in Engineering, Thessaloniki, 2003 (ZAMM **86**, No.10, 795-806 (2006)/**DOI** 10.1002/zamm.200610287)

[14] Thomas Geier, *Dynamics of Push Belt CVTs*. Fortschritt-Berichte VDI, Reihe 12, Nr. 654, VDI-Verlag Düsseldorf, 2007
[15] Chr. Glocker, *Dynamik von Starrkörpersystemen mit Reibung und Stössen*. Fortschritt-Berichte VDI, Reihe 18, Nr.182, VDI-Verlag Düsseldorf, 1995
[16] Chr. Glocker, *Set-Valued Force Laws - Dynamics of Non-Smooth Systems*. Springer Berlin, Heidelberg, New York, 2001
[17] Chr. Glocker, *Energetic consistency conditions for standard impacts Part I: Newton-type inequality impact laws and Kanes example*. Multibody Syst Dyn (2013) 29:77117
[18] Chr. Glocker, *Energetic consistency conditions for standard impacts Part II: Poisson-type inequality impact laws*. Multibody Syst Dyn (2014) 32:445509
[19] Kanehara, S., Fujii, T., Kitagawa, T., *A Study of Metal-Pushing V-Belt Type CVT - Part 3: What Forces act on Metal Blocks*, Int. Congress and Exposition Detroit, SAE Technical Paper Series, Nr. 940735, pp. 95-105, 1994
[20] R. Leine, H. Nijmeijer, *Dynamics and Bifurcations of Non-Smooth Mechanical Systems*. Springer Berlin, Heidelberg, New York, 2004
[21] Christian Lutzenberger, *Dynamik des menschlichen Ganges*. Dissertation TU-München, Lehrstuhl für Angewandte Mechanik, 2001
[22] J.J. Moreau, *Unilateral Contact and Dry Friction in Finite Freedom Dynamics*, Volume 302 of International Centre for Mechanical Sciences, Courses and Lectures. J.J. Moreau P.D. Panagiotopoulos, Springer, Vienna (1988)
[23] John G. Papastavridis, *Analytical Mechanics*. Oxford University Press, Oxford, New York, 2002
[24] F. Pfeiffer, *Mechanical System Dynamics*, Springer, Berlin, Heidelberg, 2009
[25] F. Pfeiffer, Chr. Glocker, *Multibody Dynamics with Unilateral Contacts*. John Wiley & Sons, INC., New York,1996
[26] F. Pfeiffer, Th. Schindler *Introduction to Dynamics*. Springer Berlin Heidelberg 2015
[27] F. Pfeiffer, A. Stiegelmeyr *Damping Towerlike Structures by Dry Friction*. Proc. of DETC '97, ASME Design Eng. Techn. Conf., 1997
[28] F. Pfeiffer, *Applications of Unilateral Multibody Dynamics*, Phil. Trans. of the Royal Society, Vol. 359, Nr. 1789, pp. 2609-2628, 2001
[29] F. Pfeiffer, W. Lebrecht, T. Geier, *State of the Art of CVT-Modelling*, paper 04CVT-46, CVT 2004 Congress, San Francisco (2004)
[30] F. Pfeiffer, E. König, *Modeling Normal and Hemiparetic Walking*. Digital Human Modeling Conference, Mnchen, 18.-22.6.2002, VDI-Berichte 1675, pp. 33-43, 2002
[31] Karl Popper, *Objective Knowledge*. Clarendon Press Oxford, 1972
[32] Günther Ropohl, *Eine Systemtheorie der Technik*. Hanser Verlag, München, Wien, 1979
[33] H. Sattler, *Stationäres Betriebsverhalten stufenlos verstellbarer Metallumschlingungsgetriebe*, Dissertation, Universität Hannover, 1999

[34] G. Sauer, *Grundlagen und Betriebsverhalten eines Zugketten-Umschlingungsgetriebes*, Fortschritt-Berichte VDI, Reihe 12, Nr. 293, VDI-Verlag, Düsseldorf 1996
[35] Th. Schindler, *Spatial Dynamics of Pushbelt CVTs*. Fortschritt-Berichte VDI, Reihe 12, Nr. 730, VDI-Verlag Düsseldorf 2010
[36] M. Sedlmayr, *Räumliche Dynamik von CVT-Keilkettengetrieben*. Fortschritt-Berichte VDI, Reihe 12, Nr. 558, VDI-Verlag Düsseldorf 2003
[37] M. Sedlmayr,M., Pfeiffer, F., *Spurversatz bei CVT-Ketten*, VDI-Berichte 1630 "Schwingungen in Antrieben 2001", p.117–136,2001
[38] Sedlmayr, M.; Pfeiffer, F., *Spatial Contact Mechanics of CVT Chain Drives*, 18th ASME Bien. Conf. on Mech. Vibration and Noise DETC01/VIB. 2001.
[39] Martin Sedlmayr, *Räumliche Dynamik von CVT-Keilkettengetrieben*. Fortschrittberichte VDI, Reihe 12, Nr. 558, VDI-Verlag Düsseldorf, 2003
[40] von Seherr-Toss, Hans-C., Schmelz, F., Aucktor, E., *Universal Joints and Driveshafts - Analysis, Design, Applications*. second, enlarged edtion, Springer-Verlag Berlin Heidelberg New York, 2006
[41] Ahmed Shabana *Dynamics of multibody systems*. Cambridge University Press, New York, 2005
[42] J. Srnik, *Dynamik von CVT-Keilkettengetrieben*, Fortschritt-Berichte VDI, Reihe 12, Nr. 372, VDI-Verlag Düsseldorf 1999
[43] Peter Wolfsteiner, *Dynamik von Vibrationsförderern*. Fortschrittberichte VDI,Reihe 2, Nr. 511, VDI-Verlag Düsseldorf, 1999
[44] R. Zander, H. Ulbrich, *Reference-free mixed FE-MBS approach for beam structures with constraints*, Nonlinear Dynamics, Kluwer Academic Publishers, Dordrecht Netherlands 2006

Fascination of Making Models Truth-Reality-Illusion?

Hartmut Bremer
Johannes Kepler Universität Linz, Austria

1 Introduction

Evidently, human thinking is based on an imagination of the surrounding world which one is going to explore, an abstraction, a model. Such a model, if successful, enables to understand what is really going on – reality, that is what we see and it might be the truth, but observation and imagination can also lead to confusion and illusion.

The basic aim of modeling is explanation. Explanation leads to reconstruction and prediction. The probably easiest way of reconstruction is to build a hardware model and to see what is going on under certain circumstances – like a ship model in a stormy sea pool: the Pamir [42]. This is enough to make a movie, to create an illusion, and perhaps to shed light on the one or the other reason of its accident. But is not enough to generalize these results for further applications and predictions. Here, mathematics comes into play. The hardware model then serves for experimental verification of theoretical physical modeling – like Prandtl's water channel, or wind tunnels, indispensable till today. Verification is the last step in modeling: arbiter is always nature.

Hardware modeling is sometimes not possible, like Kepler's satellite system. Then, physical modeling is pure abstraction, based on observation. The physical model states the ongoing process, but not yet its explanation. Here, Hooke comissioned Newton to calculate the assumption that gravity acts reciprocal to the square of the mass center distances – probably the best model ever – and with Euler's momentum theorem, the problem became generalizable. Physical modeling and mathematical modeling go hand in hand.

Modeling, however, contains its danger. This is, for instance, the case in politics and economics when a model, unconsciously or consciously, misguides people (how did pope Urban VIII get Galileo into prison?) But also

F. Pfeiffer and H. Bremer (eds.), *The Art of Modeling Mechanical Systems*,
CISM International Centre for Mechanical Sciences 570,
DOI 10.1007/978-3-319-40256-7_2

in technical applications when one believes on a-priori known results where they actually do not arise, like chaotic motions in a double pendulum with small amplitudes – and worse: to conclude that a double pendulum can therefore not be calculated [11]. If so, with the double pendulum as a basic robot model, this presentation would already have come to an end.

Nearer inspection of the double pendulum, however, and assumimg slender arms, leads to the question if elasticity will play a significant role.

2 Elasticity – Modeling Problems

The effect of elasticity is impressively demonstrated with a linear robot (i.e. only translational rigid body motions) where the aim (pick and place) cannot be achieved even allusively with common joint control [28].

To obtain insight into the problems of modeling, the spatial (fast) rigid body motion is decomposed into uniaxial ones. Most problems arise from rotations (linear motions are considered later). This reveals a list of model inaccuracies and/or faults in modeling elastic systems.

One common procedure to deal with elasticity is based on Hamilton's principle (which, according to Felix Klein [30], was already known to Lagrange):

$$\int_{t_o}^{t_1} [\delta T - \delta V]\, dt = 0 \tag{1}$$

(T: kinetic enery, V: potential energy).

2.1 Rotating Beams

Considering a rotating bending beam one has

$$\begin{aligned} \delta T &= \int_o^L \left(\mathbf{v}_c^T \rho A\, \delta \mathbf{v}_c + \boldsymbol{\omega}_c^T \rho \mathbf{I}\, \delta \boldsymbol{\omega}_c\right) dx \\ \delta V &= \int_o^L (EI_z v'' \delta v'' + EI_y w'' \delta w'')\, dx \end{aligned} \tag{2}$$

where ρ: mass density, A: cross sectional area, $\mathbf{I}$: tensor of area moments of inertia, $\mathbf{v}_c, \boldsymbol{\omega}_c$: translational and rotational velocity w.r.t. the mass center c of the beam element; EI_z, EI_y: bending rigidities referred to the deflection functions $v(x,t), w(x,t)$. Prime indicates differentiation: $(\ )' = \partial(\ )/\partial x$.

Longitudinal rotation. Rotation around the extensional axis x with $\dot{\alpha}(t)$ yields the velocities (up to second order)

$$\mathbf{v}_c = \begin{pmatrix} 0 \\ \dot{v} - \dot{\alpha} w \\ \dot{w} + \dot{\alpha} v \end{pmatrix}, \boldsymbol{\omega}_c = \begin{bmatrix} \left(1 - \frac{v'^2}{2} - \frac{w'^2}{2}\right) & v' & 0 \\ -v' & \left(1 - \frac{v'^2}{2}\right) & 0 \\ -w' & 0 & 1 \end{bmatrix} \begin{pmatrix} \dot{\alpha} \\ -\dot{w}' \\ \dot{v}' \end{pmatrix} \tag{3}$$

where $\mathbf{v}_c$ refers to the rotating reference frame while $\boldsymbol{\omega}_c$ is resolved in an element fixed frame due to the herein constant inertia tensor. Cardan angles (Tait-Bryan-angles) $[\alpha \to \beta = -w' \to \gamma = v']$ are chosen for description. The dot represents time derivation.

The calculation is (although a bit cumbersome) as usual: carry out variation, integrate by parts w.r.t. time

$$\begin{aligned} & \int_{t_o}^{t_1} \left[\delta T - \delta V\right] dt = \int_{t_o}^{t_1} \left[\int_{o}^{L} \{\delta\alpha[-\rho I_x \ddot{\alpha}] \right. \\ + \quad & \delta v \left[\rho A(-\ddot{v} + 2\dot{\alpha}\dot{w} + \ddot{\alpha} w + \dot{\alpha}^2 v)\right] \\ + \quad & \delta w \left[\rho A(-\ddot{w} - 2\dot{\alpha}\dot{v} - \ddot{\alpha} v + \dot{\alpha}^2 w)\right] \\ + \quad & \underline{\delta v' \left[-\rho I_z \ddot{v}' + \rho(I_z + I_y - I_x)\dot{\alpha}\dot{w}' + \rho(I_y - I_x)\dot{\alpha}^2 v' + \rho I_z w' \ddot{\alpha}\right]} \\ + \quad & \underline{\delta w' \left[-\rho I_y \ddot{w}' - \rho(I_z + I_y - I_x)\dot{\alpha}\dot{v}' + \rho(I_z - I_x)\dot{\alpha}^2 w' + \rho(I_x - I_y) v' \ddot{\alpha}\right]} \\ - \quad & \left. \underline{\underline{\delta v'' E I_z v''}} - \underline{\underline{\delta w'' E I_y w''}} \} \, dx \right] dt + \Big[\cdots \Big]_{t_o}^{t_1} = 0. \end{aligned} \tag{4}$$

Next, set the δ's at the time boundaries equal to zero. The time integral is then obsolete. Integrate by parts the underlined part once and the double underlined part twice w.r.t. space and order according to the δ's to come out with an integral part and a boundary part

$$
\begin{aligned}
\int_o^L \Big[& \delta\alpha[-\rho I_x \ddot{\alpha}] \\
+ \quad \delta v \quad & \Big\{ \rho A(-\ddot{v} + 2\dot{\alpha}\dot{w} + \ddot{\alpha} w + \dot{\alpha}^2 v) \\
& + \rho I_z \ddot{v}'' - \rho(I_z + I_y - I_x)\dot{\alpha}\dot{w}'' - \rho(I_y - I_x)\dot{\alpha}^2 v'' \underline{- \rho I_z w'' \ddot{\alpha}} \\
& -(EI_z v'')'' \Big\} \\
+ \quad \delta w \quad & \Big\{ \rho A(-\ddot{w} - 2\dot{\alpha}\dot{v} - \ddot{\alpha} v - \dot{\alpha}^2 w) \\
& + \rho I_y \ddot{w}'' + \rho(I_z + I_y - I_x)\dot{\alpha}\dot{v}'' - \rho(I_z - I_x)\dot{\alpha}^2 w'' - \rho(I_x - I_y) v'' \ddot{\alpha} \\
& - (EI_y w'')'' \Big\} \Big] \, dx,
\end{aligned} \tag{5}
$$

$$
\begin{aligned}
+ \Big[\{ \quad \delta v \quad & \big[-\rho I_z \ddot{v}' + \rho(I_z + I_y - I_x)\dot{\alpha}\dot{w}' + \rho(I_y - I_x)\dot{\alpha}^2 v' \underline{+ \rho I_z w' \ddot{\alpha}} \\
& + (EI_z v'')' \big] \} \\
+ \{ \quad \delta w \quad & \big[-\rho I_y \ddot{w}' - \rho(I_z + I_y - I_x)\dot{\alpha}\dot{v}' + \rho(I_z - I_x)\dot{\alpha}^2 w' + \rho(I_x - I_y) v' \ddot{\alpha}) \\
& + (EI_y w'')' \big] \} \\
+ \{ \quad \delta v' \quad & (-EI_z v'') \} \\
+ \{ \quad \delta w' \quad & (-EI_z w'') \} \Big]_o^L = 0.
\end{aligned} \tag{6}
$$

Assuming $\delta v, \delta w, \delta\alpha$ independent yields eq.(5) for motion equations along with eq.(6) for boundary conditions. In order to simplify matters, let $I_y = I_z = I,\ \ I_x = 2I$ (circular or square cross section) to obtain

$$
\begin{aligned}
\Big[\rho A(\ddot{v} - 2\dot{\alpha}\dot{w} \underline{- \ddot{\alpha} w} - \dot{\alpha}^2 v) - \rho I \ddot{v}'' - \rho I \dot{\alpha}^2 v'' \underline{\underline{+ \rho I w'' \ddot{\alpha}}} + (EI_z v'')'' \Big] &= 0, \\
\Big[\rho A(\ddot{w} + 2\dot{\alpha}\dot{v} \underline{+ \ddot{\alpha} v} - \dot{\alpha}^2 w) - \rho I \ddot{w}'' - \rho I \dot{\alpha}^2 w'' \underline{\underline{+ \rho I v'' \ddot{\alpha}}} + (EI_y w'')'' \Big] &= 0.
\end{aligned} \tag{7}
$$

(Notice that the sign is changed to obtain $+\rho A \ddot{v}$. Furthermore, the α-equation is decoupled from the remainder. It reads, replenished with the drive torque, $\left[\int_o^L \rho I_x dx\right] \ddot{\alpha} - M_x = A\,\ddot{\alpha} - M_x = 0$.) The result according to eq.(7) is a bit strange: same as for the (underlined) translational part one would expect skew symmetry for the (double underlined) rotational motion.

The question is: are $\delta v, \delta w, \delta \alpha$ really independent?

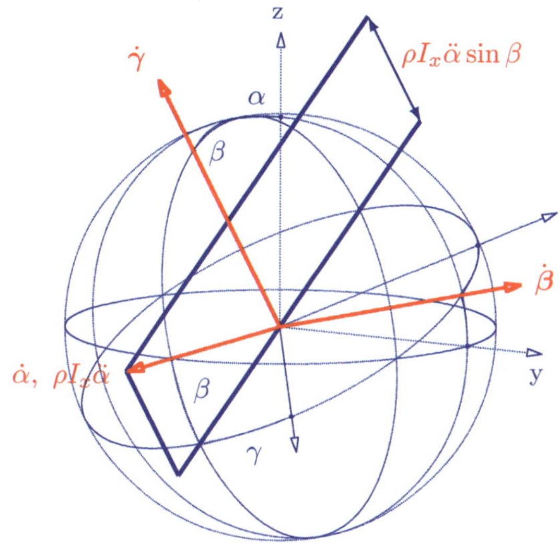

Figure 1. On orthogonality

As Figure 1 shows, the $\dot{\alpha}$-axis is not perpendicular to the $\dot{\beta}$-$\dot{\gamma}$-plane. Assuming small deflection amplitudes, one has $\rho I_x \ddot{\alpha}(\sin\beta) \simeq \rho I_x \ddot{\alpha}(-w')$ and $\rho I_x \ddot{\alpha}(\cos\beta) \simeq \rho I_x \ddot{\alpha}$. Eq.(5) has thus to be replenished with $\delta W_{add} = \int_o^L \rho I_x \ddot{\alpha}(-w')\delta v' dx = \int_o^L \rho I_x \ddot{\alpha} w'' \delta v dx - [\rho I_x \ddot{\alpha} w' \delta v]_o^L$. The underlined term in eq.(5) thus changes to $\int_o^L \delta v \left[-\rho(I_z - I_x)\ddot{\alpha} w''\right] dx$, or, for $I_z = I, I_x = 2I$, to $\int_o^L \delta v \left[+\rho I \ddot{\alpha} w''\right] dx$, yielding for eq.(7)

$$\begin{aligned}
&\left[\rho A(\ddot{v} - 2\dot{\alpha}\dot{w} - \ddot{\alpha} w - \dot{\alpha}^2 v) - \rho I \ddot{v}'' - \rho I \dot{\alpha}^2 v'' \underline{\underline{- \rho I w'' \ddot{\alpha}}} + (EI_z v'')''\right] = 0,\\
&\left[\rho A(\ddot{w} + 2\dot{\alpha}\dot{v} + \ddot{\alpha} v - \dot{\alpha}^2 w) - \rho I \ddot{w}'' - \rho I \dot{\alpha}^2 w'' \underline{\underline{+ \rho I v'' \ddot{\alpha}}} + (EI_y w'')''\right] = 0,
\end{aligned} \tag{8}$$

along with the corrected boundary terms.

Usually, Hamilton's Principle is well established for the calculation of elastic bodies, and one would never doubt the corresponding results. What is going on here, however, is due to the fact that the elastic deflections are assumed small but the rigid body motions are not. One has thus to be careful in linearization.

The beam under consideration is called a Rayleigh beam. One gets rid of such kind of difficulties when passing over to a so called Euler-Bernoulli-beam, where the inertia moments are negligible:

$$\begin{aligned} \left[\rho A(\ddot{v} - 2\dot{\alpha}\dot{w} - \ddot{\alpha}w - \dot{\alpha}^2 v) + (EI_z v'')''\right] &= 0, \\ \left[\rho A(\ddot{w} + 2\dot{\alpha}\dot{v} + \ddot{\alpha}v - \dot{\alpha}^2 w) + (EI_y w'')''\right] &= 0. \end{aligned} \tag{9}$$

However, do these equations still represent reality? *Physical model needs intuition.*

"The Swedish engineer Laval (1845–1913) was the first one to show, by practical experiments in 1889, that a shaft can rotate much faster than people at that time believed. When Laval's results came to light the reaction was disbelief everywhere" [18].

Changing notations to those usually used in rotor dynamics one can easily deduce from eq.(9) the equations for the first bending mode [$\ddot{v} \to \ddot{x}, \ddot{w} \to \ddot{y}$ etc.) along with $\dot{\alpha} = \Omega = const.$ (rotational symmetry provided)]

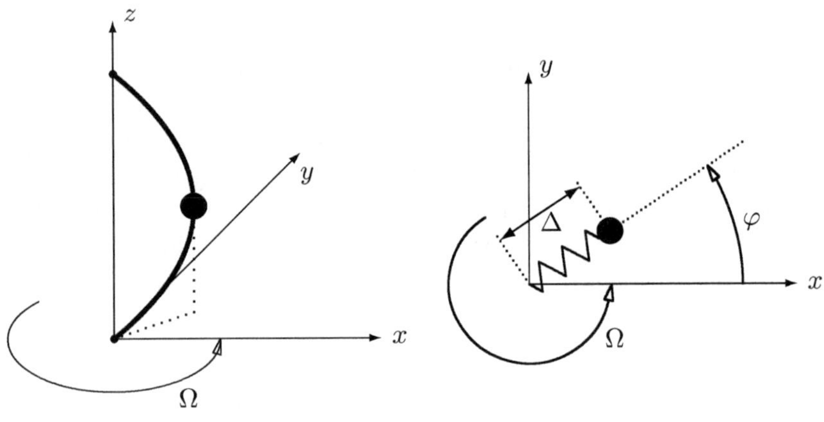

Figure 2. (a) Laval-rotor; (b) top view

$$\begin{pmatrix} \ddot{x} \\ \ddot{y} \end{pmatrix} + \begin{bmatrix} 0 & -2\Omega \\ 2\Omega & 0 \end{bmatrix} \begin{pmatrix} \dot{x} \\ \dot{y} \end{pmatrix} + \begin{bmatrix} (\omega^2 - \Omega^2) & 0 \\ 0 & (\omega^2 - \Omega^2) \end{bmatrix} \begin{pmatrix} x \\ y \end{pmatrix} = \begin{pmatrix} 0 \\ 0 \end{pmatrix}. \tag{10}$$

August Föppl (1854–1924) in a contribution from 1895 gave an explanation: W. Rankine (1820 – 1872) had used an "unfortunate model", i.e. some kind of a rotating massless guide mechanism with a point mass inside which is attached to the origin by a spring. Because the guide prevents relative velocity perpendicular to the spring elongation, the number of degrees of freedom is reduced by one, e.g. $\varphi = 0$ in Figure 2. The remaining equation $\ddot{x} + (\omega^2 - \Omega^2)x = 0$ then indicates that once Ω passes the first bending frequency ω, the motion becomes unstable. In reality, however, the Coriolis (1792 – 1843) forces stabilize motion.

Föppl's contribution had been published in a journal which was probably not well known by contemporary rotor dynamicists. In 1916, the Royal Society of London therefore commissioned Henry H. Jeffcott (1877-1937) to resolve the conflicts between theoretical investigations and practical results. Three years later, Jeffcott published his results in the Philosophical Magazine [40]. Since then, the technical term *Jeffcott-Rotor* is used in english speaking areas, while in other countries the *Laval-Rotor* is more common.

Laval's experimental setup, together with a stack of conically shaped disks inserted into the separator bowl of a milk centrifuge in around 1890 (so-called alpha-disks, patented by Clemens von Mauchenheim) was to revolutionize the dairy industry. In 2008, the Alpha-Laval-Company celebrated its 125th birthday. *Physical model needs intuition*

In autumn 1957, the former USSR launched its first statellite Sputnik. Three months later, the US Satellite Explorer successfully entered orbit. It consisted of a rigid body and some nearly massless hence negligible elastic antennas, and it was spin stabilized. One can see, in a foto from that time, how happy they were, Pickering, van Allen, von Braun. However, the satellite became unstable [35].

Considering the antimetric mode of a free rigid body in space, the equations of motion are obtained from eq.(4) by setting $\delta v, \delta w$ equal to zero and neglecting the restoring forces:

$$\begin{aligned}&\delta v' \left[\rho I_z \ddot{v}' - \rho(I_z + I_y - I_x)\dot{\alpha}\dot{w}' - \rho(I_y - I_x)\dot{\alpha}^2 v' \underline{- \rho I_z w' \ddot{\alpha}}\,\right] \\ &+ \delta w' \left[\rho I_y \ddot{w}' + \rho(I_z + I_y - I_x)\dot{\alpha}\dot{v}' - \rho(I_z - I_x)\dot{\alpha}^2 w' \underline{- \rho(I_x - I_y) v' \ddot{\alpha}}\,\right] = 0.\end{aligned}$$

The cumbersome correction for the underlined terms becomes obsolete for $\ddot{\alpha} = 0$. Then, $\delta v'$ and $\delta w'$ are independent. Changing the axes notations ($x \to z$, $y \to x$, $z \to y$, $\dot{\alpha} \to \Omega$) along with $I_y = I_z \to A$, $I_x \to C$, one obtains the equations of motion in a co-rotating frame $\mathbf{M}\ddot{\mathbf{y}} + \mathbf{G}\dot{\mathbf{y}} + \mathbf{K}\mathbf{y} = 0$ which, after dividing by A, read

$$\begin{pmatrix}\ddot\alpha\\ \ddot\beta\end{pmatrix} + (\tfrac{C}{A}-2)\Omega\begin{bmatrix}0 & +1\\ -1 & 0\end{bmatrix}\begin{pmatrix}\dot\alpha\\ \dot\beta\end{pmatrix} + (\tfrac{C}{A}-1)\Omega^2\begin{bmatrix}+1 & 0\\ 0 & +1\end{bmatrix}\begin{pmatrix}\alpha\\ \beta\end{pmatrix} = \begin{pmatrix}0\\ 0\end{pmatrix}. \tag{11}$$

Here, $\mathbf{J} = diag\{A, A, C\}$ represents the inertia tensor. Spin takes place around the z-axis and $\mathbf{y} = (\alpha,\ \beta)^T$ considers small angular deviations.

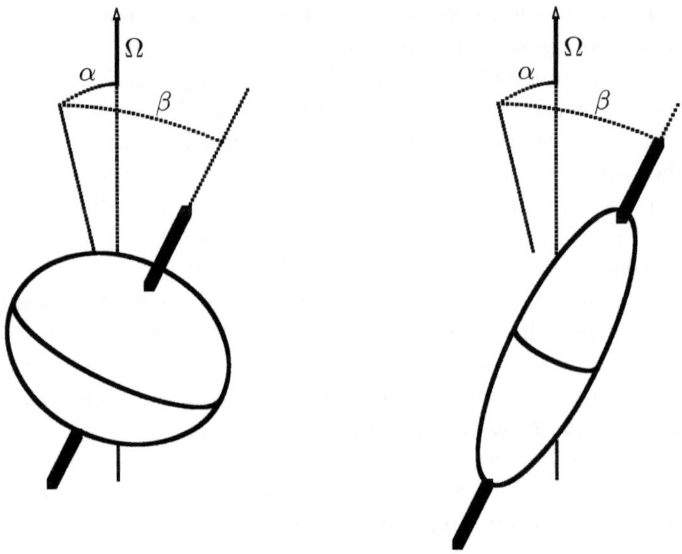

Figure 3. Spinning body, (a) $C/A > 1$; (b) $C/A < 1$

The characteristic polynomial for eq.(11) reads

$$P(\lambda) = \left[\lambda^2 + (\tfrac{C}{A}-1)\Omega^2\right]^2 + \left[\lambda(\tfrac{C}{A}-2)\Omega\right]^2 = 0 \Rightarrow \lambda = \begin{cases} \pm i(\tfrac{C}{A}-1)\Omega \\ \pm i\Omega \end{cases} . \tag{12}$$

Because $0 < \frac{C}{A} \leq 2\ \forall\ \frac{C}{A}$ holds, the motion is always stable. This is also stated by the stability theorem of W.Thomson (1824 – 1907) and P.G.Tait (1831 – 1901) from 1879: stability is warranted for $\mathbf{K} > 0, \mathbf{K} < 0$ as long as $det\mathbf{G} \neq 0$, $\parallel \mathbf{G} \parallel$ sufficient high.

When damping takes place, the equations of motion $\mathbf{M}\ddot{\mathbf{y}}+\mathbf{D}\dot{\mathbf{y}}+\mathbf{G}\dot{\mathbf{y}}+\mathbf{K}\mathbf{y}=0$ read:

$$\begin{pmatrix}\ddot{\alpha}\\ \ddot{\beta}\end{pmatrix}+(\tfrac{d}{A})\begin{pmatrix}\dot{\alpha}\\ \dot{\beta}\end{pmatrix}+(\tfrac{C}{A}-2)\Omega\begin{bmatrix}0 & +1\\ -1 & 0\end{bmatrix}\begin{pmatrix}\dot{\alpha}\\ \dot{\beta}\end{pmatrix}+(\tfrac{C}{A}-1)\Omega^2\begin{bmatrix}+1 & 0\\ 0 & +1\end{bmatrix}\begin{pmatrix}\alpha\\ \beta\end{pmatrix}=0. \tag{13}$$

Then, P.G.L. Dirichlet's (1805 – 1859) result from 1846 holds in the form: If $\mathbf{D}>0$, then $\mathbf{K}>0$ yields stability, otherwise $\mathbf{K}<0$: instability, independent from $\mathbf{G}$. This was reported by N.G. Chetaev (1902 – 1959) (english translation 1961, [12]). Hence, as $\mathbf{K}$ from eq.(13) indicates, $(C/A)<1$ reveals instability. This corresponds to the Explorer shape (Figure 3 (b)), but where should dampig come from? The breaktrough came with [38]: The reason of instability is (mainly) due to the elastic antennas with its material damping. Here, only $\mathbf{D}\geq 0$ holds, of course. Thus, separate the damping within the state equation,

$$\frac{d}{dt}\begin{pmatrix}\mathbf{y}\\ \dot{\mathbf{y}}\end{pmatrix}=\begin{bmatrix}0 & \mathbf{E}\\ -\mathbf{M}^{-1}\mathbf{K} & -\mathbf{M}^{-1}\mathbf{G}\end{bmatrix}\begin{pmatrix}\mathbf{y}\\ \dot{\mathbf{y}}\end{pmatrix}+\begin{bmatrix}0\\ -\mathbf{M}^{-1}\mathbf{D}\end{bmatrix}\dot{\mathbf{y}}, \tag{14}$$

and treat it like a control input, one comes out with

$$\mathbf{D}\geq 0 \wedge \left\{\begin{bmatrix}0 & \mathbf{E}\\ -\mathbf{M}^{-1}\mathbf{K} & -\mathbf{M}^{-1}\mathbf{G}\end{bmatrix},\begin{bmatrix}0\\ -\mathbf{M}^{-1}\mathbf{D}\end{bmatrix}\right\}\text{controllable}$$
$$\Rightarrow\begin{cases}\mathbf{K}>0: & \text{stable}\\ \mathbf{K}<0: & \text{unstable}\end{cases}. \tag{15}$$

Clearly, the material damping is very small. But, however, not negligible at all: instability means that also the smallest parameter will unevoidingly lead to amplitude growing after a short while. An advice to rotor designers: let $(C/A)>1$! *Physical model needs intuition.*

Perpendicular rotation. Considering a beam which rotates around a perpendicular axis, one has

$$[C_{\text{Hub}}+\int_o^L(\rho A x^2)]\ddot{\gamma}+\int_o^L\rho A x\ddot{v}dx=M_z,$$
$$\rho A(\ddot{v}+x\ddot{\gamma}-v\dot{\gamma}^2)+(EI_z v'')''=0, \tag{16}$$

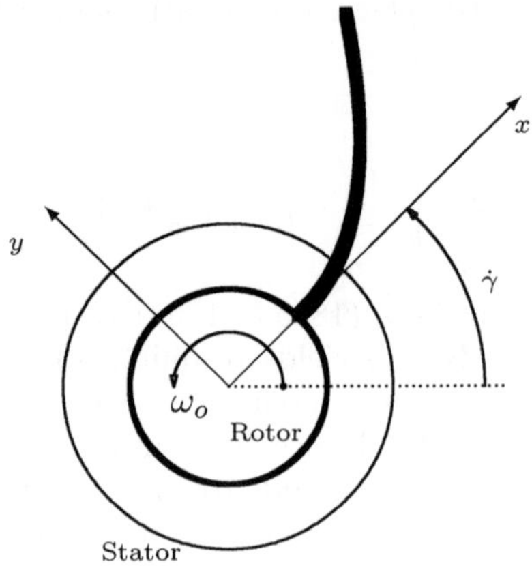

Figure 4. Rotating beam

where $v(x,t)$ denotes deflection and the prime represents differentiation w.r.t. x. In order to find a solution, one may use a Ritz ansatz [49]

$$v(x,t) = \mathbf{v}(x)^T \mathbf{q}(t) \tag{17}$$

along with Galerkin's method [20] which leads to

$$\begin{aligned} &\left[\begin{array}{cc} C^o & \int \rho A x \mathbf{v}^T dx \\ \int \rho A x \mathbf{v} dx & \int \rho A \mathbf{v}\,\mathbf{v}^T dx \end{array}\right] \begin{pmatrix} \ddot{\gamma} \\ \ddot{\mathbf{q}} \end{pmatrix} \\ + &\left[\begin{array}{cc} 0 & 0 \\ 0 & \int EI_z \mathbf{v}\,\mathbf{v}''''^T dx - \int \rho A \mathbf{v}\,\mathbf{v}^T dx \dot{\gamma}^2 \end{array}\right] \begin{pmatrix} \gamma \\ \mathbf{q} \end{pmatrix} = \begin{pmatrix} M_M \\ 0 \end{pmatrix}. \end{aligned} \tag{18}$$

We will use clamped–free beam functions for shape functions. Integrating by parts yields $\int_o^L EI_z \mathbf{v}\,\mathbf{v}''''^T dx = \int_o^L EI_z \mathbf{v}''\,\mathbf{v}''^T dx$ because of $EI_z \left[\mathbf{v}\,\mathbf{v}''' - \mathbf{v}'\,\mathbf{v}''^T\right]_o^L = 0$ due to boundary conditions.

The shape functions obbey the orthogonality conditions

$$v_i(x) = \cosh k_i x - \beta_i \sinh k_i x - \cos k_i x + \beta_i \sin k_i x, \quad \int_o^L v_i v_j dx = \delta_{ij} L, \tag{19}$$

$$\frac{v_i(x)''}{k_i^2} = \cosh k_i x - \beta_i \sinh k_i x + \cos k_i x - \beta_i \sin k_i x, \quad \int_o^L v_i'' v_j'' dx = \delta_{ij} k_i^4 L \tag{20}$$

(δ_{ij}: Kronecker's symbol), with k_i from $\cosh(k_i L)\cos(k_i L) + 1 = 0$. Hence, the only term which remains unknown in eq.(18) reads $\int_o^L \rho A x v dx$. Inserting v_i from the differential equation $v_i'''' - k_i^4 v_i = 0$ yields, with an integration by parts,

$$\int_o^L x\, v_i dx = \frac{1}{k_i^4} \int_o^L x\, v_i'''' dx = \frac{1}{k_i^4} \left[L\, v_i'''(L) + v_i''(0) - v_i''(L)\right] = \frac{2}{k_i^2}, \tag{21}$$

see eq.(20), where $v''(L) = 0, v'''(L) = 0$ due to the boundary conditions. Assuming EI and ρA constant, the equations of motion $\mathbf{M}\ddot{\mathbf{y}} + \mathbf{K}\mathbf{y} = \mathbf{Q}$ read

$$\mathbf{y} = \begin{pmatrix} \gamma \\ \hline q_1 \\ q_2 \\ \vdots \\ q_n \end{pmatrix}, \quad \mathbf{M} = m_B \left[\begin{array}{c|cccc} \frac{L^2}{3} + \frac{C_{\text{Hub}}}{m_B} & \frac{2}{k_1^2 L} & \frac{2}{k_2^2 L} & \cdots & \frac{2}{k_n^2 L} \\ \hline \frac{2}{k_1^2 L} & & & & \\ \frac{2}{k_2^2 L} & & & & \\ \vdots & & \mathbf{E} & & \\ \frac{2}{k_n^2 L} & & & & \end{array}\right], \tag{22}$$

$$\mathbf{K} = EIL \left[\begin{array}{c|ccc} 0 & 0 & 0 & \cdots \; 0 \\ \hline 0 & & & \\ 0 & & & \\ \vdots & & diag(k_i^4) - \omega_o^2 \frac{\rho A}{EI}\mathbf{E} & \\ 0 & & & \end{array} \right], \quad \mathbf{Q} = \left[\begin{array}{c} M_M \\ \hline 0 \\ 0 \\ \vdots \\ 0 \end{array} \right]. \tag{23}$$

Why to calculate shape functions analytically? Why not use a computer program? It is quite obvious that our students do not like calculus so much. A colleague of mine believes that our students today are not able to calculate the indefinite integral $\int x\,dx$ without the aid of a computer. When a told this some years ago to a young man he answered quite astonished: "How can that be achieved without computer?"

But there is a real benefit. Considering a rolling mill, for instance, which consists of several non-uniform rotating beams along with motors and gears as well as the interconnecting metal performing forces, such a system turned out to be inexecutable with common Finite Element codes. Here, Hubinger [25] reduced the computer time using the transfer matrix technique by a factor about 20 000, or, in other units, from one year to half an hour, and made the system calculable. The Computer – Ruin of science and threat to mankind [55]. Or: *Put brain in motion before starting computer.*

2.2 Curiosities

Singularity. We come back to the question: can a simple Euler-Bernoulli-beam represent, for instance, a helicopter blade? In this case, C_{Hub} does not arise in eq.(22), and the mass matrix determinant becomes

$$det\mathbf{M} = m_B \left[\frac{L^2}{3} - \sum \left(\frac{2L}{k_i^2 L^2} \right)^2 \right]. \tag{24}$$

However, with k_i from $\cosh(k_i L)\cos(k_i L) + 1 = 0$, eq.(24) turns out to be zero! It is thus necessary to retain $C_{\text{Hub}} \neq 0$ yielding

$$det\mathbf{M} = C_{\text{Hub}}. \tag{25}$$

Series truncation in eq.(24) for the shape functions (19) [or, even worse because the problem probably remains undetected, for any other shape functions] would yield a nonzero value for eq.(24) – and thus totally unreliableresults.

The reason for this is, by nearer inspection, quite clear. Consider, for instance, plane motion of a beam element represented by the deformation functions v, w. For the rotating beam one has then

$$\mathbf{p} \in I\!R^2, \ \mathbf{q} = \begin{pmatrix} v \\ w \\ \alpha \end{pmatrix} \in I\!R^3, \tag{26}$$

where $\mathbf{p}$ denotes momentum and $\mathbf{q}$ consists of the minimal coordinates. Hence, the problem is kinematically overdetermined. For correct representation one would either additionally need the angular momentum $\mathbf{L}$ (here: $\mathbf{L} \in I\!R^1$, i.e. considering C_{Hub} along with some additional assumptions). Or: $\mathbf{q} = (v \ \ w)^T \in I\!R^2$ along with $\alpha = \alpha(t)$ as a (rheonomic) constraint.

Instability. Concerning the latter we may assume $\gamma = \Omega t, \ \Omega = const.$ The equations of motion from eq.(22) et seq. are then

$$\begin{pmatrix} \ddot{q}_1 \\ \ddot{q}_2 \\ \vdots \\ \ddot{q}_n \end{pmatrix} + \left[\frac{EI}{\rho A} diag(k_i^4) - \omega_o^2 \mathbf{E} \right] \begin{pmatrix} q_1 \\ q_2 \\ \vdots \\ q_n \end{pmatrix} = \begin{pmatrix} 0 \\ 0 \\ \vdots \\ 0 \end{pmatrix}, \tag{27}$$

and the corresponding motor torque (which leads to $\Omega = const$) reads

$$M_M = m_B \left[\begin{array}{cccc} \frac{2}{k_1^2 L} & \frac{2}{k_2^2 L} & \cdots & \frac{2}{k_n^2 L} \end{array} \right] \begin{pmatrix} \ddot{q}_1 \\ \ddot{q}_2 \\ \vdots \\ \ddot{q}_n \end{pmatrix}. \tag{28}$$

$\Omega = const$ is a common assumption in rotor dynamics –mostly without asking where it comes from. The question arises wether a motor torque according to eq.(28) is possible. In 1902, Sommerfeld observed that under certain circumstances the drive torque did not longer accelerate the rotor speed but instead excited shaft bending oscillations [51].

But even if, in our case, the corresponding torque is available – is the equation of motion correct? Obviously not: Applying a constant torque leads to bending instability after a certain while.

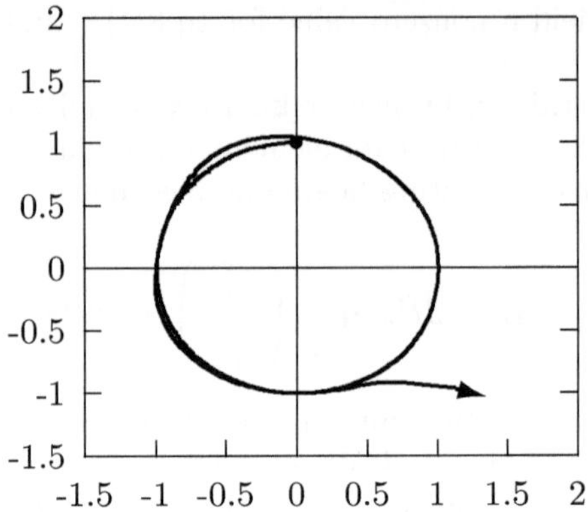

Figure 5. Unstable beam (tip deflection)

The reason is quite simple: the centrifugal force which acts on the beam element reads

$$d\mathbf{f}_{\text{cent}} = \rho A \omega_o^2 dx \begin{pmatrix} x \\ v \end{pmatrix} \tag{29}$$

while the deflection, for a linear approach, only takes the y-direction into account. In reality, however, the element does not move on a straight line but on some kind of circular arc. There is thus also a deflection in (negative) x-direction which reads $\int_o^x v'^2 dx/2$. The effect of longitudinal forces was already known to Euler [14]. It may be calculated via the corresponding virtual work (along with a Ritz ansatz)

$$\delta W_{\text{cent,x}} = -\int_o^L \left[\int_o^x v' \delta v' d\xi \right] \rho A \omega_o^2 x dx = -\delta \mathbf{q}^T \int_o^L \mathbf{v}' \mathbf{v}'^T \frac{\rho A}{2} \omega_o^2 (L^2 - x^2) dx\, \mathbf{q}. \tag{30}$$

Considering eq.(30) along with eq.(27), the motion equation reads

$$\begin{pmatrix} \ddot{q}_1 \\ \ddot{q}_2 \\ \vdots \\ \ddot{q}_n \end{pmatrix} + \left[\frac{EI}{\rho A} diag(k_i^4) - \omega_o^2 \left\{ \mathbf{E} - \frac{1}{2L} \int_o^L (L^2 - x^2) \mathbf{v}' \mathbf{v}'^T dx \right\} \right] \begin{pmatrix} q_1 \\ q_2 \\ \vdots \\ q_n \end{pmatrix} = \begin{pmatrix} 0 \\ 0 \\ \vdots \\ 0 \end{pmatrix}. \tag{31}$$

Applying a constant motor torque, the beam motion now remains stable.

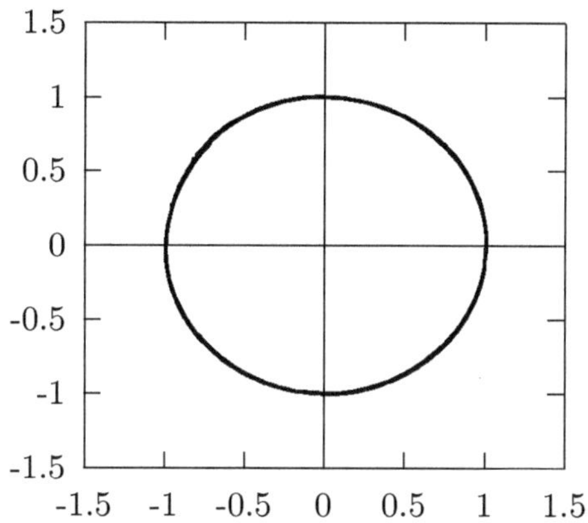

Figure 6. Rotating beam (tip deflection)

This behaviour reflects reality: a helicopter blade will never lead to instability when the rotor speed increases. Just in contrary: the blade is somehow more and more stiffened. The effect which results from second order displacment is therefore sometimes called "dynamical stiffening". Such a notation, however, should not lead to believe in a new theory. Neglecting such effects is, on the contrary, a severe mistake when linearizing the equations for small bending amplitudes. *Mathematical modeling needs interpretation.* There are more such effects which result from second order displacements. These are obtained in two steps:

- From Cauchy (fluctuating reference during motion) to Trefftz (fixed reference) where zero-order unit forces/torques are correctly applied, [26],
- Fubini type integration by parts, $\int_o^L g(x) \int_o^x h(\xi) d\xi \, dx = \int_o^L h(x) \left[\int_x^L g(\xi) d\xi \right] dx$, yields independency of load assumptions, [8].

The second order displacements [superscript (2)] are then

$$\mathbf{r}^{(2)} = -\frac{1}{2}\int_o^x \begin{pmatrix} v'^2 + w'^2 \\ -2\,[(x-\xi)\vartheta' w' - \vartheta\, w'] \\ +2\,[(x-\xi)\vartheta' v' - \vartheta\, v'] \end{pmatrix} d\xi,$$

$$\varphi^{(2)} = +\frac{1}{2}\int_o^x \begin{pmatrix} +(v'w'' - w'v'')\,d\xi \\ +2\vartheta' v'\,d\xi \\ +2\vartheta' w'\,d\xi \end{pmatrix} d\xi \qquad (32)$$

[compare $\mathbf{r}_1^{(2)}$ (first component) with eq.(30)].

First attempts to consider flexible multilink robots considered a rotating beam with a heavy end load m_E, representing the neigbouring subsystem in a first approximation [56]. Then, torsional deflection has additionally to be taken into account. The equations read

$$\mathbf{M}\ddot{\mathbf{y}} + \mathbf{K}\mathbf{y} = \mathbf{Q},\ \ \mathbf{y} = \begin{pmatrix} \gamma(t) \\ \mathbf{q}_v(t) \\ \mathbf{q}_\vartheta(t) \end{pmatrix},\ \ \mathbf{Q} = \begin{pmatrix} M_{mot} \\ 0 \\ 0 \end{pmatrix} \qquad (33)$$

where, for $m_E >> \rho AL$, (34)

$$\mathbf{K} = \int_o^L \begin{bmatrix} 0 & 0 & 0 \\ 0 & EI_z\mathbf{v}''\mathbf{v}''^T + \dot{\gamma}^2 m_E\left[L\mathbf{v}'\mathbf{v}'^T - \mathbf{v}_E\,\mathbf{v}_E{}^T\right] & 0 \\ 0 & 0 & GI_D\boldsymbol{\vartheta}'\boldsymbol{\vartheta}'^T \end{bmatrix} dx \qquad (35)$$

(index E for "end"). Here, torsion appears decoupled. However, the corresponding (first) frequency differed by around 30% from the experimental data. This cannot be explained by numerical unconciseness.

The reason is once more a second order effect: as eq.(32) shows, there is a severe coupling between bending and torsion (which was already known to Prandtl, [46]). Taking that into account, the restoring matrix becomes

$$\mathbf{K} = \int_o^L \begin{bmatrix} 0 & 0 & 0 \\ 0 & EI_z\mathbf{v}''\mathbf{v}''^T + \underline{\dot{\gamma}^2 m_E\left[L\mathbf{v}'\mathbf{v}'^T - \mathbf{v}_E\,\mathbf{v}_E{}^T\right]} & m_E g(L-x)\mathbf{v}''\boldsymbol{\vartheta}^T \\ 0 & m_E g(L-x)\boldsymbol{\vartheta}\mathbf{v}''^T & GI_D\boldsymbol{\vartheta}'\boldsymbol{\vartheta}'^T \end{bmatrix} dx. \qquad (36)$$

The centrifugal effect (underlined in eq.(36) will probably not play a significant role in robot maneuvers. However, from physical interpretation, it will not make sense to cancel the "dynamical stiffening" part and leave the divergent one. For non-permanent rotations, the whole centrifugal reaction may be neglected, coming out with a *linear* motion equation. These assumptions have to be verified, of course.

Small forces are usually neglected, but caution with nonlinear effects! A T-shaped elastic beam system for instance shows – in a certain parameter range – out-of-plane motions which correspond to parameter resonance, although resonance will of course not really take place due to limited energy: Same as in the resonance case, the beginning of motion is exponentially unstable, but then, at a certain amplitude, the effect turns the other way round. This leads to beating waves for β and q. Figure 7 shows a simplified model.

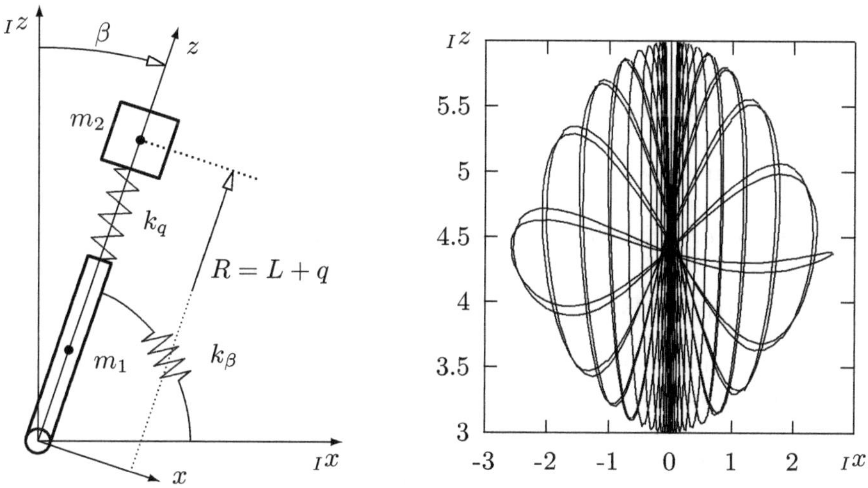

Figure 7. (a) a 2-dof oscillator; (b) tip mass motion

Physical modeling needs intuition, mathematical modeling needs interpretation.

2.3 Modeling Questions

- The-Helicopter-Blade-Example: Where is Coriolis?
  - Excursion: the tremendeous Coriolis effects,
    - e.g. weather forecast (remember hurricane Carrie and the Pamir)

    - e.g. rotor dynamics (remember Laval)
  - Caution with nonlinear effects!
    - e.g. weather forecast (see Lorenz and Chaos)
    - e.g. T-shaped beam (reveals *typical* effects as often observed)
- elastic robot slewing maneuver.
  - Experiment: Coriolis negligible? yes.
  - Experiment: Torsion a decoupled motion? mostly not.
  - Experiment: Centrifugal effects negligible? mostly yes.

What does modeling tell us – truth, reality or just illusion? To make things true one may use assumptions. The truth is, that Carthago has been destroyed; the assumption may be that it owned weapons of mass destruction to justify deletion. This corresponds to *political correctness* which has nothing in common with truth.

René Descartes – one of our philosophers' protagonists – formulated an impact rule: when a smaller body hits a bigger one, then the smaller one is reflected while the bigger remains at rest. And he continues: "if experience seems to contradict, then we must trust reason more than perception" [52]. This might be called *scientific correctness* for use in philosophy which has nothing in common with truth.

Truth may perhaps never be obtained but at least reality. What is needed here is *experimental verification*, a postulation since Galileo's days.

3 Modeling – Basic Requests

Basic request in modeling is its fundamental law *"as simple as possible, as complicated as necessary"*. The aim is to evaluate the easiest straight-on procedure(s).

The *phyical model* consists of bodies and interaction forces. As simple as possible for instance means: the hub motion of a car should not include the shell model of the cabin. The *mathematical model* may either be considered on the basis of Euler's results or on the Lagrangian approach. The angular velocity $\boldsymbol{\omega}$ hereby turns out nonholonomic – a headache for generations of scientists. To obtain $\boldsymbol{\omega}$ via Poisson's equation – as sometimes proposed – is by far too complicated. Mostly, in holonomic systems, $\boldsymbol{\omega}$ is directly inserted as a function of the minimal coordinates $\mathbf{q}$ and its derivatives – this complicates the mathematical model, even more if $\boldsymbol{\omega}$ is (as a whole) available from measurements.

The Euler approach uses the momentum and the moment of momentum theorems along with constraint forces. The Lagrangian approach (Lagrange-Principle) combines both as, so to say, a "flat rate" (all in one) but needs some explanation (how to obtain two axioms out of one?). Its evaluation needs functional matrices (Jacobians) which sometimes calculated via Poisson's equation (with an indication for numerical codes) – this is a complicated and unnecessary detour: the Jacobians are a-priori known from kinematics without any extra calculation.

The use of inertial reference frames is frequently recommended. This is by far too restrictive. Instead, moving frames adapt physical and mathematical modeling for the actual problem under consideration.

When elastic deformations take place, then Galerkin's approach is often used. Its shape functions have to fulfill the dynamical boundary conditions. The latter, however, can be considered with an extended Galerkin procedure (as often recommended). With an integration by parts, the boundary terms are shifted into the integral – which brings us back to the beginning of investigation: the direct Ritz approach (virtual work). A re-integration from Galerkin is obsolete.

If nonholonomic constraints arise: are there better suited Principles – as for instance Jourdain's one ([27], as often asserted)? This question leads to an interpretation of the Lagrangian foregoing which simultaneously builds bridges to optimization.

Considering the physical model as interacting rigid and beamlike elastic bodies, *the Euler-Bernoulli-Beam is a powerful tool.*

3.1 The Physical Model

Let us continue with physical modeling. Two typical robot assemblies are depicted in Figure 8: a motor-gear-arm unit and a telescoping arm mounted at the gear output.

Inspecting these figures shows that the physical model consists of

- rigid and elastic bodies, interconnected by
- impressed forces
  - passive: springs, dampers, ...
  - active: control

Everybody knows that a rigid body does not exist in reality. It is, however, a brilliant basic model.

3.2 The Mathematical Model

Eulerian approach.

$$\begin{aligned} \mathbf{p} = m\mathbf{v} \in I\!R^3, \;\; \dot{\mathbf{p}} + \tilde{\boldsymbol{\omega}}\mathbf{p} = \mathbf{f} \;\; &\text{(translation (1736), [13])} \\ \mathbf{L} = \mathbf{J}\boldsymbol{\omega} \in I\!R^3, \;\; \dot{\mathbf{L}} + \tilde{\boldsymbol{\omega}}\mathbf{L} = \mathbf{M} \;\; &\text{(rotation (1750), [15])} \end{aligned} \tag{37}$$

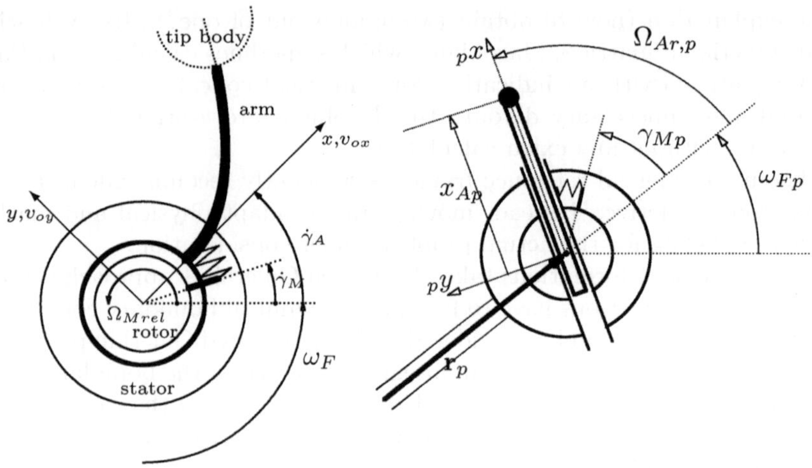

Figure 8. (a) motor-gear-arm model; (b) telescoping arm

(m: mass, $\mathbf{v}$: velocity (body-fixed frame), $\mathbf{J}$: inertia tensor (constant in body-fixed frame), $\boldsymbol{\omega}$: angular velocity (body-fixed frame), $\widetilde{\boldsymbol{\omega}}$: spin tensor, $\mathbf{f}$: force, $\mathbf{M}$: moment).

The crucial point is $\boldsymbol{\omega}$. How to obtain $\boldsymbol{\omega}$? Let $_B\mathbf{r} := \mathbf{r}$ denote a point within the considered body, represented in a body-fixed frame B. Let furthermore $\mathbf{A} = \mathbf{A}_{IB}$ be the transformation matrix which transforms $_B\mathbf{r}$ into an inertial representation: $_I\mathbf{r} = \mathbf{A}_{IB}\,{}_B\mathbf{r}$. Thereby, $\mathbf{A}_{IB}$ is considered orthonormal: $\mathbf{A}_{BI}^T = \mathbf{A}_{IB} := \mathbf{A}$, $\mathbf{A}^T\mathbf{A} = \mathbf{E}$ (unity). To obtain the absolute velocity in a body-fixed representation one has first to transform vector $\mathbf{r}$ into the inertial base, than differentiate w.r.t. time (to get the absolute measure) and eventually to transform into the (desired) body-fixed frame:

$$\mathbf{v} = \mathbf{A}^T\frac{d}{dt}[\mathbf{A}\,\mathbf{r}] = \dot{\mathbf{r}} + \left[\mathbf{A}^T\frac{d\mathbf{A}}{dt}\right]\mathbf{r}. \tag{38}$$

From eq.(38) one obtains

$$\left[\mathbf{A}^T\frac{d\mathbf{A}}{dt}\right] = \widetilde{\boldsymbol{\omega}} \tag{39}$$

where $\widetilde{\boldsymbol{\omega}} \in I\!R^{3,3}$ is the skew symmetric spin tensor assigned to $\boldsymbol{\omega} \in I\!R^3$. Eq.(39) is often called "Poisson Equation" [Simeon Poisson (1781 – 1840)]. Is eq.(39) the way to obtain $\boldsymbol{\omega}$ – as often proposed in literature? Obviously not. Once disposing of $\mathbf{A}$, the angular velocity $\boldsymbol{\omega}$ is in advance known. Let for instance $\mathbf{A}$ be

$$\mathbf{A}^T = \mathbf{A}_{BI} = \mathbf{A}_\alpha \mathbf{A}_\beta \mathbf{A}_\gamma, \tag{40}$$

i.e. consisting of consecutive elementary transformations γ (z-axis), β (y-axis), α (x-axis). The angular velocity (in body-fixed repesentation) is then simply

$$\boldsymbol{\omega} = [\ \mathbf{e}_1 \mid \mathbf{A}_\alpha \mathbf{e}_2 \mid \mathbf{A}_\alpha \mathbf{A}_\beta \mathbf{e}_3\ ] \begin{pmatrix} \dot{\alpha} \\ \dot{\beta} \\ \dot{\gamma} \end{pmatrix} \tag{41}$$

which means that the first rotation ($\dot{\gamma}\mathbf{e}_3$, $\mathbf{e}_3$: 3rd unit vector) is transformed with the following rotations β, α; the second one ($\dot{\beta}\mathbf{e}_2$, $\mathbf{e}_2$: 2nd unit vector) is transformed with the following rotation α while the last one ($\dot{\alpha}\mathbf{e}_1$, $\mathbf{e}_1$: 1st unit vector) is represented in the final (body-fixed) frame. Extracting $\boldsymbol{\omega}$ from eq.(39) is by far too tedious.

Eq.(41) reads explicitly

$$\begin{pmatrix} \omega_x \\ \omega_y \\ \omega_z \end{pmatrix} = \begin{bmatrix} 1 & 0 & -\sin\beta \\ 0 & \cos\alpha & \sin\alpha\cos\beta \\ 0 & -\sin\alpha & \cos\alpha\cos\beta \end{bmatrix} \begin{pmatrix} \dot{\alpha} \\ \dot{\beta} \\ \dot{\gamma} \end{pmatrix}. \tag{42}$$

Let $\boldsymbol{\omega}\, dt := d\boldsymbol{\pi}$. On has then, for the first component,

$$\begin{aligned} d\pi_x &= [\ 1\ ]d\alpha + [\ 0\ ]d\beta + [-\sin\beta]d\gamma \\ &= \left(\frac{\partial \pi_x}{\partial \alpha}\right) d\alpha + \left(\frac{\partial \pi_x}{\partial \beta}\right) d\beta + \left(\frac{\partial \pi_x}{\partial \gamma}\right) d\gamma \end{aligned} \tag{43}$$

which shows, with $\left(\dfrac{\partial^2 \pi_x}{\partial\beta\partial\gamma}\right) = 0$ and $\left(\dfrac{\partial^2 \pi_x}{\partial\gamma\partial\beta}\right) = -\cos\beta$, that $\boldsymbol{\omega}$ (and, consequently $\mathbf{v}$ for a noninertial representation) is not (elementary) integrable [H.A. Schwarz (1843 – 1921)].

$\boldsymbol{\omega}$ is called a nonholonomic (velocity) variable. Nonholonomicity seems somewhat rare and caused a lot of problems over decades. This might also be the reason that mostly Lagrange's (1736 – 1813) equations of the second kind [33] have been used by directly inserting eq.(42) into the kinetic energy where then $\dot{\mathbf{q}} = (\dot{\alpha}\ \dot{\beta}\ \dot{\gamma})^T$ is, of course, holonomic. Consequently one was helpless with nonholonomic constraints.

A simple example will enlighten the background (greetings from "the red baron" with his Fokker Dr.I; Dr="Dreidecker"=tri-plane): Let the plane model be a rigid body in space. Its motion is represented by eq.(37) the angular part of which (the so-called Euler Equation) reads explicitly

$$\begin{aligned} A\dot{\omega}_x - (B-C)\omega_y\omega_z &= M_x \\ B\dot{\omega}_y - (C-A)\omega_x\omega_z &= M_y \\ C\dot{\omega}_z - (A-B)\omega_x\omega_y &= M_z \end{aligned} \tag{44}$$

The Pilot applies simultaneously torques M_x (aileron), M_y (rudder), M_z (elevator) as depicted in Figure 9(a).

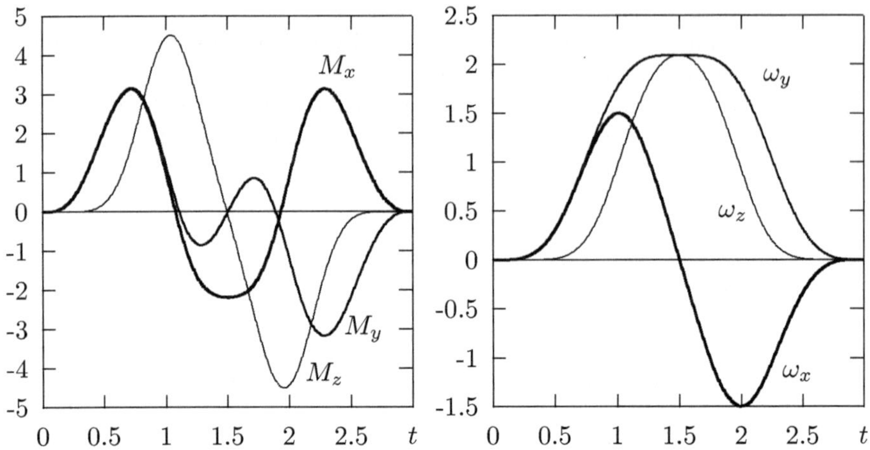

Figure 9. (a) Applied torque over time; (b) on board measurement

The aim is to hold γ equal to zero and to turn α and β around 180 degrees (Immelmann maneuver). However, as the basic gyro effects tell ("parallelism"), the w.r.t. $\dot{\gamma}$ perpendicular rotations will cause a precession around the $\mathbf{e}_3$-axis which forces the pilot to counteract. The (simultaneous) Immelmann maneuver challenges the pilot a lot.

The applied torques according to Figure 9(a) lead to angular velocities as depicted in Figure 9(b). These are on board measured by means of rate gyros for instance.

Because they are not elementary integrable a parametrization is needed, carried out with eq.(42) for instance. Resolving eq.(42) for $\dot{\mathbf{q}} = (\dot{\alpha}\ \dot{\beta}\ \dot{\gamma})^T$ leads to the ODE

$$\begin{pmatrix} \dot{\alpha} \\ \dot{\beta} \\ \dot{\gamma} \end{pmatrix} = \begin{bmatrix} 1 & \sin\alpha\tan\beta & \cos\alpha\tan\beta \\ 0 & \cos\alpha & -\sin\alpha \\ 0 & \sin\alpha/\cos\beta & \cos\alpha/\cos\beta \end{bmatrix} \begin{pmatrix} \omega_x \\ \omega_y \\ \omega_z \end{pmatrix} \tag{45}$$

which has to be solved to obtain the orientation, see Figure 10.

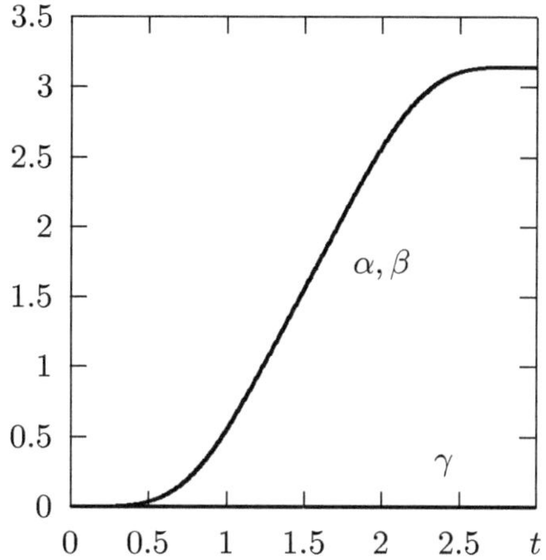

Figure 10. Resulting Orientation

Remark: Max Immelmann (1890 – 1916) flew his maneuver with a Fokker E.I. (E.="Eindecker"=mono-plane) which did not have ailerons. Immelmann could only apply wing torsion by means of cables. It is therefore quite assumable that he operated his plane not by a simultaneous maneuver but consecutively, first β ($0 \rightarrow 180^o$) and then α ($0 \rightarrow 180^o$), without any γ-motion. Such a maneuver is, of course, elementary integrable.

Lagrangian approach. The Lagrange Principle: It was in 1764 when Lagrange won the big prize of the french academy [32]. The title of his paper was *récherches sur la libration de la lune*, and on page 9 we find his principle (in modern notation)

$$\int_{(S)} (dm\ddot{\mathbf{r}} - d\mathbf{f}^e)^T \delta\mathbf{r} = 0 \tag{46}$$

which nowadays is mostly referred to d'Alembert. However, d'Alembert's principle did not have much in common with eq.(46): he inspected equilibrium conditions of the so called lost forces [24], a procedure, which is hard to apply for multi degree systems.

Lagrange concentrated his research on analytical methods; his famous second kind equations came out in 1780 (*Théorie de la libration de la lune*), and in his famous *Méchanique Analytique* from 1788 he used eq.(46) (*the general formula of dynamics*) as the basis of all his investigations.

The Flat Rate: All in One. Eq.(46) needs an inertial representation to obtain the absolute acceleration via $\ddot{\mathbf{r}}$, However, considering a rigid body, one may decompose $\mathbf{r}$ into $\mathbf{r} = {}_I\mathbf{r}_c + \mathbf{A}_{IB\,B}\mathbf{r}_p := \mathbf{r}_c + \mathbf{A}\mathbf{r}_p$ where $\mathbf{r}_c$ denotes the mass center position and $\mathbf{r}_p$ represents an arbitrary point, being constant in a body fixed representation. Then, $\mathbf{A}$ transforms $\mathbf{r}_p$ onto the inertial frame.

Along with $\dot{\mathbf{r}} = \dot{\mathbf{r}}_c + \dot{\mathbf{A}}\mathbf{r}_p \Rightarrow \delta\mathbf{r} = \delta\mathbf{r}_c + \delta\mathbf{A}\,\mathbf{r}_p$, $\mathbf{A}^T\dot{\mathbf{A}} = \widetilde{\boldsymbol{\omega}} \Rightarrow \mathbf{A}^T\delta\mathbf{A} = \delta\widetilde{\boldsymbol{\pi}}$ and $\widetilde{\mathbf{r}}_p\widetilde{\boldsymbol{\omega}}\widetilde{\boldsymbol{\omega}} + \widetilde{\boldsymbol{\omega}}\widetilde{\boldsymbol{\omega}}\widetilde{\mathbf{r}}_p + \widetilde{\boldsymbol{\omega}}\widetilde{\mathbf{r}}_p\widetilde{\boldsymbol{\omega}} = 0$, having $\int_{(B)} \mathbf{r}_p dm = 0$ (mass center) in mind, one obtains

$$\delta\boldsymbol{\pi}^T\left[\dot{\mathbf{L}} + \widetilde{\boldsymbol{\omega}}\mathbf{L} - \mathbf{M}^e\right] + \delta\boldsymbol{\rho}^T\left[\dot{\mathbf{p}} + \widetilde{\boldsymbol{\omega}}\mathbf{p} - \mathbf{f}^e\right] = 0 \tag{47}$$

where

- $\dot{\boldsymbol{\pi}} = \boldsymbol{\omega}$ (angular velocity), $\dot{\boldsymbol{\rho}} = \mathbf{v}$ (translational velocity),
- $\mathbf{L} = \mathbf{J}\boldsymbol{\omega}$ (momentum of momentum), $\mathbf{J} = \int_B \widetilde{\mathbf{r}}\widetilde{\mathbf{r}}^T dm$ (inertia tensor),
- $\mathbf{p} = m\mathbf{v}$ (momentum), m (mass),

all vector terms represented in the body fixed frame. Eq.(47) holds for a single rigid body; because it represents a scalar equation, the multi body system simply needs summation. One has thus found the "Flat Rate" [compare eq.(37)].

Consolidation. But something is seemingly going wrong. It was in 1775 (39 years after Euler's first contribution on dynamics) when Leonhard Euler proposed the linear and the rotational momentum to be independent axioms (*nova methodus motum corporum rigidorum determinandi* [17]). Then, however, the question arises how to construe these two just from one in eq.(46) – a wondrous amplification of axioms?

The answer is simple but worth to be reflected carefully: A certain number of particles (or mass elements dm) will never form a solid unless they are glued together. In a rigid body, these gluing forces do not come at sight because they are constraint forces (rigidity: they do not perform work because of missing deflection). But they are nevertheless present in order to hold the body together.

Applying the momentum of momentum theorem then yields the constraint torques which, in terms of stress components σ_{ij}, reveals a symmetric stress tensor $\sigma_{ij} = \sigma_{ji}\,\forall i \neq j$. One can argue the other way round: Assuming a symmetric stress tensor as a second axiom (so to say in the background) allows to calculate the momentum of momentum via a simple vector product which then defines the rigid body correctly. However, symmetry has then to be seen axiomatic, mostly referred to as Boltzmann's (1844–1906) axiom (the namig is possibly due to Georg Hamel (1877 – 1954) [53]).

$\boldsymbol{\omega}$ is the time derivative of the corresponding quasi-coordinate $\boldsymbol{\pi}(\mathbf{q})$. [The same holds for $\mathbf{v} = \dot{\boldsymbol{\rho}}$ where $\boldsymbol{\rho}(\mathbf{q})$]. In order to calculate eq.(47) in minimal representation one has $(\partial\boldsymbol{\pi}/\partial\mathbf{q})\delta\mathbf{q}$ and $(\partial\boldsymbol{\rho}/\partial\mathbf{q})\delta\mathbf{q}$, respectively. Because $\delta\mathbf{q}$ is arbitrary, one obtains

$$\left(\frac{\partial\boldsymbol{\pi}}{\partial\mathbf{q}}\right)^T \left[\dot{\mathbf{L}} + \widetilde{\boldsymbol{\omega}}\mathbf{L} - \mathbf{M}^e\right] + \left(\frac{\partial\boldsymbol{\rho}}{\partial\mathbf{q}}\right)^T [\dot{\mathbf{p}} + \widetilde{\boldsymbol{\omega}}\mathbf{p} - \mathbf{f}^e] = 0. \tag{48}$$

How to calculate $(\partial\boldsymbol{\pi}/\partial\mathbf{q})$ and $(\partial\boldsymbol{\rho}/\partial\mathbf{q})$? Some authors postulate to extract $\partial\widetilde{\boldsymbol{\pi}} = \mathbf{A}^T\partial\mathbf{A}$ from the Poisson-equation $\widetilde{\boldsymbol{\omega}} = \dot{\widetilde{\boldsymbol{\pi}}} = \mathbf{A}^T\dot{\mathbf{A}}$ where $\mathbf{A} = \mathbf{A}_{IB}$ consists of unit vectors ${}_I\mathbf{e}_{Bi}$, $i = 1, 2, 3$ or, in short, $\mathbf{A} = [\mathbf{e}_1|\mathbf{e}_2|\mathbf{e}_3]$. Extracting $\partial\boldsymbol{\pi}$ from the skew symmetric spin tensor then yields

$$\begin{bmatrix} 0 & -\partial\pi_z & +\partial\pi_y \\ +\partial\pi_z & 0 & -\partial\pi_x \\ -\partial\pi_y & +\partial\pi_x & 0 \end{bmatrix} = \begin{bmatrix} 0 & \mathbf{e}_1^T\partial\mathbf{e}_2 & \mathbf{e}_1^T\partial\mathbf{e}_3 \\ \mathbf{e}_2^T\partial\mathbf{e}_1 & 0 & \mathbf{e}_2^T\partial\mathbf{e}_3 \\ \mathbf{e}_3^T\partial\mathbf{e}_1 & \mathbf{e}_3^T\partial\mathbf{e}_2 & 0 \end{bmatrix}$$

$$\Rightarrow (\partial\pi_x \;\; \partial\pi_y \;\; \partial\pi_z) = \left[\mathbf{e}_3^T\partial\mathbf{e}_2 \quad \mathbf{e}_1^T\partial\mathbf{e}_3 \quad \mathbf{e}_2^T\partial\mathbf{e}_1\right]. \tag{49}$$

Considering the antimetric motions of a rotor according to Figure 3, with $\mathbf{q}^T = (\alpha, \beta)$, the transformation matrix (40) reads

$$\mathbf{A} = [\mathbf{e}_1|\mathbf{e}_2|\mathbf{e}_3] =$$

$$\left[\begin{array}{c|c|c} \cos\beta\cos\gamma & \sin\alpha\sin\beta\cos\gamma - \cos\alpha\sin\gamma & \cos\alpha\sin\beta\cos\gamma + \sin\alpha\sin\gamma \\ \cos\beta\sin\gamma & \sin\alpha\sin\beta\sin\gamma + \cos\alpha\cos\gamma & \cos\alpha\sin\beta\sin\gamma - \sin\alpha\cos\gamma \\ \sin\beta & \sin\alpha\cos\beta & \cos\alpha\cos\beta \end{array}\right]. \tag{50}$$

One obtains for instance

$$
\begin{aligned}
\partial\pi_x = \; & \big[\cos^2\alpha\sin^2\beta\cos^2\gamma + \sin\alpha\cos\alpha\sin\beta\sin\gamma\cos\gamma \\
& + \sin\alpha\cos\alpha\sin\beta\sin\gamma\cos\gamma + \sin^2\alpha\sin^2\gamma \\
& + \cos^2\alpha\sin^2\beta\sin^2\gamma - \sin\alpha\cos\alpha\sin\beta\sin\gamma\cos\gamma \\
& - \sin\alpha\cos\alpha\sin\beta\sin\gamma\cos\gamma + \sin^2\alpha\cos^2\gamma \\
& + \cos^2\alpha\cos^2\beta\big]\,\partial\alpha \\
& + \big[\sin\alpha\cos\alpha\sin\beta\cos\beta\cos^2\gamma + \sin^2\alpha\cos\beta\sin\gamma\cos\gamma \\
& + \sin\alpha\cos\alpha\sin\beta\cos\beta\sin^2\gamma - \sin^2\alpha\cos\beta\sin\gamma\cos\gamma \\
& - \sin\alpha\cos\alpha\sin\beta\cos\gamma\big]\,\partial\beta .
\end{aligned}
\tag{51}
$$

Applying the well known trigonometric formulas reduces eq.(51) to $\partial\pi_x = [\ 1\ ]\,\partial\alpha + [\ 0\ ]\,\partial\beta$ – and motivates to take a few minutes to review the procedure. At first we must state that $\boldsymbol{\pi}$ itself is not obtained but only the components of $(\partial\boldsymbol{\pi}/\partial\mathbf{q})$. But $\boldsymbol{\pi}$ does not enter the calculation anywhere, only the corresponding differentials, see eq.(48). These are, however, already known. Decomposing eq.(41) yields

$$
\boldsymbol{\omega} = \dot{\boldsymbol{\pi}} = [\ \mathbf{e}_1 \mid \mathbf{A}_\alpha\,\mathbf{e}_2] \begin{pmatrix} \dot\alpha \\ \dot\beta \end{pmatrix} + [\,\mathbf{A}_\alpha\mathbf{A}_\beta\,\mathbf{e}_3\ ]\,\dot\gamma = \left(\frac{\partial\boldsymbol{\pi}}{\partial\mathbf{q}}\right)\dot{\mathbf{q}} + \left(\frac{\partial\boldsymbol{\pi}}{\partial t}\right) \tag{52}
$$

or explicitly, along with $\dot\gamma = \Omega$

$$
\begin{pmatrix} \omega_x \\ \omega_y \\ \omega_z \end{pmatrix} = \begin{bmatrix} \underline{1} & \underline{0} \\ 0 & \cos\alpha \\ 0 & -\sin\alpha \end{bmatrix} \begin{pmatrix} \dot\alpha \\ \dot\beta \end{pmatrix} + \begin{pmatrix} -\sin\beta \\ \sin\alpha\cos\beta \\ \cos\alpha\cos\beta \end{pmatrix}\Omega \tag{53}
$$

where the underlined components refer to eq.(51). Once $\boldsymbol{\omega}$ is known, there is no additional calculation required (nor the use of a corresponding computer code as sometimes proposed to solve eq.(51)). Because the (cartesian) velocity is always linear in $\dot{\mathbf{q}}$, the requested Jacobian is nothing but the coefficient matrix w.r.t. the minimal velocities $\dot{\mathbf{q}}$: eq.(52) yields

$$
\left(\frac{\partial\boldsymbol{\pi}}{\partial\mathbf{q}}\right) = \left(\frac{\partial\boldsymbol{\omega}}{\partial\dot{\mathbf{q}}}\right) \quad \left[\text{and, analoguously } \left(\frac{\partial\boldsymbol{\rho}}{\partial\mathbf{q}}\right) = \left(\frac{\partial\mathbf{v}}{\partial\dot{\mathbf{q}}}\right)\right]. \tag{54}
$$

Thus, replacing the Jacobians in eq.(48) by eq.(54), one obtains

$$
\left(\frac{\partial\boldsymbol{\omega}}{\partial\dot{\mathbf{q}}}\right)^T \left[\dot{\mathbf{L}} + \tilde{\boldsymbol{\omega}}\mathbf{L} - \mathbf{M}^e\right] + \left(\frac{\partial\mathbf{v}}{\partial\dot{\mathbf{q}}}\right)^T [\dot{\mathbf{p}} + \tilde{\boldsymbol{\omega}}\mathbf{p} - \mathbf{f}^e] = 0 \tag{55}
$$

where the unbeloved quasi-coordinates do not appear any more.

Up to here, the calculation corresponds to a body fixed coordinate system. This is far too restrictive. Some authors prefer an inertial frame I (since then, at least the translational part is integrable). This is also far too restrictive. However, eq.(55) contains a simple access to arbitrary (orthonormal) frames: just insert a unit matrix in the form $\mathbf{A}_{RB}^T\mathbf{A}_{RB}$:
$\left(\frac{\partial\boldsymbol{\omega}}{\partial\dot{\mathbf{q}}}\right)^T \mathbf{A}_{RB}^T\mathbf{A}_{RB}\left[\dot{\mathbf{L}}+\tilde{\boldsymbol{\omega}}\,\mathbf{L}-\mathbf{M}^e\right]+\left(\frac{\partial\mathbf{v}}{\partial\dot{\mathbf{q}}}\right)^T \mathbf{A}_{RB}^T\mathbf{A}_{RB}\left[\dot{\mathbf{p}}+\tilde{\boldsymbol{\omega}}\,\mathbf{p}-\mathbf{f}^e\right]=0.$
$\mathbf{A}_{RB}^T$ can not depend on velocities, it may be shifted to $(\partial[\mathbf{A}_{RB}\,\boldsymbol{\omega}]/\partial\dot{\mathbf{q}})^T = (\partial_R\boldsymbol{\omega}/\partial\dot{\mathbf{q}})^T$. The remaining part then reads $\mathbf{A}_{RB}[\dot{\mathbf{L}}+\mathbf{A}_{BI}\,\dot{\mathbf{A}}_{IB}\,\mathbf{L}]$ which yields the representation $[_R\dot{\mathbf{L}}+\tilde{\boldsymbol{\omega}}_{IR}\,{}_R\mathbf{L}]$, and $\mathbf{A}_{RB}\,\mathbf{M}^e$ transforms to $_R\mathbf{M}^e$ (translational parts analoguously). One obtains thus, for a representation in an arbitrary coordinate system R,

$$\left(\frac{\partial\boldsymbol{\omega}}{\partial\dot{\mathbf{q}}}\right)^T\left[\dot{\mathbf{L}}+\tilde{\boldsymbol{\omega}}_{IR}\,\mathbf{L}-\mathbf{M}^e\right]+\left(\frac{\partial\mathbf{v}}{\partial\dot{\mathbf{q}}}\right)^T\left[\dot{\mathbf{p}}+\tilde{\boldsymbol{\omega}}_{IR}\,\mathbf{p}-\mathbf{f}^e\right]=0 \qquad (56)$$

where

- all vector terms represented in reference frame R
- $\boldsymbol{\omega},\mathbf{v}$: mass center velocities; $\mathbf{L}=\mathbf{J}\,\boldsymbol{\omega},\mathbf{p}=m\mathbf{v}$: momenta
- $\boldsymbol{\omega}_{IR}$: angular velocity of the reference frame w.r.t. inertial frame
  - e.g. $R=I$: inertial representation ($\boldsymbol{\omega}_{IR}=\boldsymbol{\omega}_{II}=0$),
  - e.g. $R=B$: body fixed representation ($\boldsymbol{\omega}_{IR}=\boldsymbol{\omega}_{IB}=\boldsymbol{\omega}$),
  - e.g. $R\neq I\neq B$: arbitrary representation

Moving reference frames. The advantages of moving frames are clearly demonstrated with a simple example: the edge mill (already in use in the ancient Greece, later copied by the romans, still in use today). It consists of two mill stones (Figure 11, only one is sketched).

We are interested in the reaction torque at the contact point. The use of an inertial frame then needs a post-transformation: $\mathbf{A}_\alpha\dot{\mathbf{L}}=-\mathbf{M}^{react}=$

$$\begin{pmatrix} (A\cos^2\alpha + C\sin^2\alpha)\dot\gamma^2\sin\alpha\cos\alpha - (A-C)\dot\gamma^2\sin\alpha\cos^3\alpha \\ +(A-C)\dot\gamma^2\sin^3\alpha\cos\alpha - (A\sin^2\alpha + C\cos^2\alpha)\dot\gamma^2\sin\alpha\cos\alpha \\ \\ (A\cos^2\alpha + C\sin^2\alpha)\dot\alpha\dot\gamma\cos^2\alpha - (A-C)\dot\alpha\dot\gamma\sin^2\alpha\cos^2\alpha \\ +A\dot\alpha\dot\gamma\cos^2\alpha + (A-C)\dot\alpha\dot\gamma\sin^2\alpha\cos^2\alpha - (A\sin^2\alpha + C\cos^2\alpha)\dot\alpha\dot\gamma\cos^2\alpha \\ -(A\sin^2\alpha + C\cos^2\alpha)\dot\alpha\dot\gamma\sin^2\alpha - (A-C)\dot\alpha\dot\gamma\sin^2\alpha\cos^2\alpha \\ +A\dot\alpha\dot\gamma\sin^2\alpha + (A-C)\dot\alpha\dot\gamma\sin^2\alpha\cos^2\alpha - (A\cos^2\alpha + C\sin^2\alpha)\dot\alpha\dot\gamma\sin^2\alpha \\ \\ -(A\cos^2\alpha + C\sin^2\alpha)\dot\alpha\dot\gamma\cos\alpha\sin\alpha + (A-C)\dot\alpha\dot\gamma\sin^3\alpha\cos\alpha \\ -A\dot\alpha\dot\gamma\cos\alpha\sin\alpha - (A-C)\dot\alpha\dot\gamma\sin^3\alpha\cos\alpha \\ -(A\sin^2\alpha + C\cos^2\alpha)\dot\alpha\dot\gamma\sin\alpha\cos\alpha - (A-C)\dot\alpha\dot\gamma\sin\alpha\cos^3\alpha \\ +A\dot\alpha\dot\gamma\sin\alpha\cos\alpha + (A-C)\dot\alpha\dot\gamma\sin\alpha\cos^3\alpha \\ (A\sin^2\alpha + C\cos^2\alpha)\dot\alpha\dot\gamma\cos\alpha\sin\alpha - (A\cos^2\alpha + C\sin^2\alpha)\dot\alpha\dot\gamma\sin\alpha\cos\alpha \end{pmatrix} \tag{57}$$

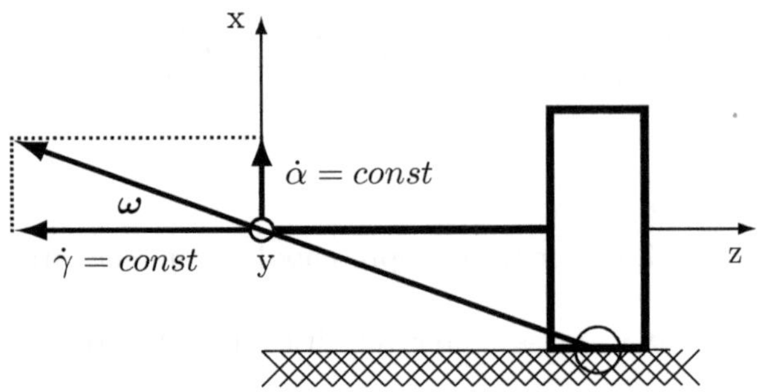

Figure 11. The edge mill

yielding, with the use of trigonometric formulas, $-\mathbf{M}^{react} = (0\ \ C\dot\gamma\dot\alpha\ \ 0)^T$. On the other hand, using the frame wich rotates with $\dot\alpha$, one has $\dot\alpha\tilde{\mathbf{e}}_1\mathbf{L} = -\mathbf{M}^{react}$ which leads directly to

$$\begin{bmatrix} 0 & 0 & 0 \\ 0 & 0 & -\dot\alpha \\ 0 & +\dot\alpha & 0 \end{bmatrix} \begin{pmatrix} A\dot\alpha \\ 0 \\ -C\dot\gamma \end{pmatrix} = \begin{pmatrix} 0 \\ C\dot\gamma\dot\alpha \\ 0 \end{pmatrix}. \tag{58}$$

Elasticity/Nonholonomicity. It is quite obvious that nor a body fixed frame nor an inertial frame will meet the aim for the general case. Considering an elastic robot, for instance, no one would use element fixed reference frames for calculation. Also, referring motions to an inertial frame is by far

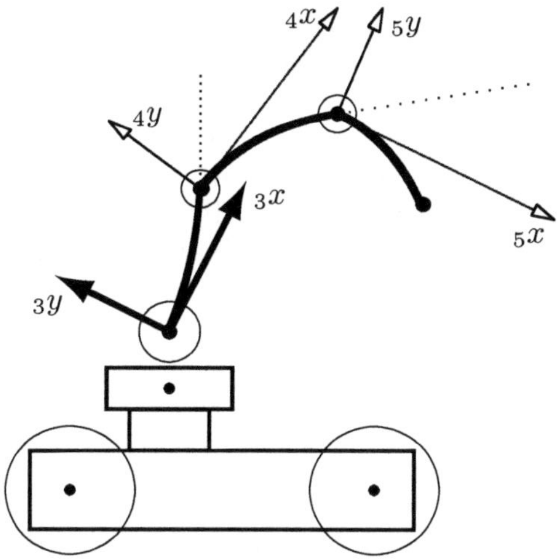

Figure 12. Sketch of a nonholonomic elastic robot

too tedious. As Figure 12 shows, the natural choice here is the undeformed moving frame for every link under consideration. We are thus coming back to elastic systems and will structurize the problem step by step.

The moving beam from eq.(8) will serve as an example for demonstration how to proceed:

Equations of Motion:

$$\left[\rho A(\ddot{v} - 2\dot{\alpha}\dot{w} - \ddot{\alpha}w - \dot{\alpha}^2 v)\right] - \left[\rho I(\ddot{v}'' + \dot{\alpha}^2 v'' + \ddot{\alpha}w'')\right] + EI_z v'''' = 0,$$

$$\left[\rho A(\ddot{w} + 2\dot{\alpha}\dot{v} + \ddot{\alpha}v - \dot{\alpha}^2 w)\right] - \left[\rho I(\ddot{w}'' + \dot{\alpha}^2 w'' - \ddot{\alpha}v'')\right] + EI_y w'''' = 0.$$

Boundary Conditions:

$$\left[\rho I\ddot{v}' + \rho I\dot{\alpha}^2 v' + \rho I\ddot{\alpha}w' - EI_z v'''\right]_o^L = 0, \quad \left[EI_z v''\right]_o^L = 0,$$

$$\left[\rho I\ddot{w}' + \rho I\dot{\alpha}^2 w' - \rho I\ddot{\alpha}v' - EI_y w''',\right]_o^L = 0, \quad \left[EI_y w''\right]_o^L = 0. \tag{59}$$

Obviously, it will (mostly) be impossible to find an analytical solution for such kind of partial differential equations (PDEs).

Galerkin – extended Galerkin – Ritz. For an approximative solution, the methods of Rayleigh-Ritz-Bubnov-Galerkin-Petrov (chronologically ordered) are at our disposal.

Boris Galerkin published his method in [20]. For a special case, the procedure had already been anticipated by I.G. Bubnov in 1913. The method is therefore sometimes referred to as the Bubnov-Galerkin method. Its original application field is mainly characterized by his original paper entitled *series solution of some problems in elastic equilibrium of rods and plates.* The procedure is, however, not restricted to linear elasticity problems but obviously reveals its main power in the field of nonlinear systems, thus extended by G.I. Petrov to nonlinear problems in gas dynamics, hydrodynamics and aerodynamics in 1940 (referred to then as the Galerkin-Petrov method). It is sometimes also called the Ritz-Galerkin method, based on a comparison with W. Ritz's procedure [49]. Ritz in his contribution expressly refers to J.W.S. Rayleigh [48] when he uses the exact (but unknown) eigenfrequency for lower bound w.r.t. the approximated one, in order to guarantee convergence. Therefore, his method is sometimes called the Rayleigh-Ritz method. However, Rayleigh approximated *one* frequency while Ritz solves the problem for a set of eigenfrequencies (46 eigenvalues and eigenvectors approximated with 49 shapefunctions for the plate problem within an accuracy of 2,2%). Although Lord Rayleigh claimed the basic idea his own literary property, it is reported that he was very astonished about Ritz's powerful procedure.

Bubnov-Galerkin's procedure makes use of a series expansion ("Ritz series") which will of course not fulfil the PDEs:

$$\begin{aligned}
&\Big[\rho A(\ddot v - 2\dot\alpha\dot w - \ddot\alpha w - \dot\alpha^2 v)\Big] - \Big[\rho I(\ddot v'' + \dot\alpha^2 v'' + \ddot\alpha w'')\Big] + EI_z v'''' \neq 0,\\
&\Big[\rho A(\ddot w + 2\dot\alpha\dot v + \ddot\alpha v - \dot\alpha^2 w)\Big] - \Big[\rho I(\ddot w'' + \dot\alpha^2 w'' - \ddot\alpha v'')\Big] + EI_y w'''' \neq 0
\end{aligned} \tag{60}$$

for $v = \mathbf{v}^T\mathbf{q}_v, \quad w = \mathbf{w}^T\mathbf{q}_w$. Hence, nullify the "weighted residua":

$$\int_o^L \mathbf{v}\Big\{ [\rho A(\ddot v - 2\dot\alpha\dot w - \ddot\alpha w - \dot\alpha^2 v)] - [\rho I(\ddot v'' + \dot\alpha^2 v'' + \ddot\alpha w'')] + EI_z v''''\Big\}dx = 0,$$

$$\int_o^L \mathbf{w}\left\{ [\rho A(\ddot{w} + 2\dot{\alpha}\dot{v} + \ddot{\alpha}v - \dot{\alpha}^2 w)] - [\rho I(\ddot{w}'' + \dot{\alpha}^2 w'' - \ddot{\alpha}v'')] + EI_y w'''' \right\} dx = 0$$

iff $v = \mathbf{v}^T \mathbf{q}_v, \quad w = \mathbf{w}^T \mathbf{q}_w$ fulfill the BCs (Galerkin-Approximation) (61)

Although correct so far, it will be nearly impossible to find shape fuctions which fulfill the BCs (see eq.(59) for instance). However, one might add the BCs which are premultiplied by $\mathbf{v}, \mathbf{v}'$ and $\mathbf{w}, \mathbf{w}'$, resp., to obtain an extended approximation scheme:

$$\begin{aligned}
&\int_o^L \mathbf{v}\left\{ [\rho A(\ddot{v} - 2\dot{\alpha}\dot{w} - \ddot{\alpha}w - \dot{\alpha}^2 v)] - [(\rho I\ddot{v}'' + \dot{\alpha}^2 v'' + \ddot{\alpha}w'')] + EI_z v'''' \right\} dx + \\
&\int_o^L \mathbf{w}\left\{ [\rho A(\ddot{w} + 2\dot{\alpha}\dot{v} + \ddot{\alpha}v - \dot{\alpha}^2 w)] - [\rho I(\ddot{w}'' + \dot{\alpha}^2 w'' - \ddot{\alpha}v'')] + EI_y w'''' \right\} dx + \\
&\left[\mathbf{v}\left[\rho I\ddot{v}' + \rho I\dot{\alpha}^2 v' + \rho I\ddot{\alpha}w' - EI_z v'''\right] + \mathbf{v}'\left[EI_z v''\right]\right]_o^L + \\
&\left[\mathbf{w}\left[\rho I\ddot{w}' + \rho I\dot{\alpha}^2 w' - \rho I\ddot{\alpha}v' - EI_y w'''\right] + \mathbf{w}'\left[EI_y w''\right]\right]_o^L = 0.
\end{aligned} \tag{62}$$

(extended Galerkin). This equation may be integrated by parts yielding

$$\begin{aligned}
\int_o^L \Big\{ &\mathbf{v}\left[\rho A(\ddot{v} - 2\dot{\alpha}\dot{w} - \ddot{\alpha}w - \dot{\alpha}^2 v)\right] + \mathbf{v}'\left[\rho I(\ddot{v}' + \dot{\alpha}^2 v' + \ddot{\alpha}w')\right] + \\
&\mathbf{w}\left[\rho A(\ddot{w} + 2\dot{\alpha}\dot{v} + \ddot{\alpha}v - \dot{\alpha}^2 w)\right] + \mathbf{w}'\left[\rho I(\ddot{w}' + \dot{\alpha}^2 w' - \ddot{\alpha}v')\right] + \\
&\mathbf{v}'' EI_z v'' + \mathbf{w}'' EI_y w'' \Big\} dx = 0.
\end{aligned} \tag{63}$$

This kind of integration is, by nearer inspection, a-priori kown as can be seen by premultiplying with $\delta\mathbf{q}_v^T, \delta\mathbf{q}_w^T$, respectively. Having

$$\begin{array}{lll} \delta\mathbf{q}_v^T\mathbf{v} = \delta v, & \delta\mathbf{q}_v^T\mathbf{v}' = \delta v', & \delta\mathbf{q}_v^T\mathbf{v}'' = \delta v'' \\ \delta\mathbf{q}_w^T\mathbf{w} = \delta w, & \delta\mathbf{q}_w^T\mathbf{w}' = \delta w', & \delta\mathbf{q}_w^T\mathbf{w}'' = \delta w'' \end{array}$$

in mind, one obtains

$$\delta W = \int_o^L \Big\{ \delta v\left[\rho A(\ddot{v} - 2\dot{\alpha}\dot{w} - \ddot{\alpha}w - \dot{\alpha}^2 v)\right] + \delta v'\left[\rho I(\ddot{v}' + \dot{\alpha}^2 v' + \ddot{\alpha}w')\right] + \delta w\left[\rho A(\ddot{w} + 2\dot{\alpha}\dot{v} + \ddot{\alpha}v - \dot{\alpha}^2 w)\right] + \delta w'\left[\rho I(\ddot{w}' + \dot{\alpha}^2 w' - \ddot{\alpha}v')\right] + \delta v''\, EI_z\, v'' + \delta w'' EI_y w'' \Big\}\, dx = 0. \tag{64}$$

Eq.(64) represents the starting point of investigation, see eq.(4) (corrected according to Figure 1): after an integration by parts w.r.t. time we carried out a subsequent integration by parts w.r.t. space in order to bring the boundary conditions at light, and these are now re-integrated. This is a totally useless calculation. The main difference in consideration hereby is that previously space was an independent variable (→ PDEs) and now it is not any more (→ ODEs). The shape functions are prechosen and have only to fulfill geometrical demands (and form an complete functional set, of course). We will call the procedure according to eq.(64) along with $v = \mathbf{v}^T\mathbf{q}_v$, $w = \mathbf{w}^T\mathbf{q}_w$ a "direct Ritz approach" and come to the following conclusion:

- Bubnov-Galerkin-Petrov's procedures drop out, because their ansatz functions have to fulfill the boundary conditions which is mostly impossible.
- Rayleigh's (original) contribution does not fit the problem.
- As Poincaré puts it: "Seulement Ritz est parvenu dans deux cas, celui du problème de Dirichlet et celui de l'élasticité" (– only Ritz succeeded in two cases: that one of Dirichlet and that one of elasticity).

Because the shape functions are prechosen, the only remaining variabe is $\mathbf{q}$. Therefore, Hamilton's priciple yields

$$\int_{t_o}^{t_1} [\delta T - \delta V]\, dt \Rightarrow \left[\frac{d}{dt}\left(\frac{\partial T}{\partial \dot{\mathbf{q}}}\right) - \left(\frac{\partial T}{\partial \mathbf{q}}\right) + \left(\frac{\partial V}{\partial \mathbf{q}}\right)\right]\delta\mathbf{q} = 0. \tag{65}$$

We are thus, for holonomic systems, left with Lagrange's equations (of the second kind). On the other hand, we have eq.(56) which yields, for moving beams,

$$\int_B \left\{ \left(\frac{\partial \omega}{\partial \dot{\mathbf{q}}}\right)^T \left[d\dot{\mathbf{L}} + \widetilde{\omega}_{IR}\, d\mathbf{L} - d\mathbf{M}^e\right] + \left(\frac{\partial \mathbf{v}}{\partial \dot{\mathbf{q}}}\right)^T \left[d\dot{\mathbf{p}} + \widetilde{\omega}_{IR}\, d\mathbf{p} - d\mathbf{f}^e\right] \right\} = 0 \tag{66}$$

(as already mentioned, eq.(56) needs, for a multi body system, summation. This summation here passes to an integral (B: body unter consideration) while the momenta $d\mathbf{L}$ and $d\mathbf{p}$, resp., refer to the mass element dm).

3.3 Are there better suited Principles?

Here, the question arises: which method will be best? and what about nonholonomicity (as Figure 12 indicates)? Up to here, the whole consideration was based on Lagrange's principle (46). Are there better suited principles – can the Lagrangean one also be used for nonholonomic systems or do we need Jourdain's Principle as has often been stated? To answer this question we will have to have a look on Lagrange's $\delta\mathbf{r}$ – often rejected as obscure (Poinsot), esoteric (Angeles), nearest to black magic (Kane).

What does $\delta\mathbf{r}$ mean? In the introduction of his paper from 1764 Lagrange refers to Johann Bernoulli and the Principle of Virtual Velocities (in statics). Lagrange reports that Bernoulli had communicated that principle to Varignon in a letter from 1717.Varignon stated its equivalence with the parallelogram of forces, thus forming a solid basis for statics, [57]. Lagrange started his investigations in 1760 with *essai d'une nouvelle méthode...*, [31], (essay on a new method...) considering the maxima and the minima of undetermined integral forms, followed one year later by *application de la méthode...* where he applied it to e.g. the brachistochrone problem. He had communicated that new method (*nouvelle méhode*) to Euler already in 1755 when he was nineteen years old, very much appreciated by Euler. Lagrange later reports (in the Miscellanea Taurinensis): "I have to emphasize, because this method needs to vary the same terms in two different ways, in order not to confuse these variations I have introduced a new characteristic δ. Thus, δZ represents a differential with respect to Z, which is not the same as dZ, but which is nevertheless calculated by the same rules, such that, if any $dZ = m\,dx$ exists, one has simultaneously $\delta Z = m\,\delta x$, and this holds for any other equation." It might be this expression which later – up to the time being – caused a lot of misunderstandig not at least due to the naming "virtual displacement" instead of "virtual velocity". It is not clear who introduced it (at the latest Ernst Mach used it already 1883, [36]) and why (may be because statics does not know velocities).

Comparing eq.(46) with dW yields

$$\delta W = \int_{(S)} (dm\ddot{\mathbf{r}} - d\mathbf{f}^e)^T \delta\mathbf{r} = 0, \quad dW = \int_{(S)} (dm\ddot{\mathbf{r}} - d\mathbf{f}^e)^T d\mathbf{r} = 0, \tag{67}$$

which states that $\delta\mathbf{r}$ has to seen infinitesimal. And it is, in this interpretation, indeed a displacement. However, what is this comparison good for? Isn't there an integration missing?

Integration yields the (real, finite) work which will be accomplished under the action of impressed forces

$$W = \int_o^{\mathbf{r}} \int_{(S)} (dm\ddot{\mathbf{r}} - d\mathbf{f}^e)^T d\mathbf{r} = \int_o^{t(\mathbf{r})} \int_{(S)} (dm\ddot{\mathbf{r}} - d\mathbf{f}^e)^T \frac{d\mathbf{r}}{dt} dt = 0. \tag{68}$$

Differentiating w.r.t. time yields the (real) work rate (or power)

$$\dot{W} = \int_{(S)} (dm\ddot{\mathbf{r}} - d\mathbf{f}^e)^T \dot{\mathbf{r}} = 0. \tag{69}$$

Looking for the (unknown) motion we are next going to replace $\dot{\mathbf{r}}$ by a comparison function:

$$\begin{aligned} \dot{\mathbf{r}} \to \dot{\mathbf{r}} + \delta\mathbf{r} : & \int_{(S)} (dm\ddot{\mathbf{r}} - d\mathbf{f}^e)^T \delta\mathbf{r} \; + \int_{(S)} (dm\ddot{\mathbf{r}} - d\mathbf{f}^e)^T \dot{\mathbf{r}} = 0 \\ & \overset{eq.(69)}{\Rightarrow} \int_{(S)} (dm\ddot{\mathbf{r}} - d\mathbf{f}^e)^T \delta\mathbf{r} = 0, \end{aligned} \tag{70}$$

and we are back with Lagrange's virtual velocity ansatz in dynamics. This goes along with Lagrange's own explanations:

1. "δZ shall express a differential w.r.t. Z which is not the same as dZ but built with the same rules" [31]
2. "The virtual velocity is the velocity which will really take place when the equilibrium is disturbed" [34].

How to calculate $\delta\mathbf{r}$? "The same rules" means that $\dot{\mathbf{r}}$ and $\delta\mathbf{r}$ belong to the same vector space. If the system under consideration is constrained via an implicit function $\mathbf{\Phi}(\mathbf{r}) = 0$, then

$$\left(\frac{\partial \mathbf{\Phi}}{\partial \mathbf{r}}\right)\dot{\mathbf{r}} = 0, \quad \left(\frac{\partial \mathbf{\Phi}}{\partial \mathbf{r}}\right)\delta\mathbf{r} = 0 \text{ and } d\mathbf{r} = \left(\frac{\partial \mathbf{r}}{\partial \mathbf{q}}\right)d\mathbf{q}, \quad \delta\mathbf{r} = \left(\frac{\partial \mathbf{r}}{\partial \mathbf{q}}\right)\delta\mathbf{q}. \tag{71}$$

$\mathbf{q} \in I\!R^f$ is totally unconstrained (minimal coordinates). Hence, $\delta\mathbf{q}$ may be looked at in a variational sense

$$\delta\mathbf{q} = \frac{\partial \mathbf{q}}{\partial \varepsilon}|_{\varepsilon=0} \cdot\varepsilon = \varepsilon \cdot \boldsymbol{\eta} \tag{72}$$

which corresponds to a Taylor expansion. Due to the linear approach in ε, the expansion is complete (no higher terms) and thus $\delta\mathbf{q}$ is arbitrary in magnitude. Then, due to eq.(71) $\delta\mathbf{r}$ is also arbitrary in magnitude. This builds bridges to optimization theory which has been started with the same approach (Lagrange, Euler, Legendre ...) until Weierstrass' new ideas (around 1900, see section 6.1).

We can draw the following conclusions:

- "Virtual displacements" (if one prefers this notation) are arbitrary tangent vectors of the constraint plane. They are well defined and by no means obscure. There is no need to assume them infinitesimal.
- There is no obstacle to impose "non-holonomic" constraints (see below). The above mentioned (very much overestimated) Jourdain-Principle is identical (and so is Gauss' [22] in its "differential" form). There is no need to add any other "Principle" to the Lagrangean one.
- The (frequent) statement *"in case of scleronomic* (i.e. time-independent) *constraints,* $d\mathbf{r}$ *and* $\delta\mathbf{r}$ *coincide"* is definitely wrong. (It led Poinsot [45] to severe mistakes).
- The (frequent) statement *"virtual displacements are infinitesimally small and happen infinitely fast"* is incomprehensible, hence rejectable.

3.4 The Euler-Bernoulli-Beam – A Powerful Model

Eq.(66) leads to several model considerations. If the (relative) elastic deflection is modeled by just one deflection function for each bending direction, then eq.(66) represents the so called Rayleigh beam. If the basic motion contains bending and shear, then the beam is called a Timoshenko beam ([54] although already established by Bresse [10]). The simplest and most powerful model is obtained when neglecting the angular momentum [first summation term in eq.(66)]. One is then left with the so called Euler–Bernoulli beam. This neglection has carefully to be considered. In a simple interpretation it refers to slender beams (rotational inertia negligible). A basic rule of thumb here indicates a L/D–ratio of about 10 (length to diameter). However, one should also have in mind that neighbouring systems (or tip masses) may have influence, such as frequency breakdown and shear deflection.

4 Solutions, Aims, Methods

Considering elastic bodies, the exact solution in Daniel Bernoulli's notation is represented by an infinite series. Infinity does not make sense in a physical interpretation. The number of requested eigenfunctions may be averaged via the influence of material damping.

Approximating the eigenfunctions needs adequate shape functions. How many? What kind? The latter is answered through convergence considerations, the first depends on the actual model (e.g. controlled system) and remains open in general.

In fast moving controlled systems, the effect of second order displacement fields are reconsidered. Reaction forces from moderately moving models may be neglected, otherwise acceleration reactions need assignment to the mass matrix (because of later inversion).

One basic task is frequency tuning. In noncontrolled systems it leads to the harmony model in occidental music (from Huygens via Bach to the time being). In controlled systems, however, it is evident for open-loop the result of which serves for closed-loop interpretations.

All these considerations go hand in hand with the mathematical method to be applied. A comparison of the up to here evaluated procedures gives already valuable hints. However, the real problem arises with nonholonomic constraints.

4.1 Exact solutions – how many eigenfunctions?

Exact solutions are obtained for a reduced class of models. These are, for instance, single Euler-Bernoulli beams with various boundary conditions als well as strings or, as a combination, tightened beams. The tightening force f_x acts at $x = L$. The corresponding elongation reads, with eq.(32), $r_x^{(2)} = -\frac{1}{2}\int_o^L w'^2 dx$ for plane motion in z-direction. Hence, f_x contributes via $\delta W_f = -\int_o^L f_x w'\delta w' dx = \int_o^L f_x w''\delta w - [f_x w'\delta w]_o^L$ which has to be considered within $\int_o^t(\delta T - \delta V + \delta W_f)dt = 0$. One obtains then

the motion equation $\quad \rho A\ddot{w} + EIw'''' - f_x w'' = 0$

$$\text{the boundary conditions} \quad \left\{ \begin{array}{lll} [w]_o^L = 0 & \vee & [f_x w' - EIw''']_o^L = 0 \\ [w']_o^L = 0 & \vee & [EIw'']_o^L = 0 \end{array} \right. \tag{73}$$

(compare eq.(5) and (6) for $\alpha = 0$ and $\rho I = 0$). The ansatz

$$w(x,t) = \sum_{n=1}^{\infty} w_n(x) q_n(t) \tag{74}$$

yields for the nth summation term of the motion equation

$$\frac{\ddot{q}_n}{q_n} = -\frac{EI}{\rho A}\left(\frac{w_n''''}{w_n}\right) + \frac{f_x}{\rho A}\left(\frac{w_n''}{w_n}\right) = -\omega_n^2 \tag{75}$$

where – because q_n depends on time and w_n on space, respectively – both terms in eq.(75) can only be fulfilled by a constant which is denoted $-\omega_n^2$. One has thus $\ddot{q}_n + \omega_n^2 q_n = 0$ for the time functions.

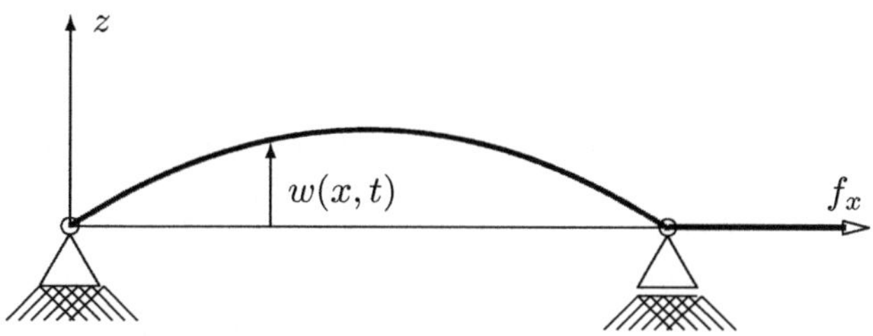

Figure 13. Tightened beam

Considering the pinned-pinned beam as representative model, the ansatz $w_n = \sin(n\pi x/L)$ yields $w_n'' = -(n\pi/L)^2 w_n$ which fulfills the boundary conditions $w(0) = w(L) = 0, w''(0) = w''(L) = 0$. Along with $w_n'''' = +(n\pi/L)^4 w_n$, one obtains from eq.(75)

$$\omega_n^2 = \frac{EI}{\rho A}\left(\frac{n\pi}{L}\right)^4 + \frac{f_x}{\rho A}\left(\frac{n\pi}{L}\right)^2 . \tag{76}$$

The nth solution term in eq.(74) then reads

$$w_n(x)q_n(t) = \sin\left(\frac{n\pi x}{L}\right)\left[q_{on}\cos\omega_n t + \frac{\dot{q}_{no}}{\omega_n}\sin\omega_n t\right] \tag{77}$$

Bernoulli's String Vibrations. The breakthrough came 1753 with D. Bernoulli's *Reflexions et Eclaircicement sur les nouvelles vibrations des cordes* (Considerations on and explanation of the new chord vibrations), [52]. Considering a string, the bending stiffness becomes negligible. The solution (77) still holds, while the frequency reduces to

$$\omega_n = \sqrt{\frac{f_x}{\rho A}}\left(\frac{n\pi}{L}\right). \tag{78}$$

It was Bernoulli who noticed the simultaneous existence of different tones which lead him to the solution in form of a (Fourier) series [see eq.(74) with eq.(77)]. His friend Euler first doubted this result but later, in 1777, he himself anticipated Fourier's results.

Considering eq.(74) leads to the question: How many modes exist? As eq.(74) shows, there are infinite modes which independently fulfill the equations of motion (73). However, this would indicate that, considering higher modes, the string in Bernoulli's solution could move faster than the speed of light. Here, modeling comes to its borders.

One possibility is to adjust modeling to the actual aim. Let for instance the basic tone of a piano string be ω_o. If one wants to model the quint w.r.t. the first octave ($\omega_5 = \frac{3}{2}(2\omega_o)$ and the third w.r.t. the second octave ($\omega_3 = \frac{5}{4}(4\omega_o)$, then $\sqrt{f_x/\rho A}(n\pi/L) \Rightarrow n = 5$ fulfills the task. This is quite trivial: one needs modeling up to the fifth overtone. Then, the vibrating basic string (Note C) with its third overtone excites the quint (Note G) from the first octave and the third (Note E) from the second octave, respectively. It is reported that this effect can be demonstrated experimentally by exciting the basic tone, such that the correspondig G- and E-strings remain resonantly vibrating for a while (only for a short while due to structural damping, and only for just tuning, see section 4.1).

Euler's beam vibrations. In real systems the always existing structural bending cannot be neglected. Introducing damping in the form $d(\omega)\dot{w}''''$ (which is quite pragmatic and corresponds to the "hypothesis of convenience") one has

$$\rho A\ddot{w} + d(\omega)\dot{w}'''' + EIw'''' = 0 \tag{79}$$

for an Euler-Beam which may serve as a span bridge model for instance. Along with eq.(73) one obtains for the pinned-pinned case

$$\frac{\ddot{q}_n}{q_n} + \frac{d(\omega)}{\rho A}\left(\frac{n\pi}{L}\right)^4\frac{\dot{q}_n}{q_n} = -\frac{EI}{\rho A}\left(\frac{n\pi}{L}\right)^4 = -\omega_n^2. \tag{80}$$

Let $d(\omega) = 2\zeta \frac{EI}{\omega_n}$. Then, eq.(74) yields for the nth summation term

$$\frac{\ddot{q}_n}{q_n} + 2\zeta\omega_n \frac{\dot{q}_n}{q_n} = -\frac{EI}{\rho A}\left(\frac{n\pi}{L}\right)^4 - = -\omega_n^2$$

$$\Rightarrow \ddot{q}_n + 2\zeta\omega_n\dot{q}_n + \omega_n^2 q_n = 0,\ \ \omega_n = \sqrt{\frac{EI}{\rho A}}\left(\frac{n\pi}{L}\right)^2,\ \ w_n = \sin\left(\frac{n\pi}{L}x\right). \tag{81}$$

The eigenfunctions $w_n(x)$ are not influenced, the Ritz-coefficients $q_n(t)$ are obtained from a damped linear oscillator where damping increases for higher modes (which reflects daily experience):

$$w(x,t) = \sum_{i=1}^{n} w_i(x)q_i(t) = \sum_{i=1}^{n} \sin\left(\frac{i\pi x}{L}\right) e^{-\zeta\omega_i t}\left[q_{oi}\cos\nu_i t + \frac{\dot{q}_{io}}{\nu_i}\sin\nu_i t\right], \tag{82}$$

$\nu_i = \omega_i\sqrt{1-\zeta^2}$. Since the eigenmodes are decoupled, one may proceed in a similar way as above: excite the structure experimentally (with a hammer, for instance) to estimate the frequencies via signal analysis. Or: look at the time history simulation for different n and estimate its minimum requested number. Adding a force f_z which acts at $x = \xi$, one obtains the corresponding virtual work via eq.(79)

$$\delta W = \int_o^L \left[\rho A\ddot{w} + d(\omega)\dot{w}'''' + EIw'''' - f_z\delta(x-\xi)\right]\delta w dx = 0 \tag{83}$$

(δ : dirac distribution). The separation statement (74) yields

$$\begin{aligned} &\int_o^L \left[\rho A\sum w_i\ddot{q}_i + d(\omega)\sum w_i''''\dot{q}_i + EI\sum w_i q_i\right] w_j dx\delta q_j \\ &= f_z \int_o^L \left[w_j\delta(x-\xi)\right] dx\delta q_j \ \ \forall\ j. \end{aligned} \tag{84}$$

($w_i = \sin(i\pi x/L)$ is a priori known and does not undergo variations any more). Hence, one has $\int w_i w_j dx = 2L$ for $j = i$ and $\int w_i w_j dx = 0$ for $j \neq i$. Regarding $w_i'''' = (i\pi/L)^4 w_i$, one obtains, along with $\int[w_j\delta(x-\xi)]dx = w_i(\xi)$

$$\begin{aligned} &\ddot{q}_i + 2\zeta\omega_i\dot{q}_i + \omega_i^2 q_n = \frac{f_z}{2m}w_i(\xi), \quad m = \rho AL \\ \Rightarrow\ &q_i(t) = w_i(\xi)e^{-\zeta\omega_i t}\frac{p}{2m\nu_i}\sin(\nu_i t), \quad \nu_i = \omega_i\sqrt{1-\zeta^2} \end{aligned} \tag{85}$$

if f_z is modeled as $p\delta(t - to)$ (impulse at time $t = t_o = 0$). The solution

$$w(x,t) = \sum_{i=1}^{n} e^{\zeta\omega_i t}\frac{p}{2m\nu_i}\sin(\nu_i t)\sin[(i\pi/L)\xi]\sin[(i\pi/L)x] \qquad (86)$$

shows that every eigenmode $w_i(x)$ will take part as long as $w_i(\xi) \neq 0$ (controllability). However, exciting higher modes needs much energy. To give an impression: Figure 14 is initially calculated with $n = 50$. A recalculation with $n = 20$ does not show perceivable differences.

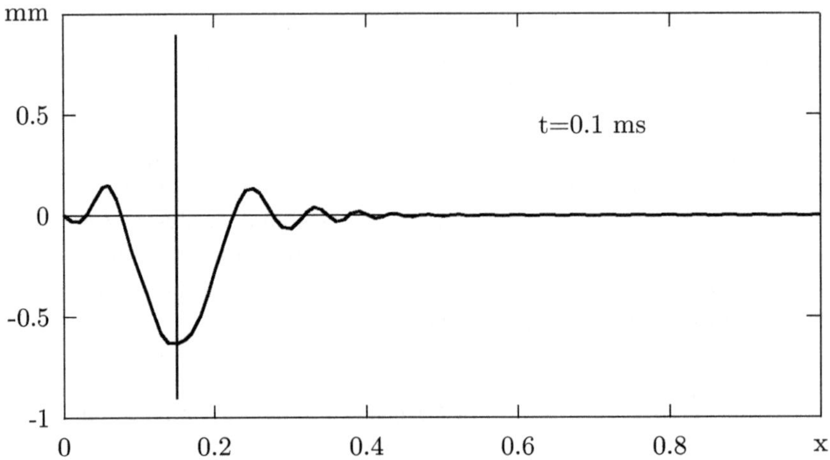

Figure 14. Time response after impact at $x = 0,15$

The piano chord – a tightened beam. In reality, a piano chord is more a beam than a string, especially for the lower tones. Then the modes are calculated with eq.(76): Let the string frequency (78) be ω_{sn}. One has then

$$\frac{\omega_n}{\omega_{s1}} = \sqrt{\left[\frac{EI\left(\frac{n\pi}{L}\right)^2}{f_x} + 1\right]} \cdot n$$

To obtain the string overtones $(\omega_{sn}/\omega_{s1}) = n, \;\; n = 1,2,3\cdots$ one needs quite high forces f_x. In a piano, these range up to 1,6 kN.

The Euler-Bernoulli-beam is a powerful model

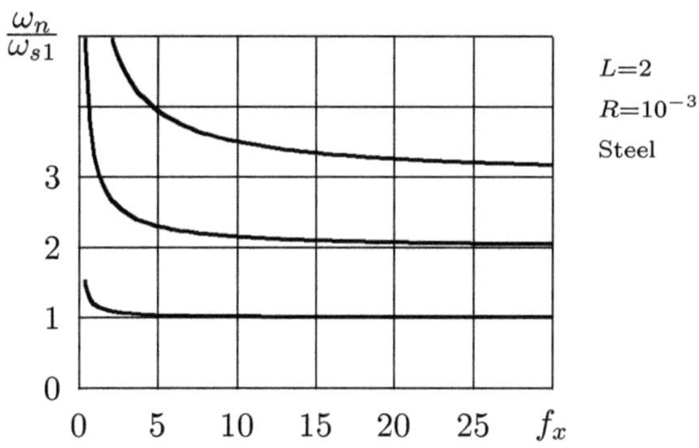

Figure 15. Tightened beam frequencies

4.2 Approximative solutions: how many shape functions?

Once the *number* of requested eigenfunctions is estimated, the eigenfunctions themselves remain unknown (for the general case). One may ask oneself how to represent these by a series expansion (the Ritz ansatz in this context). This task refers to expand an unknown function $f(x)$ in terms of $\mathbf{a}^T\mathbf{g}(x)$ with known functions $g_i(x)$, for instance to expand a periodic function into a Fourier series. The well known procedure is to minimize the (quadratic) error $\Delta = \frac{1}{2}\int[\mathbf{a}^T\mathbf{g}(x) - f(x)]^2 dx$ w.r.t. $\mathbf{a}$ yielding $\int[\mathbf{a}^T\mathbf{g}(x) - f(x)]\mathbf{g}(x)^T dx = 0$, thus

$$\left[\int \mathbf{g}(x)\mathbf{g}(x)^T dx\right]\mathbf{a} = \int f(x)\mathbf{g}(x)dx. \tag{87}$$

Using orthogonal beam functions $g_i(x) = v_i(x)$ with $v_i'''' - k_i^4 v_i = 0$, one obtains with an integration by parts

$$a_i = \int_o^L f(x)\frac{v_i''''}{k_i^4}dx = \frac{1}{k_i^4}\left[(f''v_i' - f'v_i'' + fv_i''')\,|_o^L - \int_o^L f'''v_i' dx\right]. \tag{88}$$

For free-free boundaries, $v_i''\,|_o^L$ and $v_i'''\,|_o^L$ vanish, hence

$$a_i = \frac{1}{k_i^3}\left[f''\left(\frac{v_i'}{k_i}\right)_o^L - \int_o^L f'''\left(\frac{v_i'}{k_i}\right)dx\right] \le \kappa_i \;\Rightarrow a_i \le \frac{\kappa_i}{k_i^3} \tag{89}$$

since $(v_i'/k_i) \in \{-2, +2\}$ and f'', f''' are assumed finite and steady. Because $k_i \simeq (i + \frac{1}{2})\pi$, the series $\mathbf{a}^T \mathbf{v}$ converges with $\simeq (1/i^3)$ for increasing i to $f(x)$ as a lower bound. An expansion of $f(x,y) = \mathbf{a}_{ij}^T u_i(x) v_j(y)$ with beam functions vor $u_i(x), i = 1 \cdots n$, and $v_j(y), j = 1 \cdots m$, resp., led Ritz to his series representation for a plate with free borders which converges rapidly, with $(1/(ij)^3)$, to its lower bound the existence of which is warranted with Rayleigh's quotient.

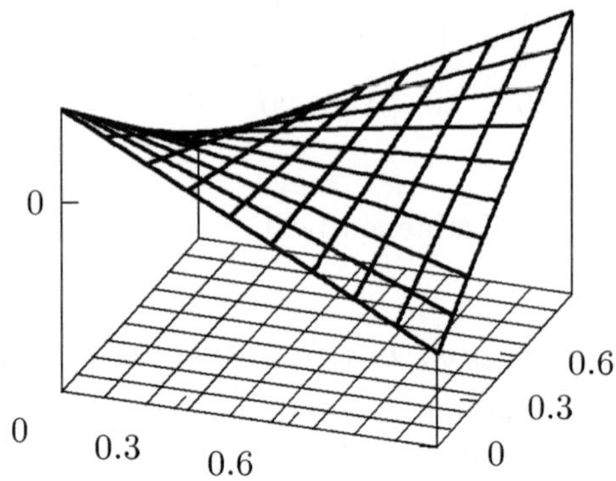

Figure 16. 1st plate eigenfunction

Let us consider Figure 17(a) which corresponds to Figure 8(a) with a stiff gear. The restoring matrix $\mathbf{K} = 0$ is known with eq.(36). We adress to the mass matrix calling attention to modeling aspects.

We assume the tip tensor of inertia $\mathbf{J}_{\text{tip}}$ negligible. Along with $\rho I_y \to 0$ and $\rho I_z \to 0$ (Euler beam) one obtains the mass matrix for a rotating beam

$$\mathbf{M} = \begin{bmatrix} \int \frac{\partial m}{\partial x} x^2 dx & \int x \mathbf{v}^T \frac{\partial m}{\partial x} dx & 0 \\ \int x \mathbf{v} \frac{\partial m}{\partial x} dx & \int \mathbf{v}\,\mathbf{v}^T \frac{\partial m}{\partial x} dx & 0 \\ 0 & 0 & \int \rho I_x \boldsymbol{\vartheta}\,\boldsymbol{\vartheta}^T dx \end{bmatrix} \tag{90}$$

where the sequence of coordinates is

$$\mathbf{y} = (\gamma \quad \mathbf{q}_v^T \quad \mathbf{q}_\vartheta^T)^T. \tag{91}$$

Is the Euler-Bernoulli beam really a powerful model? C_{Hub} does not arise anywhere which in eq.(24) led to sigularity. However, inserting

$$\frac{\partial m}{\partial x} = m_{\text{hub}} \delta(x - 0) + m_{\text{tip}} \delta(x - x|_e) + \rho A \tag{92}$$

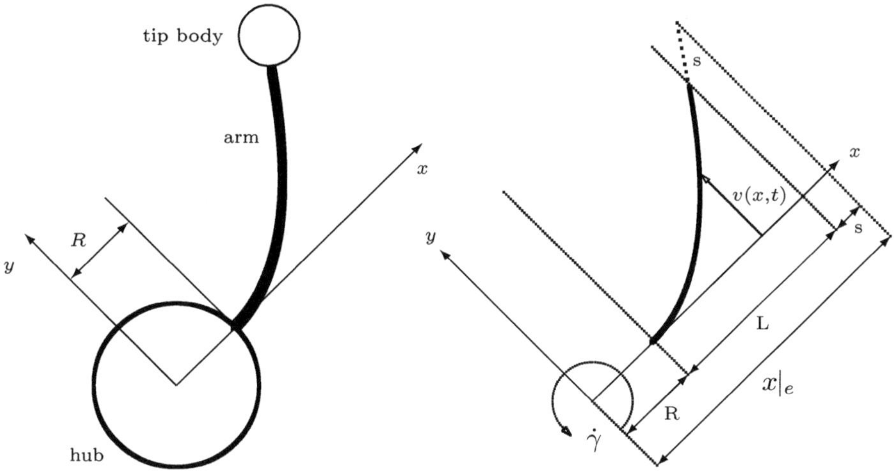

Figure 17. (a) beam model, (b) coordinates

(δ: Dirac distribution) where, see Figure 17(b)

$$x|_e = (R + L + s), \quad \mathbf{v}|_e = (\mathbf{v}_L + \mathbf{v}'_L s), \quad \mathbf{v}'|_e = \mathbf{v}'_L \tag{93}$$

yields

$$\mathbf{M} = \begin{bmatrix} m_{\text{beam}}\frac{L^2}{3} + m_{\text{tip}} x|_e^2 & \int (x\mathbf{v}^T)\rho A dx + (x\mathbf{v}^T)|_e\, m_{\text{tip}} & 0 \\ \int (x\mathbf{v})\rho A dx + (x\mathbf{v})|_e\, m_{\text{tip}} & \int (\mathbf{v}\,\mathbf{v}^T)\rho A dx + (\mathbf{v}\,\mathbf{v}^T)|_e\, m_{\text{tip}} & 0 \\ 0 & 0 & \int \rho I_x \boldsymbol{\vartheta}\,\boldsymbol{\vartheta}^T dx \end{bmatrix}, \tag{94}$$

thus $\mathbf{M} > 0$. The foregoing holds as long as $m_{tip} x|_e^2 >> C_{beam} + C_{Hub} + C_{tip}$. Otherwise, the Rayleigh beam comes into play. *Modeling needs careful interpretation.*

At a first glance, it suggests itself to use beam eigenfunctions for shape functions since convergence is warranted. For a beam with tip mass, for instance, the series $\mathbf{a}^T\mathbf{v}$ with $\mathbf{v} \in I\!R^n$ will converge to the n-th bending form $v_{n,eigen}$ with $\sim (1/n)^3$, see eq.(89). The number of requested shape functions will remain transparent. However, convergence here depends on the beam bending functions with its fourth order differential equation. The situation will change dramatically for torsional deflections with its second order differential equation.

Quality of shape functions – increasing the convergence. Let us consider a torsional shaft with disks attached at its ends. The shaft shall be clamped at $x = 0$ ($\gamma = const.$)

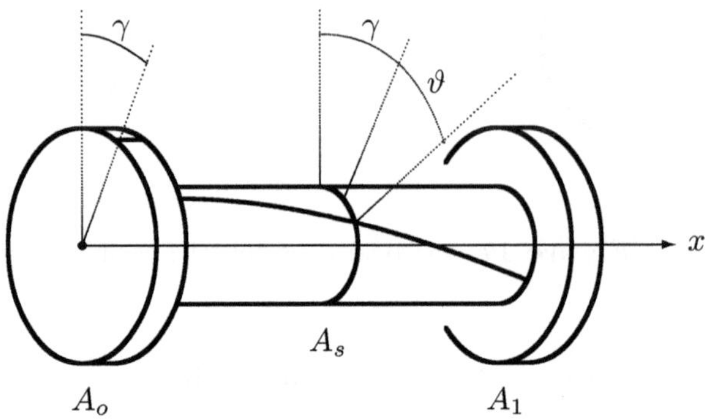

Figure 18. torsional shaft

The eigenfunctions are known with $\vartheta = \sin\left(\frac{k_i x}{L}\right)$ where $\tan k_i = \frac{A_s}{A_1}\frac{1}{k_i}$ with $A_s = \rho I_x L$. (They are easily obtained from Hamilton's princple for instance). The first three eigenfunctions are sketched in Figure 19(a). From eq.(87) we obtain

$$\mathbf{a} = \left[\int \mathbf{g}(x)\mathbf{g}(x)^T dx\right]^{-1} \int f(x)\mathbf{g}(x)dx. \tag{95}$$

Using *orthogonal* beam functions $g_i(x) = \vartheta_{bi}(x)$ with $\vartheta''_{bi} + k_i^2\vartheta_{bi} = 0$, one obtains with an integration by parts

$$a_i = -\int_o^L f(x)\frac{\vartheta''_{bi}}{k_i^2}dx = \frac{1}{k_i}\left[-f\left(\frac{\vartheta'_{bi}}{k_i}\right)|_o^L + \int_o^L f'\left(\frac{\vartheta'_{bi}}{k_i}\right)dx\right] \leq \frac{\kappa}{k_i}. \tag{96}$$

Since $(\vartheta'_{bi}/k_i) = \cos(k_i x) \in \{-1, +1\}$, $k_i \simeq (i - \frac{1}{2})\pi$, and f, f' are assumed finite and steady, the series $\mathbf{a}^T\boldsymbol{\vartheta}_b$ converges *slowly* with $\simeq (1/i)$ for increasing i.

Figure 19(b) shows the right end (magnified) of the second eigenfunction ϑ_{2e}, approximated with n=50, 100, 500 beam functions $\boldsymbol{\vartheta}_b^T = \left(\sin\left(\frac{\pi}{2}x\right), \sin\left(\frac{3\pi}{2}x\right), \cdots \sin\left(\frac{(2n-1)\pi}{2}x\right)\right)$. It seems obvious that a very high number of shape functions is needed, since none of these show a slope at the right end. This leads to the well known effect that the approximation forms some kind of oscillation around the true solution.

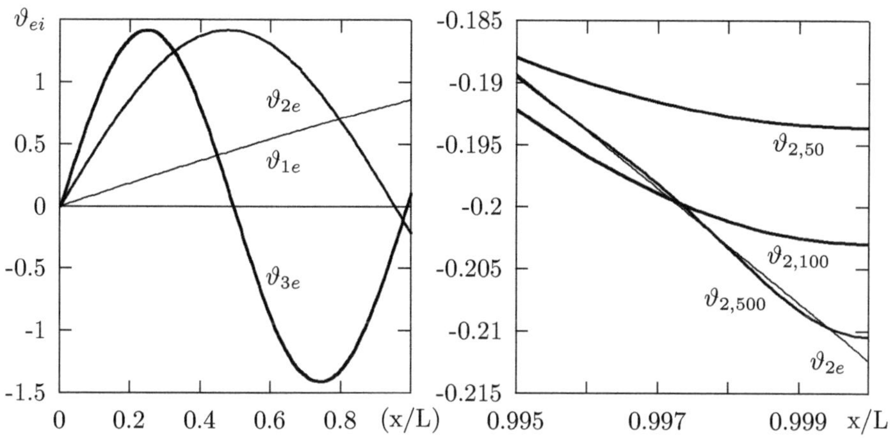

Figure 19. (a) EF (exact), (b) 2.EF approximations (magnified)

Being aware of this, one might add an auxiliary function with a slope at the right end: $\boldsymbol{\vartheta}_{ab}^T = \left(\underline{\sin(\pi x)}, \sin\left(\frac{\pi}{2}x\right), \sin\left(\frac{3\pi}{2}x\right), \cdots \sin\left(\frac{(2n-1)\pi}{2}x\right)\right)$. These functions are *not orthogonal*, thus the inverse in eq.(95) enters the consideration and the convergence is not so easy to check. However, the result shown in Figure 20(b) is tremendeous. Here, only four shape functions are used. Figure 20(a) depicts the first four approximated eigenfunctions for this case; the deviations become apparent with the highest (the fourth) eigenfunction.

How to construe auxiliary functions? Here, a physical interpretation will be helpful. Considering rapidly moving higher modes, the inertia of the disk comes more and more into play and will act in the limiting case as a clamped support (notice that $\sin(\pi x)$ is the first eigenmode for this case). This tendency can already be seen in Figure 19 (a) where the higher modes tend more and more to zero at the right end.

One has thus also the recipe at hand how to treat the bending beam with

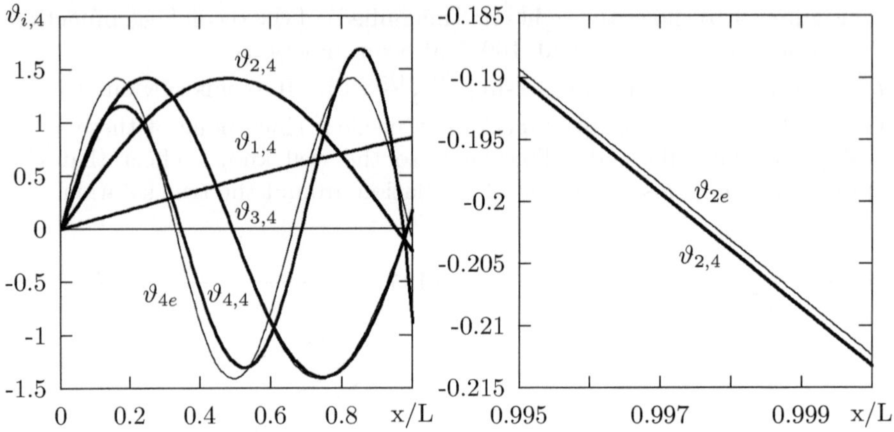

Figure 20. (a) EF (approximated), (b) 2.EF approximations (magnified)

a tip mass acording to Figure 17(a): for a point mass (rotational inertia neglected), the auxiliary function corresponds to the clamped-pinned case, for a rigid body (with rotational inertia), it is represented by the clamped-clamped case.

Being aware of the quality of shape functions, one could start to calculate the motions of a rotating beam, see Figure 17, for instance. The mass matrix is known from eq.(94) and the restoring matrix from eq.(36), respectively, which contains the influences of the second order elastic displacements. However, the latter need reconsideration.

Second order field effects. A simple case will give insight into the problem: the "linear robot" (only linear guided motions, no rotation) according to Figure 21.

Same as in the case of centrifugal reaction forces $df_x = \rho A \omega^2 dx$ which led to

$$\int_o^L \int_x^L \rho A(\ \omega^2 \xi \ ) d\xi \ \mathbf{v}' \mathbf{v}'^T dx \ \mathbf{q} \tag{97}$$

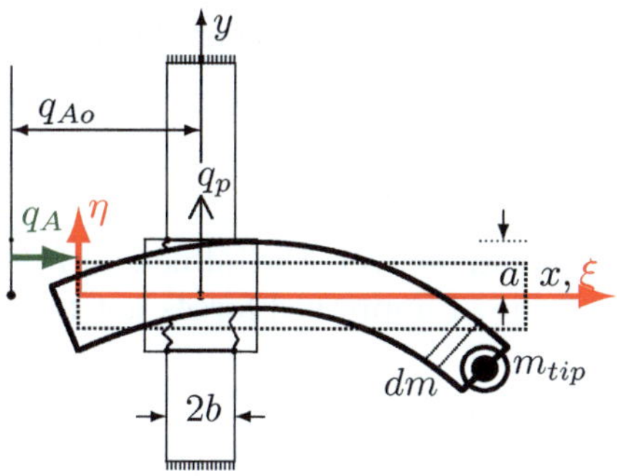

Figure 21. A "linear robot"

$$\Rightarrow \mathbf{K}_{cent}\mathbf{y} = \begin{bmatrix} 0 & 0 \\ 0 & \int_o^L \rho A \left[\frac{1}{2}\omega_o^2(L^2{-}x^2)\right] \mathbf{v}'\mathbf{v}'^T dx \end{bmatrix} \begin{pmatrix} \gamma \\ \mathbf{q} \end{pmatrix} \tag{98}$$

(see eq.(30)), we now have a reaction force $df_x = -\rho A\, \ddot{q}_A dx$ which leads to

$$-\int_o^L \int_x^L \rho A(\ \ddot{q}_A \xi\ ) d\xi\ \mathbf{v}'\mathbf{v}'^T dx\ \mathbf{q} \tag{99}$$

and consequently

$$\mathbf{K}_{acc}\mathbf{y} = \begin{bmatrix} 0 & 0 & 0 \\ 0 & 0 & 0 \\ 0 & 0 & \int_o^L \rho A \ddot{q}_A (x-L) \mathbf{v}'\mathbf{v}'^T dx \end{bmatrix} \begin{pmatrix} q_p \\ q_A \\ \mathbf{q} \end{pmatrix} \tag{100}$$

However, eq.(100) may be reorganized to

$$\mathbf{M}_{acc}\ddot{\mathbf{y}} = \begin{bmatrix} 0 & 0 & 0 \\ 0 & 0 & 0 \\ 0 & \int_o^L \rho A (x-L) \mathbf{v}'\mathbf{v}'^T dx\, \mathbf{q} & 0 \end{bmatrix} \begin{pmatrix} \ddot{q}_p \\ \ddot{q}_A \\ \ddot{\mathbf{q}} \end{pmatrix} \tag{101}$$

This is essential. $\mathbf{M}_{acc}$ has to be assigned to the mass matrix which later needs inversion! We can thus draw the following conclusions:

- reaction forces:
  - accelerations to mass matrix (mandatory)

- consequently coriolis and centrifugals to gyroscopic matrix (not necessary)
- impressed forces
  - to restoring matrix

In case of moderate maneuvers the influence of the reaction forces may be negligible, but that one of impressed forces not (see buckling, for instance).

4.3 Frequency tuning

Frequency tuning plays an importat role in modeling. It starts already with the above mentioned piano: the harmony model is based on the string overtones which might be achieved with strongly tightened beams. Thereby, at least for occidential ears, the third ($\omega/\omega_o = 5/4$) and the quint ($\omega/\omega_o = 3/2$) are the most harmonic sequences. Then, after a certain sequence of quints, one shoud meet the basic tone (or rather one of its octaves). However, with 12 quints one has $(3/2)^{12} = 129,75$, while 7 octaves yield $2^7 = 128$. Hence, since the sequence of octaves is mandatory, one has to adjust the tone intervals anyhow. This busied a lot of scientists, from Huygens and Euler up to d'Alembert [1], just to mention a few of them. Some people believe that the great J.S. Bach gave a crytographic rule on the first page of his welltempered clavier – nobody knows, it remains a mistery.

Misteries are not allowed in mechanical modeling. Setting $q_A = const.$ in Figure 21, one arrives, along with stiff gears, to a gantry robot model, or the crane problem, Figure 22. Here, an open loop control can easily be modeled: Once the beam frequency ω_{beam} is known, then a desired q–motion for an end position q_e with maneuver time T can be calculated as

$$\begin{aligned} q(t)_d &= \tfrac{q_e}{2\pi}(\Omega t - \sin\Omega t) \\ \text{where } \Omega &= \tfrac{2\pi}{T} = \tfrac{1}{2}\omega_{\text{beam}}. \end{aligned} \tag{102}$$

The payload motion results then in

$$y(t) = \frac{q_e}{6\pi}\left[\frac{1}{2}\sin\omega_{\text{beam}}t - \sin\Omega t\right] \tag{103}$$

where the vanishing slope at the beginning and at the end of the maneuver yield a soft placement of the payload. The force f_q which is needed for q is then proportional to $\sin\Omega t$. This scenario is typical for any point-to-point maneuver: at the beginning, the payload motion y stays behind the driver's position as seen from the driver's cabin. After a while (at half the maneuver time in the present), the driver changes from acceleration to deceleration.

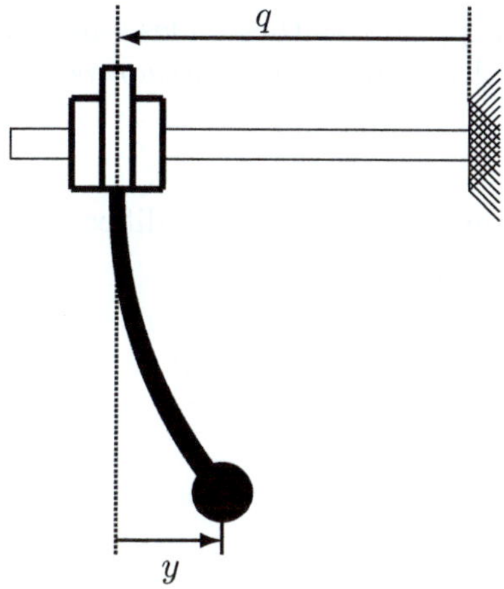

Figure 22. The crane model

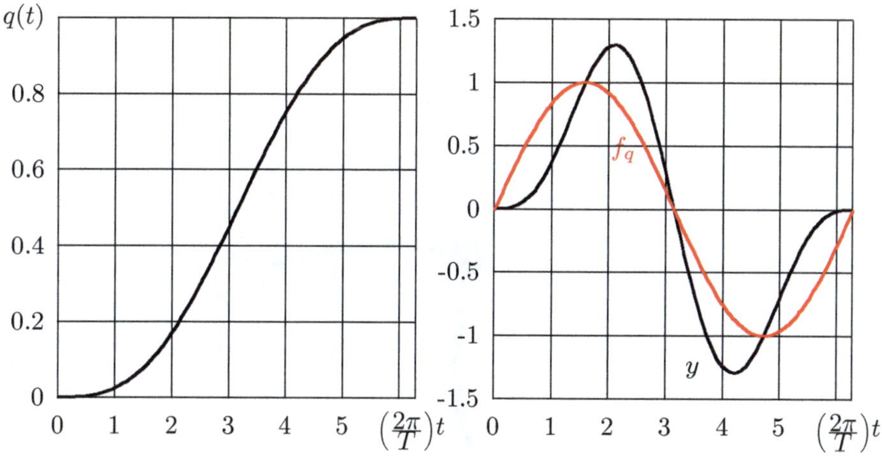

Figure 23. (a) driver motion (b) force and payload motion

The payload gets ahead such that at the end of the maneuver it comes smoothly to rest, see Figure 23.

This is, in principal, the behaviour of any point-to-point maneuver, also in closed loop controlled system. The crane problem consideration is helpful to interpret the results. Also the slewing maneuver of the Truckenbrodt model (see eq.(33) et seq.) obbeys these rules; any model assumptions there have been experimentally verified. *Verification is the last step in modeling.*

4.4 Procedures

Having a look on the fore arm (partial differential) equation of a plane elastic double pendulum (see Figure 24),

$$
\begin{aligned}
&\underline{\frac{\partial m_2}{\partial x_2}\ddot{v}_2 - \frac{\partial}{\partial x_2}\left[\frac{\partial J_2}{\partial x_2}\ddot{v}_2'\right] + \frac{\partial^2}{\partial x_2^2}\left[(EI)_2 v_2''\right]} \\
&+\frac{\partial m_2}{\partial x_2}\Big\{ \left[L_1\cos\gamma_2 - (L_1 v_{1_L}' - v_{1_L})\sin\gamma_2 + x_2\right]\ddot{\gamma}_1 + x_2\ddot{\gamma}_2 \\
&+ \left[L_1 \sin\gamma_2 + (L_1 v_{1_L}' - v_{1_L})\cos\gamma_2 - v_2\right]\dot{\gamma}_1^2 - v_2\dot{\gamma}_2^2 \\
&+ \left[2\dot{v}_{1_L}\sin\gamma_2 - 2v_2\dot{\gamma}_2\right]\dot{\gamma}_1 + g\left[\sin(\gamma_1+\gamma_2) + v_{1_L}'\cos(\gamma_1+\gamma_2)\right] \\
&+ (\ddot{v}_{1_L}\cos\gamma_2 + x_2\ddot{v}_{1_L}'')\Big\} - \frac{\partial}{\partial x_2}\left[(\ddot{\gamma}_1+\ddot{\gamma}_2+\ddot{v}_{1_L}')\frac{\partial J_2}{\partial x_2}\right] \\
&-\frac{\partial}{\partial x_2}\left[\int_{x_2}^{L_2}\left[g\cos(\gamma_1+\gamma_2) - L_1\ddot{\gamma}_1\sin\gamma_2 + L_1\dot{\gamma}_1^2\cos\gamma_2 + \xi(\dot{\gamma}_1+\dot{\gamma}_2)^2\right]\frac{\partial m_2}{\partial \xi}d\xi\, v_2'\right] \\
&= 0,
\end{aligned}
\tag{104}
$$

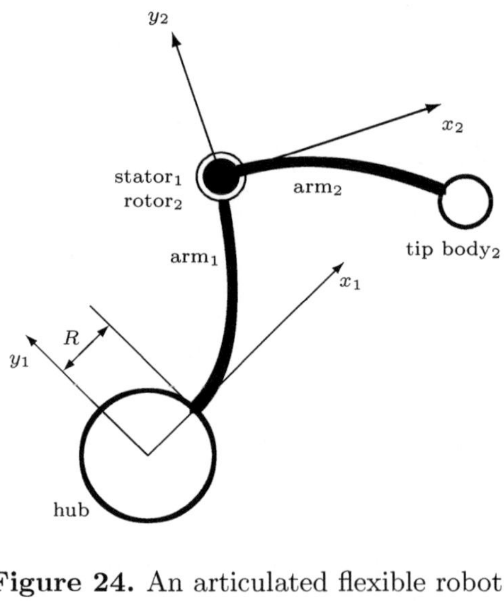

Figure 24. An articulated flexible robot

it becomes obvious that effective procedures are needed. Without going into detail, one interprets the first line of eq.(104) as the Rayleigh beam equation, and the remainder comes from gravity and dynamical reactions from the upper arm [9]. To find a solution, Galerkin's procedure drops out due to the corresponding boundary conditions. Considering the boundary conditions within the approximation procedure (the "extended Galerkin") is tedious. But integrating these one obtains the "direct Ritz" as pointed out in eq.(64). This was the very first step in direction of effectivity.

Eq.(64) considers the elastic rotor which will serve for demonstration: One has simply to set $v(x,t) = \mathbf{v}(x)^T \mathbf{q}_v(t)$, $\delta v(x,t) = \delta \mathbf{q}_v(t)^T \mathbf{v}(x)$ etc. If one concentrates on the $\ddot{\alpha}$-depending terms,

$$\begin{aligned} \delta W &= \int_o^L \ddot{\alpha}\rho \Big\{ -\delta v\, Aw \underline{+\delta v'\, Iw'} + \delta w\, Av - \delta w'\, Iv' \Big\}\, dx + \cdots \\ &= (\delta \mathbf{q}_v^T\, \delta \mathbf{q}_w^T) \int_o^L \ddot{\alpha}\rho \begin{bmatrix} -(A\mathbf{v}\mathbf{w}^T - I\mathbf{v}'\mathbf{w}'^T)\mathbf{q}_w \\ +(A\mathbf{w}\mathbf{v}^T - I\mathbf{w}'\mathbf{v}'^T)\mathbf{q}_v \end{bmatrix} dx + \cdots = 0, \end{aligned} \tag{105}$$

one obtains the **N**-matrix (nonconservative restoring matrix) which reads

$$\mathbf{N}\,\mathbf{q} = \ddot{\alpha} \int_o^L \rho \begin{bmatrix} 0 & -[A\mathbf{v}\mathbf{w}^T - I\mathbf{v}'\mathbf{w}'^T] \\ +[A\mathbf{w}\mathbf{v}^T - I\mathbf{w}'\mathbf{v}'^T] & 0 \end{bmatrix} dx \begin{pmatrix} \mathbf{q}_v \\ \mathbf{q}_w \end{pmatrix}. \tag{106}$$

Notice, however, that eq.(105) needed a tedious correction (underlined in eq.(105)).

On the other hand, we have eq.(66) for deriving the motion equations. We define v and w as functions w.r.t. the undeformed system (moving frame). The mass center velocities are, for the moving frame representation,

$$\mathbf{v}_c = \begin{pmatrix} 0 \\ \dot{v} - \dot{\alpha} w \\ \dot{w} + \dot{\alpha} v \end{pmatrix}, \quad \boldsymbol{\omega}_c = \begin{pmatrix} \dot{\alpha} \\ -\dot{w}' \\ +\dot{v}' \end{pmatrix}. \tag{107}$$

The inertia tensor for a moving frame representation reads $d\mathbf{J} = [\mathbf{E} - \widetilde{\boldsymbol{\varphi}}]^T \rho \mathbf{I} dx\, [\mathbf{E} - \widetilde{\boldsymbol{\varphi}}]$ where $\mathbf{I} = I\, diag\{2,\, 1,\, 1\}$ (circular cross section). The bending angles are $\boldsymbol{\varphi} = (0 \quad -w' \; v')^T$. Thus, the first order momenta read

$$d\mathbf{p} = \begin{pmatrix} 0 \\ \dot{v} - \dot{\alpha} w \\ \dot{w} + \dot{\alpha} v \end{pmatrix} \rho A dx, \quad d\mathbf{L} = \begin{pmatrix} 2\dot{\alpha} \\ \dot{\alpha} v' - \dot{w}' \\ \dot{\alpha} w' + \dot{v}' \end{pmatrix} \rho I dx. \tag{108}$$

Inserting the Ritz ansatz into eq.(107) yields

$$\mathbf{v}_c = \begin{bmatrix} 0 & 0 \\ +\mathbf{v}^T & 0 \\ 0 & +\mathbf{w}^T \end{bmatrix} \dot{\mathbf{q}} + \begin{bmatrix} 0 & 0 \\ 0 & -\dot{\alpha}\mathbf{w}^T \\ +\dot{\alpha}\mathbf{v}^T & 0 \end{bmatrix} \mathbf{q}, \tag{109}$$

$$\boldsymbol{\omega}_c = \begin{bmatrix} 0 & 0 \\ 0 & -\mathbf{w}'^T \\ +\mathbf{v}'^T & 0 \end{bmatrix} \dot{\mathbf{q}} + \begin{pmatrix} \dot{\alpha} \\ 0 \\ 0 \end{pmatrix}, \tag{110}$$

the functional matrices (Jacobians) are thus

$$\left(\frac{\partial \mathbf{v}_c}{\partial \dot{\mathbf{q}}}\right) = \begin{bmatrix} 0 & 0 \\ +\mathbf{v}^T & 0 \\ 0 & +\mathbf{w}^T \end{bmatrix}, \left(\frac{\partial \boldsymbol{\omega}_c}{\partial \dot{\mathbf{q}}}\right) = \begin{bmatrix} 0 & 0 \\ 0 & -\mathbf{w}'^T \\ +\mathbf{v}'^T & 0 \end{bmatrix}. \tag{111}$$

Along with $\boldsymbol{\omega}_R = (\dot{\alpha}\ 0\ 0)$, the motion equation reads

$$\begin{aligned} &\int_o^L \begin{bmatrix} 0 & 0 & +\mathbf{v}' \\ 0 & -\mathbf{w}' & 0 \end{bmatrix} \begin{pmatrix} 2\ddot{\alpha} \\ -\ddot{w}' - \dot{\alpha}^2 w' + \ddot{\alpha} v' \\ +\ddot{v}' + \dot{\alpha}^2 v' + \ddot{\alpha} w' \end{pmatrix} \rho I dx + \\ &\int_o^L \begin{bmatrix} 0 & +\mathbf{v} & 0 \\ 0 & 0 & +\mathbf{w} \end{bmatrix} \begin{pmatrix} 0 \\ \ddot{v} - 2\dot{\alpha}\dot{w} - \dot{\alpha}^2 v - \ddot{\alpha} w \\ \ddot{w} + 2\dot{\alpha}\dot{v} - \dot{\alpha}^2 w + \ddot{\alpha} v \end{pmatrix} \rho A dx. \end{aligned} \tag{112}$$

Once more concentrating on $\mathbf{N}$, one has

$$\begin{aligned} \mathbf{N}\,\mathbf{q} = &\int_o^L \begin{bmatrix} 0 & 0 & +\mathbf{v}' \\ 0 & -\mathbf{w}' & 0 \end{bmatrix} \begin{pmatrix} 0 \\ \ddot{\alpha} v' \\ \ddot{\alpha} w' \end{pmatrix} \rho I dx + \\ &\int_o^L \begin{bmatrix} 0 & +\mathbf{v} & 0 \\ 0 & 0 & +\mathbf{w} \end{bmatrix} \begin{pmatrix} 0 \\ -\ddot{\alpha} w \\ +\ddot{\alpha} v \end{pmatrix} \rho A dx. \end{aligned} \tag{113}$$

One obtains automatically the correct signs. The procedure (66) is easy (matrix-vector multiplications, no directional derivatives any more). After all, eq.(66) shows clear cut advantages, but how to deal with nonholonomic constraints?

5 Nonholonomicity and Redundancy

Including nonholonomic constraints leads to the Projection Equation as most powerful and easy to handle procedure. A fair comparison of methods is obtained through the Central Equation of Dynamics. Focusing on modeling aspects, the decomposition into subsystems leads to their topology, the interchangeability of submodels, and their interactions. The corresponding equations are either obtained in minimal form or as a recursive scheme which saves a tremendeous amount of CPU time. Disposing of modern computers: should one confine oneself with rigid body segmentation (or lumped mass) models? A simple example: to obtain the first mode of a clamped beam with a certain accuracy needs one hundred spring interconnected rigid links – this is not convincing. Or the other way round: Why not use commercial FE-codes, also for beamlike (sub-) structures? An example: the motions of a rolling mill turned out inexecutable. Reconsideration saved the CPU time with a factor of one year to half an hour – isn't this convincing?

5.1 Analytical methods

In case of nonholonomic variables s_m one can not conclude from

$$\sum_n \left[\frac{d}{dt}\frac{\partial T}{\partial \dot{q}_n} - \frac{\partial T}{\partial q_n} - \mathbf{Q}_n \right] \delta q_n = 0 \tag{114}$$

on the Lagrange equations because δq_n is not arbitrary any more. Hence

$$\sum_{n=1}^{f} \left[\frac{d}{dt}\frac{\partial T}{\partial \dot{q}_n} - \frac{\partial T}{\partial q_n} - \mathbf{Q}_n \right] \sum_{m=1}^{g} \left(\frac{\partial q_n}{\partial s_m} \right) \delta s_m = 0 \tag{115}$$

(Maggi, [37]). Hereby, the relationship $(\partial \mathbf{q}/\partial \mathbf{s})$ is always known: Assume $\mathbf{q} = \mathbf{q}(\mathbf{s})$. Then $\dot{\mathbf{q}} = (\partial \mathbf{q}/\partial \mathbf{s})\dot{\mathbf{s}}$ yields by differentiation w.r.t. $\dot{\mathbf{s}}$ simply $(\partial \dot{\mathbf{q}}/\partial \dot{\mathbf{s}}) = (\partial \mathbf{q}/\partial \mathbf{s})$ (same as in eq.(52) et seq.). In eq.(115), f denotes the positional degree of freedom and g the velocity degree of freedom, $g = f$ for the holonomic case. Notice that the Lagrangean part in eq.(115) needs a full range calculation and, afterwards, replacement of $\dot{\mathbf{q}} \to \dot{\mathbf{s}}$ etc. This is is extremely laborious.

Hamel [23] derived his famous *Euler-Lagrange-Equations* which in its variational form reads (for derivation see for instance eq.(128))

$$\sum_n \left[\frac{d}{dt}\frac{\partial T}{\partial \dot{s}_n} - \frac{\partial T}{\partial s_n} - \mathbf{Q}_n \right] \delta s_n + \sum_\nu \frac{\partial T}{\partial \dot{s}_\nu} \left[\frac{d\delta s_\nu - \delta d s_\nu}{dt} \right] = 0. \tag{116}$$

Of course, the kinetic energy T does not depend on s_n directly but $T = T(\mathbf{q}, \dot{\mathbf{s}})$. What is meant by $(\partial T/\partial s_n)$ is simply $(\partial T/\partial \mathbf{q})(\partial \mathbf{q}/\partial s_n)$ with known $(\partial \mathbf{q}/\partial s_n)$ (see above). Hamel says "I found them [the equations] when I asked myself: which equations replace the Lagrangean ones, when

I insert, besides the n position variables $\mathbf{q}$, some other n independent linear combinations of $d\mathbf{q}/dt$?" Thus, at first, nonholonomic constraints are not considered (same number n). In the first term of eq.(116), δs_n is factored out while in the second it is somehow hidden. But along with $\delta s_\nu = \sum_i(\partial s_\nu/\partial q_i)\delta q_i \Rightarrow d\delta s_\nu = \sum_{i,k}(\partial^2 s_\nu/\partial q_i\partial q_k)\delta q_i dq_k + \sum_i(\partial s_\nu/\partial q_i)d\delta q_i$ one obtains

$$\frac{d\delta s_\nu - \delta ds_\nu}{dt} = \sum_{i,k}\left[\frac{\partial^2 s_\nu}{\partial q_i\partial q_k} - \frac{\partial^2 s_\nu}{\partial q_k\partial q_i}\right]\delta q_i\dot{q}_k + \sum_i \frac{\partial s_\nu}{\partial q_i}(d\delta q_i - \delta dq_i). \quad (117)$$

Using a variational approach $\delta\mathbf{q} = (\partial\mathbf{q}/\partial\varepsilon)|_{\varepsilon=0}\cdot\varepsilon = \varepsilon\cdot\boldsymbol{\eta}$ one obtains $d\delta\mathbf{q} - \delta d\mathbf{q} = \epsilon\cdot d\boldsymbol{\eta} - \epsilon\cdot d\boldsymbol{\eta} = 0$: Then (and only then), the last term in eq.(117) vanishes and the remainder [the brackets to be more precise] represents Schwarz' integrability rule – which is, preconditionally, not fulfilled for nonholonomic variables. However, $\delta q_i = \sum_n(\partial q_i/\partial s_n)\delta s_n$ and $\dot{q}_k = \sum_\mu(\partial q_k/s_\mu)\dot{s}_\mu$ are, of course, known. Insertion into eq.(117) and eq.(116), δs_n is also factored out for the second term. Since δs_n is arbitrary, one obtains the nth motion equation

$$\sum_n\left[\frac{d}{dt}\frac{\partial T}{\partial\dot{s}_n} - \frac{\partial T}{\partial s_n} - \mathbf{Q}_n\right] + \sum_n\sum_{\nu,\mu}\left[\frac{\partial T}{\partial\dot{s}_\nu}\dot{s}_\mu\beta_\nu^{\mu,n}\right] = 0$$
$$\text{where } \beta_\nu^{\mu,n} = \sum_{i,k}\frac{\partial q_k}{\partial s_\mu}\frac{\partial q_i}{\partial s_n}\left(\frac{\partial^2 s_\nu}{\partial q_i\partial q_k} - \frac{\partial^2 s_\nu}{\partial q_k\partial q_i}\right). \quad (118)$$

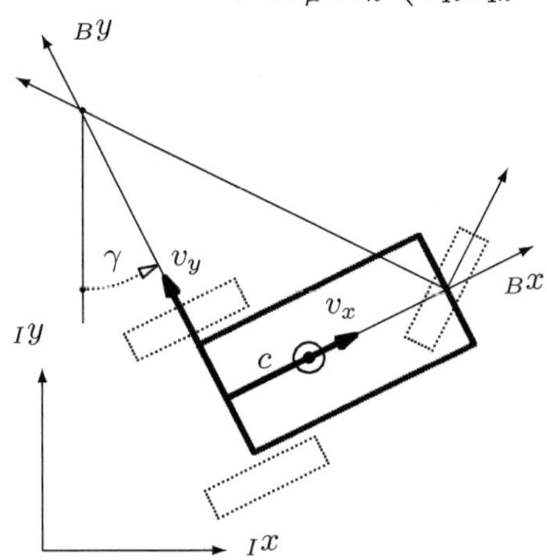

Figure 25. Carriage

Example. Carriage: The reference frame origin lies on the rear axle; the rear wheels are not allowed to slide sideways (nonholonomic constraint). The car is able to reach every point within the inertial x-y-plane and orient itself arbitrarily (angle γ). Using the inertial coordinates $\mathbf{q}^T = (x\ y\ \gamma)$, the rear axle velocities (represented in the body fixed reference frame) are

$$\dot{\mathbf{s}} = \begin{pmatrix} v_x \\ v_y \\ \omega_z \end{pmatrix} = \begin{bmatrix} \cos\gamma & \sin\gamma & 0 \\ -\sin\gamma & \cos\gamma & 0 \\ 0 & 0 & 1 \end{bmatrix} \begin{pmatrix} \dot{x} \\ \dot{y} \\ \dot{\gamma} \end{pmatrix} := \left(\frac{\partial \mathbf{s}}{\partial \mathbf{q}}\right) \dot{\mathbf{q}} \in I\!R^{f=3}. \tag{119}$$

The kinetic energy refers to the mass center velocities

$$T = \frac{m}{2}\left[v_x^2 + (v_y + c\omega_z)^2\right] + \frac{C^c}{2}\omega_z^2 = \frac{m}{2}\left(v_x^2 + v_y^2\right) + \underline{mcv_y\omega_z} + \frac{C^o}{2}\omega_z^2. \tag{120}$$

Since $(\partial\mathbf{s}/\partial\mathbf{q}) \in I\!R^{3,3}$ is orthonormal for this example, its inverse $(\partial\mathbf{q}/\partial\mathbf{s})$ is simply the transpose. The Hamel-coefficients (118) are

$$\begin{aligned}
\beta_1^{1,2} &= 0 \\
\beta_1^{1,3} &= -\frac{\partial q_1}{\partial s_1}\frac{\partial^2 s_1}{\partial q_1 \partial q_3} - \frac{\partial q_2}{\partial s_1}\frac{\partial^2 s_1}{\partial q_2 \partial q_3} = (-\cos\gamma)(-\sin\gamma) - (\sin\gamma)(\cos\gamma) = 0 \\
\beta_1^{2,3} &= -\frac{\partial q_1}{\partial s_2}\frac{\partial^2 s_1}{\partial q_1 \partial q_3} - \frac{\partial q_2}{\partial s_2}\frac{\partial^2 s_1}{\partial q_2 \partial q_3} = (\sin\gamma)(-\sin\gamma) - (\cos\gamma)(\cos\gamma) = -1.
\end{aligned} \tag{121}$$

Analoguously, one obtains for $\nu = 2$: $\beta_2^{1,2} = 0, \beta_2^{1,3} = 1, \beta_2^{2,3} = 0$ etc. Hence, the motion equations read

$$\left[\frac{d}{dt}\frac{\partial T}{\partial v_x} + \underbrace{\beta_2^{3,1}}_{-1}\dot{\gamma}\,\frac{\partial T}{\partial v_y} - Q_1\right]\delta s_1 = \left[m\dot{v}_x - \underline{mc\dot{\gamma}^2 - m\dot{\gamma}v_y} - f_x\right]\delta s_1 = 0$$

$$\left[\frac{d}{dt}\frac{\partial T}{\partial v_y} + \underbrace{\beta_1^{3,2}}_{+1}\dot{\gamma}\,\frac{\partial T}{\partial v_x} - Q_2\right]\delta s_2 = \left[\underline{m\dot{v}_y} + mc\ddot{\gamma} + m\dot{\gamma}v_x - f_y\right]\delta s_2 = 0$$

$$\left[\frac{d}{dt}\frac{\partial T}{\partial \dot{\gamma}} + \underbrace{\beta_1^{2,3}}_{-1}v_y\frac{\partial T}{\partial v_x} + \underbrace{\beta_2^{1,3}}_{+1}v_x\frac{\partial T}{\partial v_Q} - Q_3\right]\delta s_3 =$$

$$\left[C^o\ddot{\gamma} + \underline{mc\dot{v}_y + msv_x\dot{\gamma}} - M_z\right]\delta s_3 = 0 \tag{122}$$

Considering the nonholonomic constraint $\dot{s}_2 = v_y = 0$ yields $\delta s_2 = 0$ (null!), while δs_1 und δs_3 are arbitrary. One has thus

$$\Rightarrow \begin{bmatrix} m & 0 \\ 0 & C^o \end{bmatrix}\begin{pmatrix} \dot{v}_x \\ \ddot{\gamma} \end{pmatrix} + \begin{bmatrix} 0 & -mc\dot{\gamma} \\ +mc\dot{\gamma} & 0 \end{bmatrix}\begin{pmatrix} v_x \\ \dot{\gamma} \end{pmatrix} = \begin{pmatrix} f_x \\ M_z \end{pmatrix} \tag{123}$$

Also Hamel's equations need a full range calculation at first (f=3). The nonholonomic constraint has to be considered afterwards. Setting $v_y = 0$ already in T [eq.(120), underlined] would yield a loss of information because then, due to the required partial derivatives, the underlined terms in eq.(122) vanish and the gyroscopic matrix in eq.(123) gets lost. The procedure is laborious. (We should, however, have in mind that our scientific ancestors were much more trained in calculus – and perhaps in thinking.)

5.2 Synthetical method(s)

On the other hand we have eq.(66) (where $\boldsymbol{\omega} = \dot{\boldsymbol{\pi}} \bigwedge \boldsymbol{\pi}$ is a quasicoordinate – so what? And $\mathbf{v} = \dot{\boldsymbol{\rho}} \bigwedge \boldsymbol{\rho}$ is a quasicoordinate – so what?) Replacement of $(\partial\boldsymbol{\rho}/\partial\mathbf{q}) = (\partial\mathbf{v}/\partial\dot{\mathbf{q}})$ and $(\partial\boldsymbol{\pi}/\partial\mathbf{q}) = (\partial\boldsymbol{\omega}/\partial\dot{\mathbf{q}})$ makes the quasicoordinates disappear and simultaneously shows that the Jacobians are simply coefficient matrices which are known from kinematics, obtained thus as a by-product without any extra calculation. This holds also if we introduce new minimal velocities $\dot{\mathbf{s}} = \mathbf{H}(\mathbf{q})\dot{\mathbf{q}}$ where $\mathbf{s}$ is a quasicoordinate – so what? We call this final result

$$\sum_{i=1}^{N} \left[\left(\frac{\partial \mathbf{v}_i}{\partial \dot{\mathbf{s}}}\right)^T \left(\frac{\partial \boldsymbol{\omega}_i}{\partial \dot{\mathbf{s}}}\right)^T \right] \begin{pmatrix} \dot{\mathbf{p}} + \widetilde{\omega}_R \mathbf{p} - \mathbf{f}^e \\ \dot{\mathbf{L}} + \widetilde{\omega}_R \mathbf{L} - \mathbf{M}^e \end{pmatrix}_i = 0 \qquad (124)$$

the *Projection Equation* since it projects the momentum balances into the unconstrained directions. (Other projections are possible, $\dot{\mathbf{s}} \to \dot{\mathbf{z}}$, which characterizes either a congruence transformation ($\dot{\mathbf{z}} \in I\!R^g$) or any other but constrained vector space ($\dot{\mathbf{z}} \in I\!R^n$, $n > g$) which needs to consider the corresponding constraint forces, of course). Eq.(124) refers to moving frames indicated by $\boldsymbol{\omega}_R$. However, it enables to choose different coordinate systems for different bodies, yet different systems for momentum and momentum of momentum of the same body. This makes the procedure very flexible. There are no directional derivatives any more; the equation holds for nonholonomic variables and nonholonomic constraints: $\dot{\mathbf{s}} \in I\!R^g$ considers the constraints in advance.

Example. Consider once more Figure 25. The rear axle velocities are $(v_x \;\; v_y \;\; \omega_z)$. The nonholonomic constraint $v_y = 0$ is in advance taken into account:

$$\dot{\mathbf{s}} = \begin{pmatrix} v_x \\ \omega_z \end{pmatrix} = \begin{bmatrix} \cos\gamma & \sin\gamma & 0 \\ 0 & 0 & 1 \end{bmatrix} \begin{pmatrix} \dot{x} \\ \dot{y} \\ \dot{\gamma} \end{pmatrix} \in I\!R^{g=2}. \qquad (125)$$

The mass center velocities $(v_{cx}\ v_{cy}\ \omega_{cz})$ are

$$\begin{pmatrix} v_{cx} \\ v_{cy} \\ \omega_{cz} \end{pmatrix} = \begin{bmatrix} 1 & 0 \\ 0 & c \\ 0 & 1 \end{bmatrix} \begin{pmatrix} v_x \\ \omega_z \end{pmatrix} \Rightarrow \begin{pmatrix} (\partial v_{cx}/\partial \dot{\mathbf{s}}) \\ (\partial v_{cy}/\partial \dot{\mathbf{s}}) \\ (\partial \omega_{cz}/\partial \dot{\mathbf{s}}) \end{pmatrix} = \begin{bmatrix} 1 & 0 \\ 0 & c \\ 0 & 1 \end{bmatrix}. \quad (126)$$

Along with the momenta $(p_x\ p_y\ L_z) = (mv_{cx}\ mv_{cy}\ C^c\omega_{cz})$ one obtains

$$\begin{bmatrix} 1 & 0 & 0 \\ 0 & c & 1 \end{bmatrix} \left[\begin{bmatrix} m & 0 \\ 0 & mc \\ 0 & C^c \end{bmatrix} \begin{pmatrix} \dot{v}_x \\ \dot{\omega}_z \end{pmatrix} + \begin{bmatrix} 0 & -\omega_z & 0 \\ +\omega_z & 0 & 0 \\ 0 & 0 & 0 \end{bmatrix} \begin{bmatrix} m & 0 \\ 0 & mc \\ 0 & C^c \end{bmatrix} \begin{pmatrix} v_x \\ \omega_z \end{pmatrix} - \begin{pmatrix} f_x \\ f_y \\ M_z \end{pmatrix} \right] \quad (127)$$

which leads directly to eq.(123) without any detour.

5.3 The horn of plenty?

Collecting procedures (Euler, Hamel, Hamilton, Kane, Lagrange, Maggi, Mangeron, Tzenoff ...) reveals the beautiful variety of methods, the horn of plenty – really? Kane's method for instance turns out to be a restrictive special case, along with misleading abbreviations, although Kane ("the man who mastered motion", [47]) is keen enough to call Lagrange's results black magic. And Poinsot called Lagrange's δ obscure [45]. Are the procedures interconnected – and if, how are they connected? It is time to bring some order into the variety of methods, just to enable a fair comparison and a motivated choice.

The Central Equation of Dynamics. Lagrange is indeed sometimes a bit vague: "because this method needs to vary the same terms in two different ways ... I have introduced a new characteristic δ." And he was probably not aware of the independency of the momentum and the momentum of momentum theorems. The background is (hopefully) clarified with the above interpretation of $\delta\mathbf{r}$ and with Boltzmann's axiom in the background. So let us continue with Lagrange's $\int dm\ddot{\mathbf{r}}^T\delta\mathbf{r} - \delta W = 0$ and extract one time derivative (which needs a correction term due to the product rule): $\int dm\ddot{\mathbf{r}}^T\delta\mathbf{r} = \frac{d}{dt}\left[\int dm\dot{\mathbf{r}}^T\delta\mathbf{r}\right] - \int dm\dot{\mathbf{r}}^T\frac{d}{dt}\delta\mathbf{r} = 0$. Within the last term, the d- and the δ-operation may be interchanged: Consider eq.(117) for $s_\nu \to r_\nu$. Along with the variational ansatz for δq_i, the last term vanishes and one is left with Schwarz' rule which for r_ν has to be fulfilled just from modeling assumptions. One has then $\int dm\dot{\mathbf{r}}^T\delta\frac{d\mathbf{r}}{dt} = \delta\int\frac{1}{2}\dot{\mathbf{r}}^T dm\dot{\mathbf{r}}$. Using the same quadratic form for the penultimate term yields $\frac{d}{dt}\int\frac{\partial}{\partial\dot{\mathbf{r}}}\left[\frac{1}{2}\dot{\mathbf{r}}^T dm\dot{\mathbf{r}}\right]\delta\mathbf{r}$. From $\mathbf{r} = \mathbf{r}(\mathbf{q}(\mathbf{s})) := \mathbf{r}(\mathbf{s})$ one has $\delta\mathbf{r} = (\partial\mathbf{r}/\partial\mathbf{s})\delta\mathbf{s}$. Same as above $(\partial\mathbf{r}/\partial\mathbf{s}) =$

$(\partial\dot{\mathbf{r}}/\partial\dot{\mathbf{s}})$ holds, yielding $\frac{d}{dt}\int\frac{\partial}{\partial\dot{\mathbf{r}}}\left[\frac{1}{2}\dot{\mathbf{r}}^T dm\dot{\mathbf{r}}\right]\delta\mathbf{r} = \frac{d}{dt}\int\frac{\partial}{\partial\dot{\mathbf{r}}}\left[\frac{1}{2}\dot{\mathbf{r}}^T dm\dot{\mathbf{r}}\right]\frac{\partial\dot{\mathbf{r}}}{\partial\dot{\mathbf{s}}}\delta\mathbf{s} = \frac{d}{dt}\frac{\partial}{\partial\dot{\mathbf{s}}}\left[\int\frac{1}{2}\dot{\mathbf{r}}^T dm\dot{\mathbf{r}}\right]\delta\mathbf{s}$ (chain rule). Along with $T = \int\frac{1}{2}\dot{\mathbf{r}}^T dm\dot{\mathbf{r}}$ (kinetic energy) one obtains *the Central Equation of Dynamics* [4]:

$$\frac{d}{dt}\left[\left(\frac{\partial T}{\partial\dot{\mathbf{s}}}\right)\delta\mathbf{s}\right] - \delta T - \delta W = 0. \tag{128}$$

Methodology. From eq.(128) one obtains by elementary calculations a considerable body of methods, see Table 1. If the aim is to derive the motion equations – and this is the aim in the present context – then the choice of procedure is without any doubt the Projection Equation. This is simply because here any step of calculation is carrried out as far as possible such that one only has to insert the problem specific quantities.

5.4 Structurizing the problem

Subsystems. Let the total system consist of a number of subsystems N_{sub} each of which contains N_n bodies. Splitting the sum in eq.(124) into a double sum, and inserting $\dot{\mathbf{y}}_n$ which describes the nth subsystem yields, along with the chain rule,

$$\sum_{n=1}^{N_{sub}}\sum_{i=1}^{N_n}\left(\frac{\partial\dot{\mathbf{y}}_n}{\partial\dot{\mathbf{s}}}\right)^T\left\{\left[\left(\frac{\partial\mathbf{v}_c}{\partial\dot{\mathbf{y}}_n}\right)^T\left(\frac{\partial\boldsymbol{\omega}_c}{\partial\dot{\mathbf{y}}_n}\right)^T\right]\left[\begin{array}{c}\dot{\mathbf{p}}+\tilde{\boldsymbol{\omega}}_{IR}\mathbf{p}-\mathbf{f}^e\\ \dot{\mathbf{L}}+\tilde{\boldsymbol{\omega}}_{IR}\mathbf{L}-\mathbf{M}^e\end{array}\right]\right\}_i = 0. \tag{129}$$

Shifting the inner sum,

$$\sum_{n=1}^{N_{sub}}\left(\frac{\partial\dot{\mathbf{y}}_n}{\partial\dot{\mathbf{s}}}\right)^T\sum_{i=1}^{N_n}\left\{\left[\left(\frac{\partial\mathbf{v}_c}{\partial\dot{\mathbf{y}}_n}\right)^T\left(\frac{\partial\boldsymbol{\omega}_c}{\partial\dot{\mathbf{y}}_n}\right)^T\right]\left[\begin{array}{c}\dot{\mathbf{p}}+\tilde{\boldsymbol{\omega}}_{IR}\mathbf{p}-\mathbf{f}^e\\ \dot{\mathbf{L}}+\tilde{\boldsymbol{\omega}}_{IR}\mathbf{L}-\mathbf{M}^e\end{array}\right]\right\}_i = 0, \tag{130}$$

yields, in a well known notation for the subsystems,

$$\sum_{n=1}^{N_{sub}}\left(\frac{\partial\dot{\mathbf{y}}_n}{\partial\dot{\mathbf{s}}}\right)^T\left[\mathbf{M}\ddot{\mathbf{y}}+\mathbf{G}\dot{\mathbf{y}}-\mathbf{Q}\right]_n = 0. \tag{131}$$

Flexible subsystem are obtained with the limitig case

$$\sum_{n=1}^{N_{sub}}\int_{B_n}\left(\frac{\partial\dot{\mathbf{y}}_n}{\partial\dot{\mathbf{s}}}\right)^T\left\{\left[\left(\frac{\partial\mathbf{v}_c}{\partial\dot{\mathbf{y}}_n}\right)^T\left(\frac{\partial\boldsymbol{\omega}_c}{\partial\dot{\mathbf{y}}_n}\right)^T\right]\left[\begin{array}{c}d\dot{\mathbf{p}}+\tilde{\boldsymbol{\omega}}_{IR}\,d\mathbf{p}-d\mathbf{f}^e\\ d\dot{\mathbf{L}}+\tilde{\boldsymbol{\omega}}_{IR}\,d\mathbf{L}-d\mathbf{M}^e\end{array}\right]\right\}_n = 0. \tag{132}$$

(1) $$T + V = H$$

(2) $$\left(\frac{\partial G}{\partial \ddot{\mathbf{s}}}\right)^T = \mathbf{Q}$$

(3) $$\delta \int_{t_o}^{t_1} (T - V)dt = 0$$

(4) $$\dot{\mathbf{p}}^T = -\left(\frac{\partial H}{\partial \mathbf{q}}\right); \quad \dot{\mathbf{q}}^T = \left(\frac{\partial H}{\partial \mathbf{p}}\right)$$

(5) $$\frac{d}{dt}\frac{\partial T}{\partial \dot{\mathbf{q}}} - \frac{\partial T}{\partial \mathbf{q}} - \mathbf{Q}^T = 0$$

(6) $$\frac{1}{m}\frac{\partial T^m}{\partial \mathbf{q}^m} - \frac{m+1}{m}\frac{\partial T}{\partial \mathbf{q}} - \mathbf{Q}^T = 0$$

(7) $$\left[\frac{d}{dt}\frac{\partial T}{\partial \dot{\mathbf{q}}} - \frac{\partial T}{\partial \mathbf{q}} - \mathbf{Q}^T\right]\left(\frac{\partial \dot{\mathbf{q}}}{\partial \dot{\mathbf{s}}}\right) = 0$$

(8) $$\left[\frac{d}{dt}\frac{\partial T}{\partial \dot{\mathbf{s}}} - \frac{\partial T}{\partial \mathbf{s}} - \mathbf{Q}^T\right]\delta\mathbf{s} + \frac{\partial T}{\partial \dot{\mathbf{s}}}\left(\frac{d\delta\mathbf{s} - \delta d\mathbf{s}}{dt}\right) = 0$$

Central- $$\frac{d}{dt}\left[\left(\frac{\partial T}{\partial \dot{\mathbf{s}}}\right)\delta\mathbf{s}\right] - \delta T - \delta W = 0$$ Equation

---

(9) $$\sum_{i=1}^{N}\left\{\left(\frac{\partial \mathbf{v}}{\partial \dot{\mathbf{s}}}\right)^T(\dot{\mathbf{p}} + \widetilde{\boldsymbol{\omega}}_R\mathbf{p} - \mathbf{f}^e) + \left(\frac{\partial \boldsymbol{\omega}}{\partial \dot{\mathbf{s}}}\right)^T(\dot{\mathbf{L}} + \widetilde{\boldsymbol{\omega}}_R\mathbf{L} - \mathbf{M}^e)\right\}_i = 0$$

Table 1. Methodology

The special structure of equations in this case is

$$\sum_{n=1}^{N_{sub}} \int_{B_n} \left(\frac{\partial \dot{\mathbf{y}}_n}{\partial \dot{\mathbf{s}}}\right)^T [d\mathbf{M}\ddot{\mathbf{y}} + d\mathbf{G}\dot{\mathbf{y}} - d\mathbf{Q}]_n = 0. \tag{133}$$

Shifting the integral (like the sum in eq.(130)) is not yet possible because $\mathbf{y}_n$ contains for instance the deflections and its spatial derivatives while $\mathbf{s}$ refers to the deflections only. The way out here is to define spatial operators which eventually leads to the partial differential equations along with the boundary conditions, see [9]. But this is clearly not the aim of investigation here. To obtain the ordinary (approximative) equations one has to insert the Ritz ansatz $\dot{\mathbf{y}}(\mathbf{x},t) = \mathbf{\Psi}(\mathbf{x})^T \dot{\mathbf{y}}_R(t)$ and make use of the chain rule once more:

$$\sum_{n=1}^{N_{sub}} \int_{B_n} \left(\frac{\partial \dot{\mathbf{y}}_{Rn}}{\partial \dot{\mathbf{s}}}\right)^T \left(\frac{\partial \dot{\mathbf{y}}_n}{\partial \dot{\mathbf{y}}_{Rn}}\right)^T [d\mathbf{M}\ddot{\mathbf{y}} + d\mathbf{G}\dot{\mathbf{y}} - d\mathbf{Q}]_n = 0. \tag{134}$$

Since then $(\partial \dot{\mathbf{y}}_{Rn}/\partial \dot{\mathbf{s}})$ depends on time and $(\partial \dot{\mathbf{y}}_n/\partial \dot{\mathbf{y}}_{Rn}) = \mathbf{\Psi}^T$ on space only, the integral may be shifted, and, inserting the Ritz ansatz for $\mathbf{y}$, $\dot{\mathbf{y}}$, $\ddot{\mathbf{y}}$, one obtains

$$\sum_{n=1}^{N_{sub}} \left(\frac{\partial \dot{\mathbf{y}}_{Rn}}{\partial \dot{\mathbf{s}}}\right)^T \left[[\int_B \mathbf{\Psi}\, d\mathbf{M}\, \mathbf{\Psi}^T]\, \ddot{\mathbf{y}}_R + [\int_B \mathbf{\Psi}\, d\mathbf{G}\, \mathbf{\Psi}^T]\, \dot{\mathbf{y}}_R - \int_B \mathbf{\Psi}\, d\mathbf{Q} \right]_n = 0 \tag{135}$$

which has the same structure as eq.(131). The general subsystem representation results in (with $N_{sub} \rightarrow N$ for simplicity)

$$\left[\left(\frac{\partial \dot{\mathbf{y}}_1}{\partial \dot{\mathbf{s}}}\right)^T \left(\frac{\partial \dot{\mathbf{y}}_2}{\partial \dot{\mathbf{s}}}\right)^T \cdots \left(\frac{\partial \dot{\mathbf{y}}_N}{\partial \dot{\mathbf{s}}}\right)^T \right] \begin{bmatrix} \mathbf{M}_1\ddot{\mathbf{y}}_1 + \mathbf{G}_1\dot{\mathbf{y}}_1 - \mathbf{Q}_1 \\ \mathbf{M}_2\ddot{\mathbf{y}}_2 + \mathbf{G}_2\dot{\mathbf{y}}_2 - \mathbf{Q}_2 \\ \vdots \\ \mathbf{M}_N\ddot{\mathbf{y}}_N + \mathbf{G}_N\dot{\mathbf{y}}_N - \mathbf{Q}_N \end{bmatrix} = 0. \tag{136}$$

The kinematic chain. Starting with the first subsystem, one has $\dot{\mathbf{y}}_1 = \mathbf{F}_1\dot{\mathbf{s}}_1$. Next, calculate the velocities at the coupling point and transform the result into the successor frame to obtain the guidance velocity of the successor subsystem in the form $\mathbf{T}_{21}\dot{\mathbf{y}}_1$. (Notice that $\dot{\mathbf{y}}_i$ refers to the corresponding frame origin). This guidance velocity is superimposed by the relative velocity which the neighbouring subsystem contributes, hence $\dot{\mathbf{y}}_2 = \mathbf{T}_{21}\dot{\mathbf{y}}_1 + \mathbf{F}_2\dot{\mathbf{s}}_2$, etc. One obtains, with p as predecessor of i, $\dot{\mathbf{y}}_i = \mathbf{T}_{ip}\dot{\mathbf{y}}_p + \mathbf{F}_i\dot{\mathbf{s}}_i$ (where $\mathbf{F}_i$ are the local functional matrices, or Jacobians) explicitly

$$\begin{pmatrix} \dot{\mathbf{y}}_1 \\ \dot{\mathbf{y}}_2 \\ \vdots \\ \dot{\mathbf{y}}_N \end{pmatrix} = \begin{bmatrix} \mathbf{F}_1 & & & \\ \mathbf{T}_{21}\mathbf{F}_1 & \mathbf{F}_2 & & \\ \vdots & \vdots & \ddots & \\ \mathbf{T}_{N1}\mathbf{F}_1 & \mathbf{T}_{N2}\mathbf{F}_2 & \cdots & \mathbf{F}_N \end{bmatrix} \begin{pmatrix} \dot{\mathbf{s}}_1 \\ \dot{\mathbf{s}}_2 \\ \vdots \\ \dot{\mathbf{s}}_N \end{pmatrix} := \mathbf{F}\dot{\mathbf{s}}. \quad (137)$$

$\mathbf{F}$ is the global functional matrix (nonmarked elements are zero).

The equations of motion. Using eq.(137) for eq.(136) yields

$$\begin{bmatrix} \mathbf{F}_1^T & \mathbf{F}_1^T\mathbf{T}_{21}^T & \cdots & \mathbf{F}_1^T\mathbf{T}_{N1}^T \\ & \mathbf{F}_2^T & \cdots & \mathbf{F}_2^T\mathbf{T}_{N2}^T \\ & & \ddots & \vdots \\ & & & \mathbf{F}_N^T \end{bmatrix} \begin{bmatrix} \mathbf{M}_1\ddot{\mathbf{y}}_1 + \mathbf{G}_1\dot{\mathbf{y}}_1 - \mathbf{Q}_1 \\ \mathbf{M}_2\ddot{\mathbf{y}}_2 + \mathbf{G}_2\dot{\mathbf{y}}_2 - \mathbf{Q}_2 \\ \vdots \\ \mathbf{M}_N\ddot{\mathbf{y}}_N + \mathbf{G}_N\dot{\mathbf{y}}_N - \mathbf{Q}_N \end{bmatrix} = 0. \quad (138)$$

This gives either the access to the minimal form or to a recursive algorithm. The minimal form is

$$\mathbf{M}\ddot{\mathbf{s}} + \mathbf{G}\dot{\mathbf{s}} - \mathbf{Q} = 0 \in I\!R^g,$$

$$\begin{aligned} \mathbf{M} &= \mathbf{F}^T \mathit{blockdiag}\{\mathbf{M}_1 \cdots \mathbf{M}_N\}\mathbf{F}, \\ \mathbf{G} &= \mathbf{F}^T \mathit{blockdiag}\{\mathbf{G}_1 \cdots \mathbf{G}_N\}\mathbf{F} + \mathbf{F}^T \mathit{blockdiag}\{\mathbf{M}_1 \cdots \mathbf{M}_N\}\dot{\mathbf{F}}, \\ \mathbf{Q} &= \mathbf{F}^T[\mathbf{Q}_1^T \cdots \mathbf{Q}_N^T]^T. \end{aligned} \quad (139)$$

with $\mathbf{F}$ from eq.(137), while the recursive form is already given with eq.(138): solve in a Gaussian sense.

Example. Consider Figure 26: at distance a from the carriage frame origin (index 1) a robot arm is mounted (index 2). Its motion in z-direction (perpendicular to the drawing plane) is decoupled from the remainder and will not be considered.

The carriage (see eq.(125)/eq.(123)):

$$\dot{\mathbf{s}}_1 = \begin{pmatrix} v_x \\ \omega_z \end{pmatrix}_1 = \begin{bmatrix} \cos\gamma_1 & \sin\gamma_1 & 0 \\ 0 & 0 & 1 \end{bmatrix} \begin{pmatrix} \dot{x} \\ \dot{y} \\ \dot{\gamma} \end{pmatrix}_1 = \mathbf{H}_1(\mathbf{q}_1)\dot{\mathbf{q}}_1. \quad (140)$$

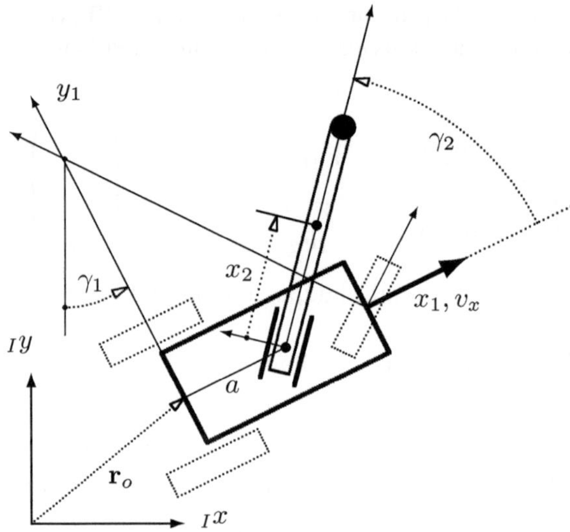

Figure 26. Nonholonomic robot model, top view

$$\mathbf{M}_1 = \begin{bmatrix} m & 0 \\ 0 & C^o \end{bmatrix}_1, \mathbf{G}_1 = \begin{bmatrix} 0 & -mc\dot{\gamma} \\ +mc\dot{\gamma} & 0 \end{bmatrix}_1. \tag{141}$$

The robot arm: mass center velocities in terms of the "describing velocities" $\dot{\mathbf{y}}$ (frame origin o)

$$\begin{pmatrix} v_{cx} \\ v_{cy} \\ \omega_{cz} \end{pmatrix}_2 = \begin{bmatrix} 1 & 0 & 0 & 1 \\ 0 & 1 & x & 0 \\ 0 & 0 & 1 & 0 \end{bmatrix}_2 \begin{pmatrix} v_{ox} \\ v_{oy} \\ \omega_{oz} \\ \dot{x} \end{pmatrix}_2 := \overline{\mathbf{F}}_2 \, \dot{\mathbf{y}}_2 \tag{142}$$

(Notice that $\overline{\mathbf{F}}$ depends on time due to x_2.) Along with $\overline{\mathbf{M}} = diag\{m, m, C^c\}$ one obtains the mass matrix $\mathbf{M} = \overline{\mathbf{F}}^T \overline{\mathbf{M}} \overline{\mathbf{F}}$ and the gyroscopic matrix $\mathbf{G} = \overline{\mathbf{F}}^T \overline{\mathbf{M}} \dot{\overline{\mathbf{F}}} + \overline{\mathbf{F}}^T \widetilde{\boldsymbol{\omega}}_R \overline{\mathbf{M}} \overline{\mathbf{F}}$ which read

$$\mathbf{M}_2 = \begin{bmatrix} m & 0 & 0 & m \\ 0 & m & mx & 0 \\ 0 & mx & C^o & 0 \\ m & 0 & 0 & m \end{bmatrix}_2, \mathbf{G}_2 = \begin{bmatrix} 0 & -m & -mx & 0 \\ m & 0 & -0 & 2m \\ mx & 0 & 0 & 2mx \\ 0 & -m & -mx & 0 \end{bmatrix}_2 \omega_{oz2}, \tag{143}$$

($\omega_{oz2} = \dot{\gamma}_1 + \dot{\gamma}_2$). The kinematic chain: the cariage needs $\dot{\mathbf{y}}_1 = (v_x \; \omega_z)_1^T = \dot{\mathbf{s}}_1$. Thus, $\mathbf{F}_1$ is simply the unit matrix. The arm frame origin is guided with the velocity at distance a: For a representation in frame 1 one has $(v_x \;\; a\dot{\gamma}_1 \;\; 0)$. Transformed into reference frame 2 and superimposed with the relative angular velocity $\dot{\gamma}_2$ one has

$$\begin{pmatrix} v_{ox} \\ v_{oy} \\ \omega_{oz} \end{pmatrix}_2 = \begin{pmatrix} \cos\gamma_2 & \sin\gamma_2 & 0 \\ -\sin\gamma_2 & \cos\gamma_2 & 0 \\ 0 & 0 & 1 \end{pmatrix} \begin{bmatrix} 1 & 0 \\ 0 & a \\ 0 & 1 \end{bmatrix} \begin{pmatrix} v_x \\ \dot{\gamma}_1 \end{pmatrix} + \begin{bmatrix} 0 \\ 0 \\ 1 \end{bmatrix} \dot{\gamma}_2. \tag{144}$$

To obtain $\dot{\mathbf{y}}_2$, eq.(144) has simply to be augmented with $\dot{x}_2$:

$$\dot{\mathbf{y}}_2 = \begin{pmatrix} v_{ox} \\ v_{oy} \\ \omega_{oz} \\ \dot{x} \end{pmatrix}_2 = \underbrace{\begin{bmatrix} \cos\gamma_2 & a\sin\gamma_2 \\ -\sin\gamma_2 & a\cos\gamma_2 \\ 0 & 1 \\ 0 & 0 \end{bmatrix}}_{\mathbf{T}_{21}\mathbf{F}_1} \underbrace{\begin{pmatrix} v_x \\ \dot{\gamma}_1 \end{pmatrix}}_{\dot{\mathbf{s}}_1} + \underbrace{\begin{bmatrix} 0 & 0 \\ 0 & 0 \\ 0 & 1 \\ 1 & 0 \end{bmatrix}}_{\mathbf{F}_2} \underbrace{\begin{pmatrix} \dot{x}_2 \\ \dot{\gamma}_2 \end{pmatrix}}_{\dot{\mathbf{s}}_2}, \tag{145}$$

the (global) functional matrix thus reads

$$\mathbf{F} = \begin{bmatrix} \mathbf{F}_1 & 0 \\ \mathbf{T}_{21}\mathbf{F}_1 & \mathbf{F}_2 \end{bmatrix} = \left[\begin{array}{cc|cc} 1 & 0 & 0 & 0 \\ 0 & 1 & 0 & 0 \\ \hline \cos\gamma_2 & a\sin\gamma_2 & 0 & 0 \\ -\sin\gamma_2 & a\cos\gamma_2 & 0 & 0 \\ 0 & 1 & 0 & 1 \\ 0 & 0 & 1 & 0 \end{array}\right]. \tag{146}$$

From eq.(139) one eventually obtains, for $\dot{\mathbf{s}}^T = \begin{pmatrix} v_x & \dot{\gamma}_1 & \dot{x}_2 & \dot{\gamma}_2 \end{pmatrix}$,

$$\mathbf{M} = \begin{bmatrix} m_1 + m_2 & -m_2 x_2 \sin\gamma_2 & m_2\cos\gamma_2 & -m_2 x_2 \sin\gamma_2 \\ -m_2 x_2 \sin\gamma_2 & \begin{matrix} C_1^o + C_2^o + \\ m_2(a^2 + 2a x_2 \cos\gamma_2) \end{matrix} & m_2 a \sin\gamma_2 & \begin{matrix} C_2^o + \\ m_2 a x_2 \cos\gamma_2 \end{matrix} \\ m_2 \cos\gamma_2 & m_2 a \sin\gamma_2 & m_2 & 0 \\ -m_2 x_2 \sin\gamma_2 & \begin{matrix} C_2^o + \\ m_2 a x_2 \cos\gamma_2 \end{matrix} & 0 & C_2^o \end{bmatrix}, \tag{147}$$

$$\mathbf{G}=\begin{bmatrix} 0 & \begin{matrix}-m_2a\omega_1-m_1c_1\omega_1\\ -m_2x_2\omega_2\cos\gamma_2\end{matrix} & -2m_2\omega_2\sin\gamma_2 & -m_2x_2\omega_2\cos\gamma_2 \\ \begin{matrix}m_2x_2\omega_1\cos\gamma_2+\\ m_2a\omega_1+m_1c_1\omega_1\end{matrix} & -m_2ax_2\dot\gamma_2\sin\gamma_2 & \begin{matrix}2m_2x_2\omega_2+\\ 2m_2a\omega_2\cos\gamma_2\end{matrix} & -m_2ax_2\omega_2\sin\gamma_2 \\ m_2\omega_1\sin\gamma_2 & \begin{matrix}-m_2x_2\omega_2\\ -m_2a\omega_1\cos\gamma_2\end{matrix} & 0 & -m_2x_2\omega_2 \\ m_2x_2\omega_1\cos\gamma_2 & m_2ax_2\omega_1\sin\gamma_2 & 2m_2x_2\omega_2 & 0\end{bmatrix}. \tag{148}$$

The state vector reads $\mathbf{x}^T=(\mathbf{q}^T\ \ \dot{\mathbf{s}}^T)$ where $\dot q_1=\dot x=v_x\cos\gamma_1$, $\dot q_2=\dot y=v_x\sin\gamma_1$ (the remainder is holonomic). The state equations need to invert eq.(147). This is, with $\mathbf{M}\in I\!R^{4,4}$, of course unproblematic. However, concerning large systems, the Gaussian form gains in importance. For a plane chain with twenty links, the time saving factor is already more then ten when solving the equations with an elimination process instead of eq.(139).

It was about thirty years ago when Brandl et.al. proposed a recursive procedure ("order-n-algorithm") for the numerical simulation of multi body systems without inversion of the (entire) mass matrix [3]. The probably first attempt in this field is already due to Vereshchagin [58] as pointed out by Schwertassek [50]; quite an amount of procedures followed in the eighties till the late nineties and sporadicly till today, see e.g. [41]. This fact clearly demonstrates its general interest and importance concerning large (rigid) multibody systems.

These papers look at single rigid bodies for subsystems and pursue the idea to *eliminate* the (generalized) constraint forces step by step, starting with the last body, along with successively applying Euler's cut principle. Clearly, this procedure leads to quite a big amount of non-structured equations/abbreviations. Over the time, several authors tried to simplify the equations, e.g. [59]. Bremer and Pfeiffer extended the procedure to elastic bodies (along with a Ritz series expansion) [6]. Such a procedure results in an extreme analytical effort which is susceptible to careless mistakes.

In the late nineties we decided to abandon the *Elimination*-process in favor of a *Projection*-process [7]. Elimination means to write down the momentum equations and then to eliminate line by line the (generalized) constraint forces (somehow comparable to d'Alembert's principle, [2]). Projection, on the other hand, means to project the momentum equations into the unconstrained ("minimal") space (as an application of Lagrange's principle, or "virtual work"). Its mathematical evaluation then leads to the Gauss-form (139) where simultaneously, as a generalization, instead of single bodies subsystems enter consideration along with nonholonomic variables and constraints. The simplest subsystem hereby is, of course, the single body itself but in general it characterizes an assembly group (like in Figure 8 for instance) . This foregoing yields clear cut advantages not only concerning modeling aspects but also due to the clear structured representation of the basic dynamics. Easy subsystem interchangeability without restarting the whole calculation is one of its advantages. (The "order-n-evaluation" needs three steps: kinematics from bottom to top, kinetics from top to bottom, resolution of minimal accelerations from bottom to top; for details see see e.g. [8]).

5.5 Redundancy

The robot model from Figure 26 reveals another aspect: the motion z_E of the end effector is somehow overdetermined. The position is well defined with $z_E = \overline{x}_1 + b + \overline{x}_2$ (if $\overline{x}_1 = x$ and $\overline{x}_2 = x_2 \cos\gamma_2$ are known, "forward kinematics") but, on the other hand, if z_E is known, how much does $\overline{x}_1$ contribute, and how much $\overline{x}_2$ ("inverse kinematics")?

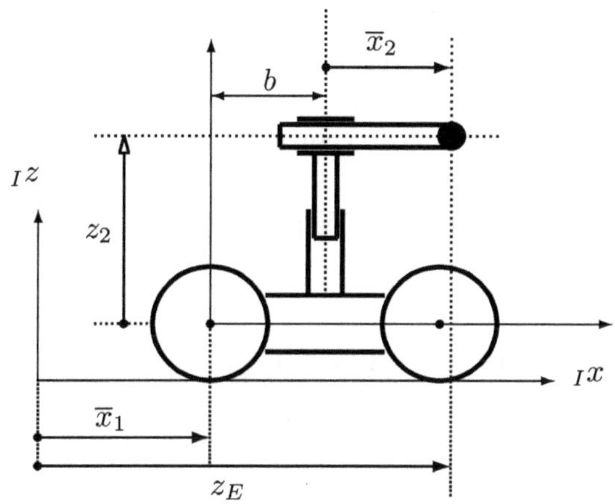

Figure 27. Nonholonomic robot model, side view

The problem reads in mathematical notation (for $\dot{\mathbf{s}} \rightarrow \dot{\mathbf{q}}$, holonomic)

$$\dot{\mathbf{z}}_E = \left[\begin{array}{c} \left(\dfrac{\partial \mathbf{v}_E}{\partial \dot{\mathbf{q}}}\right) \\ \left(\dfrac{\partial \boldsymbol{\omega}_E}{\partial \dot{\mathbf{q}}}\right) \end{array} \right] \dot{\mathbf{q}} := \mathbf{J}_E \dot{\mathbf{q}} \in I\!R^6, \quad \dot{\mathbf{q}} \in I\!R^{f>6}. \tag{149}$$

The Jacobian $\mathbf{J} := \mathbf{J}_E \in I\!R^{6,f}$ is not invertible, at least not directly (index E supressed for simplicity; not to be confused with the end effector inertia tensor). How to solve for $\dot{\mathbf{q}}$? The answer, however, is simple and exciting: $f > 6$ offers the possibility to additionally impose conditions on the system. This has for instance masterfully been done since 1989 in walking machines [44]. How should one construe effective walking? Effectivity may have been detected by nature with its long time evolution. Hence, assuming minimum hip muscle tension (why should one apply torques while standing on a non-slippy ground?) and minimum bending potential in the legs (why should one load the legs unnecessarily?) etc. – and then contact biologists in order to find out if these model assumptions match reality.

Let the additional condition (on the velocity level) for example be

$$\frac{1}{2}\dot{\mathbf{q}}\mathbf{W}\dot{\mathbf{q}} \quad \rightarrow \quad min. \tag{150}$$

Minimization with eq.(149) as a constraint leads to the Lagrange-function

$$L(\boldsymbol{\lambda}, \dot{\mathbf{q}}) = \frac{1}{2}\dot{\mathbf{q}}\mathbf{W}\dot{\mathbf{q}} + \boldsymbol{\lambda}^T(\dot{\mathbf{z}}_E - \mathbf{J}\dot{\mathbf{q}}). \tag{151}$$

The solution reads

$$\dot{\mathbf{q}} = \mathbf{W}^{-1}\mathbf{J}^T \left[\mathbf{J}\,\mathbf{W}^{-1}\mathbf{J}^T\right]^{-1} \dot{\mathbf{z}}_E := \mathbf{J}^+ \dot{\mathbf{z}}_E \tag{152}$$

where $\mathbf{J}^+ \in I\!R^{f,6}$ is the weighted pseudo-inverse. (Notice $\mathbf{J}\mathbf{J}^+ = \mathbf{E} \in I\!R^{6,6}$ (unit matrix) while $\mathbf{J}^+\mathbf{J} \in I\!R^{f,f}$ is not regular). In the last step, one obtains $\mathbf{q}$ by elementary integration. Eq.(152) distributes thus the presribed $z_{Ei}(t)$ over the minimal coordinates q_j. It enables the robot according to Figure 27 for instance to catch a ball.

To fulfill the task $\mathbf{z}_E$, the system contains redundant minimal coordinates. One may preselect these within $\mathbf{q}_N$ to identify the null space. Eq.(150) is therefore changed to

$$\frac{1}{2}(\dot{\mathbf{q}} - \dot{\mathbf{q}}_N)\mathbf{W}(\dot{\mathbf{q}} - \dot{\mathbf{q}}_N) \quad \rightarrow \quad min \tag{153}$$

with solution

$$\dot{\mathbf{q}} = \mathbf{J}^{+}\dot{\mathbf{z}}_E + \left[\mathbf{E} - \mathbf{J}^{+}\mathbf{J}\right]\dot{\mathbf{q}}_N, \quad \mathbf{E} \in I\!R^{f,f}, \tag{154}$$

[41]. Premultiplying eq.(154) by $\mathbf{J}$ yields eq.(149) and shows, with $\mathbf{J}\mathbf{J}^{+} = \mathbf{E}$, that $\mathbf{z}_E$ is not influenced by $\mathbf{q}_N$. The null space dynamics can thus be operated separately and be used e.g. for collision avoidance.

Considering nonholonomic systems, the foregoing is almost the same: Instead of $\dot{\mathbf{z}}_E = \mathbf{J}\dot{\mathbf{q}}$ one has $\dot{\mathbf{z}}_E = \mathbf{J}\dot{\mathbf{s}} = \mathbf{J}\mathbf{H}(\mathbf{q})\dot{\mathbf{q}}$.

Example. Put the robot from Figure 26 on a straight rail and fix the coordinates $(y, \gamma_1, \gamma_2) = 0$ to come out with Figure 27 for $\overline{x}_1 = x := x_1$, $\overline{x}_2 = x_2$ (holonomic) where

$$\mathbf{q} = \begin{pmatrix} x_1 \\ x_2 \end{pmatrix}, \quad \mathbf{M} = \begin{bmatrix} m_1 + m_2 & m_2 \\ m_2 & m_2 \end{bmatrix}, \quad \mathbf{G} = 0. \tag{155}$$

Eq.(152) yields, for $\mathbf{W} = diag\{w_1, w_2\}$,

$$\dot{z}_E = [1 \quad 1]\begin{pmatrix} \dot{x}_1 \\ \dot{x}_2 \end{pmatrix} \Rightarrow \begin{pmatrix} \dot{x}_1 \\ \dot{x}_2 \end{pmatrix} = \frac{1}{w_1 + w_2}\begin{pmatrix} w_2 \\ w_1 \end{pmatrix}\dot{z}_E. \tag{156}$$

There exits a lot of solutions; $w_1 = w_2 = 1$ yields for instance $x_1 = x_2 = z_E/2$ which is, considering the mass distribution, not a very good decision. Intuitively one would rather set $w_1 = m_1$, $w_2 = m_2$ since m_2 is much smaller than m_1 just to balance the energetic flow etc.

After all one must state that all these considerations give valuable hints but need a lot of modeling intuition. Moreover, kinematics is not the whole story. To move the system needs kinetics. Prescribing the end effector motion in a desired manner and conclude on the corresponding forces belongs to open loop control – the success of which is not at all warranted when disturbances occur. We will therefore have a short look on control.

6 Control

The fundamental (historical) control scheme has not changed so much, nor did the (underlying) optimization process. Some basic considerations show problems especially in elastic modeling. Nonlinearity has carefully to be investigated. A path correction procedure for the open-loop part which had already been postulated 25 years ago [29] is still a powerful tool – with the exception that the formerly off-line calculated part can now on-line be computed, due to modern computers. The corresponding result is impressively demonstrated with a "linear elastic robot" (or gantry robot).

A (computer oriented) automated procedure to get rid of nonlinearities can not be recommended if the nonlinear part has stabilizing character – *the art of modeling* needs interpretation. Moreover, friction is critical. The latter can be overcome with a (high gain) observer. The electronic circuit for such procedures is quite simple. Being aware of the fact that, in view of separately considered motor units, the nonlinearities contain the disturbances as well as all the interaction forces leads to a decentralized control concept with the computer as a "black box". This impressive technology does (nearly) not need any model assumption, hence one may shift the whole consideration towards the beautiful world of Big Data.

However, although computational data collection and statistical evaluation without any intermediate "physical law" may be (at least economically) successful in some applications, it will fail (and has already failed) in mechanical egineering. Concentrating on separated motors for control is indeed promising, but not without a mechanical model. Modeling remains indispensable. The computer cannot overtake thinking.

6.1 Classical Methods

One of the perhaps most instructive examples is from early rocket control (e.g. Fieseler Fi 103, a predecessor of the A4 (till A10) which formed the basis of astronautics/kosmonautics after WWII). Its centerpiece is a displacement gyro which holds its direction in space (air driven with an internal feedback via small pendulums: once the gyro deviates, the pendulums open small holes which act like nozzles to restore the gyro – patented by Elmer Ambrose Sperry (1860-1930), a Edison-like inventor who did not care much about patent holders and their patents). The outer gimbal is subjected to an external torque (electric motor) which comes from a precomputed flight program. The resulting gyro movement then leads compressed air via cam disks to pistons which operate the pitch elevator and the yaw rudder. This procedure represents the open loop control $\mathbf{u}_o$ in Fig.28.

As well known from daily life, nothing is perfect. Hence, it is necessary to compare the actual path $\mathbf{q}$ with the desired one $\mathbf{q}_d$ and feed the difference back (gain P, "proportional", already achieved with the cam discs in the present). Doing so, however, one risks unstable oscillations (flutter) which is avoided by superimposing damping (gain D, "differential") which is achieved with rate gyros. We have thus arrived at the classical closed loop control.

The desired path $\mathbf{q}_d$ is sometimes to be reconsidered along with a certain correction model. This will be done later. For the moment the question arises how to chose P, D adequately. Besides some rough rules of thumb

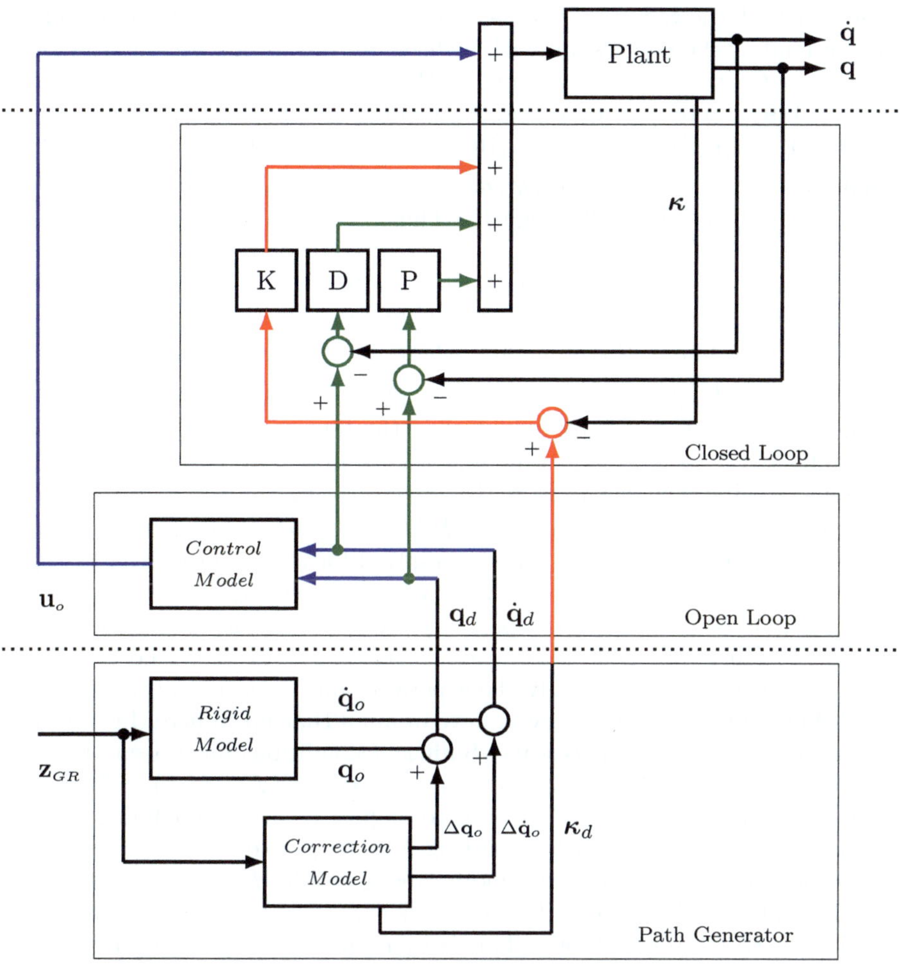

Figure 28. Classical control concept

one is often interested in an optimal choice. The corresponding criterion is then usually performed as a functional where optimality simultaneously needs to fulfill some constraints. A classical example is Dido's problem: Dido, a phoenician princess, had to flee from her brother Pygmalion after he had killed her husband Sychaeus (9th century BC). When she reached

the coast of (nowadays) Tunesia, the inhabitants there promised her as much land as she could span with a sliced cowhide. This became the foundation of Carthago [destroyed in 146 BC, remember Marcus Porcius Cato (234 - 149); "ceterum censeo carthaginem esse delendam"]. The criterion is $J = \int_{x_o}^{x_1} f(y, y')\, dx \to max$. (Here, $f = y$: as much ground as possible) along with constraint $\int_{x_o}^{x_1} g(y, y')\, dx = const.$ (which, with $g = (1 + y'^2)^{1/2}$, represents the cowhide slice length). Thus,
$\int_{x_o}^{x_1} [(\partial f/\partial y)\delta y + (\partial f/\partial y')\delta y']dx = 0 \bigwedge \int_{x_o}^{x_1} [(\partial g/\partial y)\delta y + (\partial g/\partial y')\delta y']dx = 0$ have simulatenously to be fulfilled. Lagranges brilliant idea is to combine $\int_{x_o}^{x_1} \{(\partial f/\partial y)\delta y + (\partial f/\partial y')\delta y'] + \lambda[(\partial g/\partial y)\delta y + (\partial g/\partial y')\delta y']\}dx = 0$
$= \int_{x_o}^{x_1} [(\partial L/\partial y)\delta y + (\partial L/\partial y')\delta y']dx$ [where $L = (f + \lambda g)$: *Lagrange-function*] which yields $\int_{x_o}^{x_1} [(\partial L/\partial y) - \frac{d}{dx}(\partial L/\partial y')]\delta y\, dx = 0$ (integration by parts). Then, δy is arbitrary if, by means of λ, the term $[(\partial L/\partial y) - \frac{d}{dx}(\partial L/\partial y')]$ is nullified (*Euler-equation*, fundamental lemma). Moreover, since in the present L is independent from x, the corresponding *Hamilton-function* $H = (\partial L/\partial y')y' - L$ is constant because $dH/dx = [\frac{d}{dx}(\partial L/\partial y') - (\partial L/\partial y)]y' = 0$ (see Euler-equation); H can then be resolved for y' and integrated by separating the variables to come out with $(x - x_o)^2 + (y - y_o)^2 = \lambda^2 := R^2$. The biggest plot is obtained with (part of) a circle – as one would intuitively suggest. Hence, did Dido make such calculations? Probably not. But the above keywords sketch already the general results in optimization as shown in Table 2. One should, however, have in mind that up to here the optimality conditions are only necessary. Sufficient conditions have been obtained by Weierstrass (1815 - 1897): he concentrated on $(J - J_{min}) > 0 \; \forall \; J$ to come out with his *Excess-function* E to be positive (semi-) definite – which is hard to check. Nevertheless, setting $\boldsymbol{\lambda} = \boldsymbol{\lambda}_{opt}$ (and $\mathbf{x} = \mathbf{x}_{opt}$) leads the condition $(\partial H/\partial \mathbf{u}) = 0$ to the more precise result $H_{opt} = H_{max}$ w.r.t. $\mathbf{u}$ – which also holds for bounded $\mathbf{u} \in \{U\}$ (*Maximum-principle*, Pontryagin and co-workers around 1950). This is by no means trivial: an unrestricted time optimal control would of course yield $\mathbf{u} \to \infty$. However, one should still have in mind that this condition is only necessary. A numerical check will be unevoidable – which is of course automatically included in (usually performed) simulations. Here the computer comes into play.

6.2 PD-Control

PD-Control usually refers to small deviations, e.g. $\mathbf{q}_d - \mathbf{q}$ and $\dot{\mathbf{q}}_d - \dot{\mathbf{q}}$ in case of the rocket control (Figure 28), which is superimposed to the open loop control $\mathbf{u}_o$. Thus, the state equation $\dot{\mathbf{x}} = \mathbf{a}(\mathbf{x}, \mathbf{u})$ is subjected to $\mathbf{u} \to \mathbf{u}_o + \mathbf{u}$ which yields by linearization $\mathbf{x} \to \mathbf{x}_o + \mathbf{x}$ the open loop part

$\dot{\mathbf{x}} = \mathbf{a}(\mathbf{x}, \mathbf{u})$. Determine $\mathbf{u}$ such that $J = \int f(\mathbf{x}, \mathbf{u})dt \to min$	
$H = \boldsymbol{\lambda}^T \mathbf{a}(\mathbf{x}, \mathbf{u}) - f$ (Lagrange) (Hamilton)	$E = -H(\mathbf{x}, \mathbf{u}, \boldsymbol{\lambda}) + \mathbf{a}^T(\boldsymbol{\lambda} - \boldsymbol{\lambda}_{opt})$
necessary	sufficient
$H(\mathbf{x}_{opt}, \mathbf{u}_{opt}, \boldsymbol{\lambda}_{opt}) = 0$ "Canonical Equations" (Euler): $+\left(\frac{\partial H}{\partial \boldsymbol{\lambda}}\right) = \dot{\mathbf{x}}^T$ $-\left(\frac{\partial H}{\partial \mathbf{x}}\right) = \dot{\boldsymbol{\lambda}}^T$ $\pm\left(\frac{\partial H}{\partial \mathbf{u}}\right) = 0$	"Excess Function" (Weierstrass) : $E > 0$
$H_{opt} = \max_{u \in U} H$	$\Leftarrow$

Table 2. Results from classical optimization theory, e.g. [8]

$\dot{\mathbf{x}}_o = \mathbf{a}_o(\mathbf{x}_o, \mathbf{u}_o)$ and the linear plant

$$\dot{\mathbf{x}} = \mathbf{A}\mathbf{x} + \mathbf{B}\mathbf{u}. \tag{157}$$

Let the control be optimized w.r.t.

$$J = \frac{1}{2} \int_0^\infty (\mathbf{x}^T \mathbf{Q} \mathbf{x} + \mathbf{u}^T \mathbf{R} \mathbf{u}) dt \to min \tag{158}$$

$[(\mathbf{A}, \mathbf{B})$ controllable, $\mathbf{Q} \geq 0 \bigwedge (\mathbf{A}, \mathbf{Q})$ observable, $\mathbf{R} > 0]$. The canonical equations from Table 2 yield the optimal PD-control

$$\mathbf{u} = -\mathbf{R}^{-1} \mathbf{B}^T \mathbf{P}_{\mathrm{R}} \mathbf{x} = -[\mathbf{P} \;\; \mathbf{D}] \begin{pmatrix} \mathbf{q} \\ \dot{\mathbf{q}} \end{pmatrix} \tag{159}$$

($\dot{\mathbf{q}} \to \dot{\mathbf{s}}$ for the nonholonomic case) with $\mathbf{P}_{\mathrm{R}}$ being solution of the Riccati-type equation (J. F. Riccati (1676 – 1754))

$$\dot{\mathbf{P}}_{\mathrm{R}} + \mathbf{A}^T\mathbf{P}_{\mathrm{R}} + \mathbf{P}_{\mathrm{R}}\mathbf{A} - \mathbf{P}_{\mathrm{R}}\mathbf{B}\mathbf{R}^{-1}\mathbf{B}^T\mathbf{P}_{\mathrm{R}} + \mathbf{Q} = 0. \tag{160}$$

Several procedures exist for solving eq.(160), see e.g. [5]. Here, the computer comes into play. However, the optimal feedback (159) refers to a linear plant according to eq.(157).

Example. Consider once more a robot according to Figure 26 but with an elastic arm. The carriage is hold at a certain position $\mathbf{q}_c = const.$ and the arm is clamped at its hub. Along with $\mathbf{q} = (\gamma\ \mathbf{q}_{\mathrm{Ritz}}^T)^T$, the motion equations read $\mathbf{M}\ddot{\mathbf{q}} + \mathbf{K}\mathbf{q} = \mathbf{b}u$.

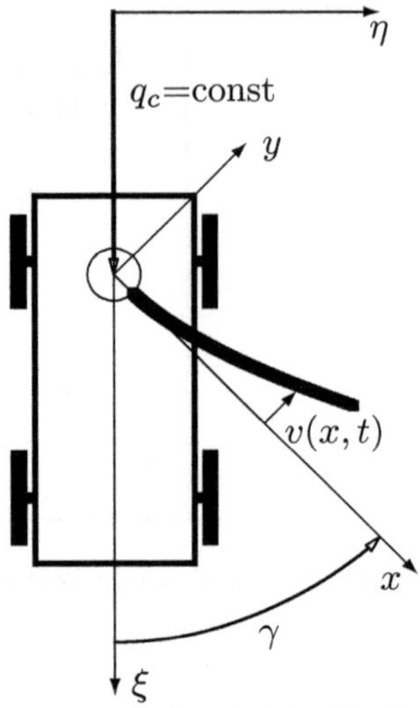

Figure 29. Carriage, elastic arm

Taking the second order field (32) into account, one has

$$\mathbf{M} = \int_o^L \begin{bmatrix} \frac{C^o}{L} & \rho A x \mathbf{v}^T \\ \rho A x \mathbf{v} & \rho A \mathbf{v}\,\mathbf{v}^T \end{bmatrix} dx,$$

$$\mathbf{K} = \int_o^L \begin{bmatrix} 0 & 0 \\ 0 & EI\mathbf{v}''\,\mathbf{v}''^T - \dot{\gamma}^2\rho A\left[\mathbf{v}\,\mathbf{v}^T - \frac{1}{2}(L^2 - x^2)\mathbf{v}'\mathbf{v}'^T\right] \end{bmatrix} dx, \tag{161}$$

$$\mathbf{b}^T = [1 \quad 0],\ u = M_\gamma.$$

The arm can only be influenced by the hub motor $u = M_\gamma$, but controllability is fulfilled. We consider a 90°-slewing manoeuvre.

Due to the centrifugal effects in $\mathbf{K}$, the plant is nonlinear. However, linearization w.r.t. the end point $\mathbf{q} = \mathbf{q}_e + \mathbf{y}$ with $\mathbf{q}_e = (\gamma_e \; 0^T)^T$ yields a linear equation $\mathbf{M}_e\ddot{\mathbf{y}} + \mathbf{K}_e\mathbf{y} = \mathbf{b}_e u$ where

$$\mathbf{M}_e = \int_o^L \begin{bmatrix} \frac{C^o}{L} & \rho A x \mathbf{v}^T \\ \rho A x \mathbf{v} & \rho A \mathbf{v}\,\mathbf{v}^T \end{bmatrix} dx, \mathbf{K}_e = \int_o^L \begin{bmatrix} 0 & 0 \\ 0 & EI\mathbf{v}''\,\mathbf{v}''^T \end{bmatrix} dx, \mathbf{b}_e = \begin{bmatrix} 1 \\ 0 \end{bmatrix}. \tag{162}$$

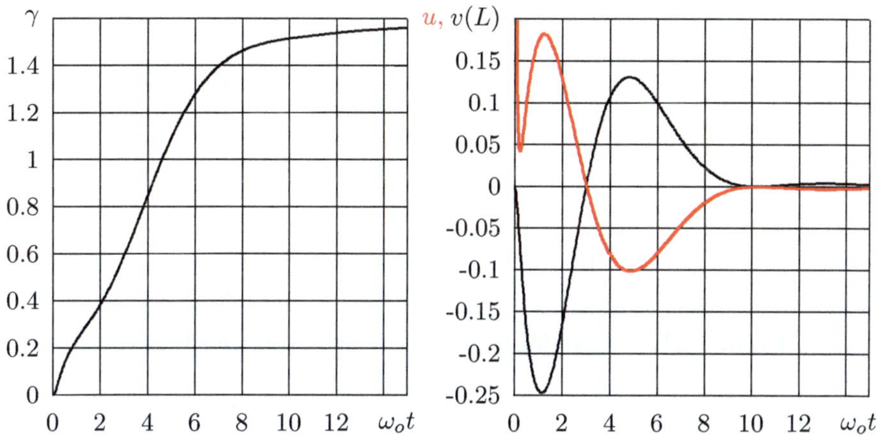

Figure 30. (a) $\gamma(t)$, (b) tip deflection $v(L,t)$ and control $u(t)$

Specifying the control torque as $u = M_\gamma = -P(\gamma - \gamma_e) - D(\dot{\gamma} - \dot{\gamma}_e)$ drives the arm asymptotically to the requested endposition, see Figure 30 (the abscissa is normalized with the first beam bending frequency ω_o).

The result looks quite good at a first glance. Control u and tip deflection $v(L)$ show the desired behaviour (compare Figure 23). However, the main drawback here is due to $u(0) = P\,\gamma_e$: the systems reacts like a step response ($\gamma_e = \pi/2$; in more general systems, $\mathbf{q}_e$ may be very far away from the initial state). This may perhaps be tolerated in rigid systems, but not at all in elastic ones. Such a step response causes jerks which initiate a lot of bending modes. These are not to be seen in Figure 30 which only takes the first bending mode into account.

One can thus not do without an open loop control, for instance by means of a "computed torque"

$$\mathbf{M}\ddot{\mathbf{q}}_d + \mathbf{K}\mathbf{q}_d = \mathbf{u}_{oL} \tag{163}$$

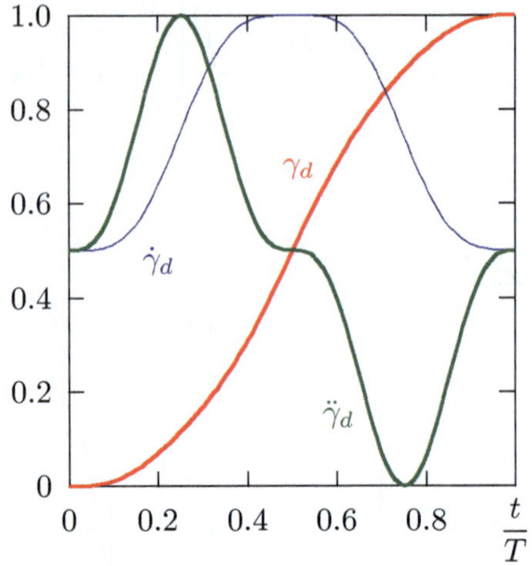

Figure 31. Sinusoidal transition trajectories (normalized)

e.g. using $\ddot{\gamma}_d = \left(\frac{3\pi}{T^2}\right)\sin^3\left(\frac{2\pi}{T}t\right)$ ("sinusoidal") leading to γ_d at least being four times steady and steadily differentiable as a basic request (Figure 31). The open loop part smoothes the trajectory, and the closed loop control then needs less shape functions for simulation. This is due to its pervasive damping character (controllability). Nevertheless, we are once more left with a nonlinear plant. How to get rid of the nonlinearities?

6.3 Flatness based control

There exists an elegant procedure. Consider the nonlinear plant (single input u, single output y for simplicity):

$$\dot{\mathbf{x}} = \mathbf{a}(\mathbf{x}) + \mathbf{b}\,u \in I\!R^n, \tag{164}$$

$$y = \mathbf{c}^T\mathbf{x}. \tag{165}$$

Consecutive time differentiation yields

$$
\begin{aligned}
\dot{y} &= \frac{\partial y}{\partial \mathbf{x}}(\mathbf{a}(\mathbf{x})+\mathbf{b}u) &&= \mathbf{c}^T\mathbf{a}(\mathbf{x}) &&:= \mathbf{c}^T\mathbf{a}_1 &&= 0 \text{ if } &&\mathbf{c}^T\mathbf{b}=0\\
\ddot{y} &= \frac{\partial \dot{y}}{\partial \mathbf{x}}(\mathbf{a}(\mathbf{x})+\mathbf{b}u) &&= \mathbf{c}^T\frac{\partial \mathbf{a}_1}{\partial \mathbf{x}}\mathbf{a}(\mathbf{x}) &&:= \mathbf{c}^T\mathbf{a}_2 &&= 0 \text{ if } &&\mathbf{c}^T\frac{\partial \mathbf{a}_1}{\partial \mathbf{x}}\mathbf{b}=0\\
\dddot{y} &= \frac{\partial \ddot{y}}{\partial \mathbf{x}}(\mathbf{a}(\mathbf{x})+\mathbf{b}u) &&= \mathbf{c}^T\frac{\partial \mathbf{a}_2}{\partial \mathbf{x}}\mathbf{a}(\mathbf{x}) &&:= \mathbf{c}^T\mathbf{a}_3 &&= 0 \text{ if } &&\mathbf{c}^T\frac{\partial \mathbf{a}_2}{\partial \mathbf{x}}\mathbf{b}=0\\
& && \vdots
\end{aligned}
\tag{166}
$$

If, for the kth derivative, $\mathbf{c}^T(\partial \mathbf{a}_{k-1}/\partial \mathbf{x})\mathbf{b} \neq 0$, then

$$y^k = \mathbf{c}^T\frac{\partial \mathbf{a}_{k-1}}{\partial \mathbf{x}}[\mathbf{a}(\mathbf{x})+\mathbf{b}u] := \overline{a}(\mathbf{x}) + \overline{b}u = 0. \tag{167}$$

If $k = n$, y is called a "flat output". Let $u = u_{comp} + u_{CL}$ where $u_{comp} = -\frac{\overline{a}}{\overline{b}}$ compensates the nonlinearities ("exact linearization"). The closed loop part

$$u_{CL} = \frac{1}{\overline{b}}\left[\frac{d^n y_d}{dt^n} + \sum_{i=0}^{n-1} k_i\left(\frac{d^i y_d}{dt^i} - \frac{d^i y}{dt^i}\right)\right] \tag{168}$$

yields, with $z = y - y_d$, the Frobenius form

$$\frac{d}{dt}\begin{pmatrix} z\\ \dot{z}\\ \ddot{z}\\ \vdots\\ z^{n-1}\end{pmatrix} = \begin{bmatrix} 0 & 1 & 0 & \cdots & 0\\ 0 & 0 & 1 & \cdots & 0\\ \vdots & & & \ddots & \vdots\\ 0 & & & & 1\\ -k_o & -k_1 & -k_2 & \cdots & -k_{n-1}\end{bmatrix}\begin{pmatrix} z\\ \dot{z}\\ \ddot{z}\\ \vdots\\ z^{n-1}\end{pmatrix} \tag{169}$$

[Georg Frobenius (1849 – 1917)] where k_i are the characteristic coefficients from $\lambda^n + k_{n-1}\lambda^{n-1} + \cdots + k_o = 0$ which can be used for pole assignment.

The differentiation process is somewhat tedious and will, of course, fail in case of backlash and friction. Moreover: The example according to eq.(161) contains a nonlinearity which stabilizes flexural oscillation. Should one really compensate positive effects with such an immense effort? Of course not. Looking at "exact linearization" procedures goes along with a strict warning not to rely on automated procedures without physical interpretation. In case of controlled systems, modeling refers to physical reality on the one hand side (which is needed for simulation) and to a control model which is not the same but reduced to minimal requirements (– *the art of modeling!*) We are thus left with the linear plant according to eq.(162) for control calculation. Its result has to be tested by simulation and, in the last step, verified in hardware.

6.4 Path correction

Considering a rigid articulated robot, the desired gripper position and orientation $\mathbf{z}_{\mathrm{Gd}} \in I\!R^6$ is obvious. In case of an elastic articulated robot, the control model may be the purely rigid one, i.e. neglecting all structural and gear elasticities. The "real" elastic model then represents the simulation model. However, $\mathbf{z}_{\mathrm{Gd}}$ does not more characterize the desired aim. Correcting the path is done in two steps.

Let $\mathbf{q} = (\mathbf{q}_M^T\ \mathbf{q}_A^T\ \mathbf{q}_e^T)^T$ (M: motor, A: arm, e: elastic (Ritz-coefficients)) and define $\mathbf{q}_M = \mathbf{q}_R$ (reference path). Along with $\mathbf{y}$ for deviation one has

$$\mathbf{q} = \mathbf{q}_o + \mathbf{y} = \begin{pmatrix} \mathbf{q}_R \\ \mathbf{q}_R \\ \mathbf{0} \end{pmatrix} + \begin{pmatrix} \mathbf{y}_M \\ \mathbf{y}_A \\ \mathbf{q}_e \end{pmatrix}, \quad \| \mathbf{y} \| << \| \mathbf{q} \| . \tag{170}$$

1. **Reference motion control.**
Decompose $\mathbf{M}(\mathbf{q})\ddot{\mathbf{q}} + \mathbf{g}(\mathbf{q}, \dot{\mathbf{q}}) = \overline{\mathbf{B}}(\mathbf{u}_{ol} + \mathbf{u}_{cl} + \mathbf{u}^c)$ and linearize,

$$\begin{bmatrix} \mathbf{M}_{MM} & \mathbf{M}_{MA} & \mathbf{M}_{Me} \\ \mathbf{M}_{MA}^T & \mathbf{M}_{AA} & \mathbf{M}_{Ae} \\ \mathbf{M}_{Me}^T & \mathbf{M}_{Ae}^T & \mathbf{M}_{ee} \end{bmatrix} \begin{pmatrix} \ddot{\mathbf{q}}_R \\ \ddot{\mathbf{q}}_R \\ 0 \end{pmatrix} + \begin{pmatrix} \mathbf{g}_{0M} \\ \mathbf{g}_{0A} \\ \mathbf{g}_{0e} \end{pmatrix} - \begin{pmatrix} \mathbf{u}_{ol} \\ \mathbf{0} \\ \mathbf{0} \end{pmatrix}$$

$$+\mathbf{M}_o\ddot{\mathbf{y}} + \mathbf{P}_o\dot{\mathbf{y}} + \mathbf{Q}_o\mathbf{y} = \begin{pmatrix} \mathbf{u}_{cl} + \mathbf{u}^c \\ \mathbf{0} \\ \mathbf{0} \end{pmatrix}, \tag{171}$$

set $\mathbf{y}$ and $\mathbf{u}_{cl} + \mathbf{u}^c$ zero, formulate the virtual work for this case via $\delta\mathbf{q}^T = \delta\mathbf{q}_R^T\,[\,\mathbf{E}\ \ \mathbf{E}\ \ 0\,]$ to obtain the open loop control $\mathbf{u}_{ol}$,

$$\mathbf{u}_{ol} = \left[\mathbf{M}_{MM} + \mathbf{M}_{MA} + \mathbf{M}_{MA}^T + \mathbf{M}_{AA}\right]\ddot{\mathbf{q}}_R + (\mathbf{g}_{0M} + \mathbf{g}_{0A}). \tag{172}$$

Insert $\mathbf{u}_o$ (and $\mathbf{u}_{cl} = 0$) into eq.(171),

$$\begin{bmatrix} -\mathbf{f}_R \\ +\mathbf{f}_R \\ +\mathbf{f}_{eR} \end{bmatrix} + \mathbf{M}_o\ddot{\mathbf{y}} + \mathbf{P}_o\dot{\mathbf{y}} + \mathbf{Q}_o\mathbf{y} = \begin{pmatrix} \mathbf{u}^c \\ \mathbf{0} \\ \mathbf{0} \end{pmatrix}, \tag{173}$$

$$\text{where} \quad \begin{bmatrix} -\mathbf{f}_R \\ +\mathbf{f}_R \\ +\mathbf{f}_{eR} \end{bmatrix} = \begin{bmatrix} -([\,\mathbf{M}_{MA}^T + \mathbf{M}_{AA}\,]\ddot{\mathbf{q}}_R + \mathbf{g}_{0A}) \\ +([\,\mathbf{M}_{MA}^T + \mathbf{M}_{AA}\,]\ddot{\mathbf{q}}_R + \mathbf{g}_{0A}) \\ +([\,\mathbf{M}_{Me}^T + \mathbf{M}_{Ae}^T\,]\ddot{\mathbf{q}}_R + \mathbf{g}_{0e}) \end{bmatrix}.$$

From eq.(173) one obtains the special "quasistatic" equilibrium $\mathbf{y}_{qs}^T = \left(\mathbf{0}^T \mathbf{y}_A^T \mathbf{q}_e^T\right)_{qs}$:

$$\begin{bmatrix} -\mathbf{f}_R \\ +\mathbf{f}_R \\ +\mathbf{f}_{eR} \end{bmatrix} + \mathbf{Q}_o \mathbf{y}_{qs} = \begin{pmatrix} \mathbf{u}^c \\ \mathbf{0} \\ \mathbf{0} \end{pmatrix} \Rightarrow \begin{cases} \mathbf{u}^c = [\mathbf{Q}_{MA} \mathbf{Q}_{Me}] \begin{pmatrix} \mathbf{y}_A \\ \mathbf{q}_e \end{pmatrix}_{qs} \\ \begin{pmatrix} \mathbf{y}_A \\ \mathbf{q}_e \end{pmatrix}_{qs} = -\begin{bmatrix} \mathbf{Q}_{AA} & \mathbf{Q}_{Ae} \\ \mathbf{Q}_{eA} & \mathbf{Q}_{ee} \end{bmatrix}^{-1} \begin{bmatrix} \mathbf{f}_R \\ \mathbf{f}_{eR} \end{bmatrix} \end{cases} \tag{174}$$

where $\mathbf{Q}_o$ is decomposed into submatrices $\mathbf{Q}_{ij}, \{i,j\} \in \{M, A, e\}$ like in eq.(171). $\mathbf{u}^c$ is a constraint force which holds $\mathbf{y}_M$ zero within $\mathbf{y}_{qs}$. *Presuming that* $\mathbf{u}_o$ *follows* $\mathbf{q}_R$ *in an ideal manner such that* $\mathbf{y}_M$ *vanishes all the time,* $\mathbf{u}^c$ *is implicitly taken into account by* $\mathbf{u}_o$.

2. **Path matching.**
Calculate the resulting gripper deviation $\Delta\mathbf{z}_G$, and the deviation $\Delta\mathbf{z}_{GR}$ which would be obtained by modifying the reference path $\mathbf{q}_R$ (Taylor expansion)

$$\Delta\mathbf{z}_G = \underbrace{\begin{bmatrix} \dfrac{\partial \mathbf{v}_G}{\partial \dot{\mathbf{q}}_A} & \dfrac{\partial \mathbf{v}_G}{\partial \dot{\mathbf{q}}_e} \\ \dfrac{\partial \boldsymbol{\omega}_G}{\partial \dot{\mathbf{q}}_A} & \dfrac{\partial \boldsymbol{\omega}_G}{\partial \dot{\mathbf{q}}_e} \end{bmatrix}}_{\mathbf{F}_G} \begin{pmatrix} \mathbf{y}_A \\ \mathbf{q}_e \end{pmatrix}, \quad \Delta\mathbf{z}_{GR} = \underbrace{\begin{bmatrix} \dfrac{\partial \mathbf{v}_G}{\partial \dot{\mathbf{q}}_R} \\ \dfrac{\partial \boldsymbol{\omega}_G}{\partial \dot{\mathbf{q}}_R} \end{bmatrix}_R}_{\mathbf{F}_{GR}} \Delta\mathbf{q}_R. \tag{175}$$

$(\Delta\mathbf{z}_G + \Delta\mathbf{z}_{GR}) = 0 \Rightarrow \Delta\mathbf{q}_R$ would compensate the deviation $\Delta\mathbf{z}_G$:

$$\Delta\mathbf{q}_R = -\mathbf{F}_{GR}^+ \mathbf{F}_G \begin{pmatrix} \mathbf{y}_A \\ \mathbf{q}_e \end{pmatrix} := \left[-\mathbf{F}_{\mathrm{A,corr}} - \mathbf{F}_{\mathrm{e,corr}}\right] \begin{pmatrix} \mathbf{y}_A \\ \mathbf{q}_e \end{pmatrix} \tag{176}$$

where $(\ )^+$ is the generalized inverse. (Notice that the gripper motion does not depend on the "inner" coordinates $\mathbf{q}_M$ (motor coordinates) when defined according to Figure 8).

However, eq.(176) needs the solution of a differential equation [i.e. eq.(173) along with eq.(174): $\mathbf{M}_o\ddot{\mathbf{y}} + \mathbf{P}_o\dot{\mathbf{y}} + \mathbf{Q}_o(\mathbf{y} - \mathbf{y}_{qs}) = \mathbf{0}$]. This is almost tedious and time consuming. Thus, assuming $\ddot{\mathbf{y}}$ and $\dot{\mathbf{y}}$ negligible to calculate eq.(176), one may use eq.(174) to come out with

$$\Delta\mathbf{q}_R = \left[-\mathbf{F}_{\mathrm{A,corr}} - \mathbf{F}_{\mathrm{e,corr}}\right] \begin{pmatrix} \mathbf{y}_A \\ \mathbf{q}_e \end{pmatrix}_{qs}. \tag{177}$$

As a result one has a corrected path $\mathbf{q}_R + \Delta\mathbf{q}_R$. One could think of using the new (corrected) path and restart calculations.

However, the correction may also be shifted to the deviations which means to retain the old reference and to define a new target

$$\mathbf{q} = \begin{pmatrix} \mathbf{q}_R + \Delta\mathbf{q}_R \\ \mathbf{q}_R + \Delta\mathbf{q}_R \\ \mathbf{0} \end{pmatrix} + \begin{pmatrix} \mathbf{0} \\ \mathbf{y}_A \\ \mathbf{q}_e \end{pmatrix} \rightarrow \begin{pmatrix} \mathbf{q}_R \\ \mathbf{q}_R \\ \mathbf{0} \end{pmatrix} + \begin{pmatrix} \Delta\mathbf{q}_R \\ \mathbf{y}_A + \Delta\mathbf{q}_R \\ \mathbf{q}_e \end{pmatrix} \tag{178}$$

yielding the correction model

$$\mathbf{y}_d = \begin{bmatrix} -\mathbf{F}_{\mathrm{A,corr}} & -\mathbf{F}_{\mathrm{e,corr}} \\ \mathbf{E} - \mathbf{F}_{\mathrm{A,corr}} & -\mathbf{F}_{\mathrm{e,corr}} \\ 0 & \mathbf{E} \end{bmatrix} \begin{pmatrix} \mathbf{y}_{\mathrm{A}} \\ \mathbf{q}_{\mathrm{e}} \end{pmatrix}_{qs} \tag{179}$$

to build a closed loop control $\mathbf{u}_{cl}$, compare Figure 28 (lower part).

Thereby, the elastic deflecture itself can not be nullified itself via the hub motor (think of bending due to gravity, for instance). The deformation $\mathbf{q}_e$ which leads to the desired gripper position is, however, obtained from the correction model. It is thus recommendable to also feed $\mathbf{q}_e$ back, preferably with a curvature (output) control $\boldsymbol{\kappa} = (\partial\boldsymbol{\kappa}/\partial\mathbf{q}_e)\mathbf{q}_e$ which is easily accomplished with strain gauge measurements.

A lot of assumptions have been made to obtain the correction model. The result has therefore first to be checked by numerical simulations before going into hardware realization.

The outlined procedure was experimentally verified for an "elastic linear robot" (or gantry robot) with an overwhelming success, [28].

6.5 Observers

As already mentioned, friction plays an important role. An exact linearization according to eq.(166) ff. will fail due to the required differentiation process. This problem can be overcome with an observer.

A (state) observer is, in principle, nothing but an computerized reconstruction of the plant, but without any knowledge on the initial conditions. Even if these were known one would have a severe numerical drift-off after just a short time. Hence, if the real output $\mathbf{y} = \mathbf{C}\mathbf{x}$ differs from the observer signal $\hat{\mathbf{y}} = \mathbf{C}\hat{\mathbf{x}}$, then the difference is fed back with an adequate gain matrix $\mathbf{L}_x$ to assure zero phase between the models.

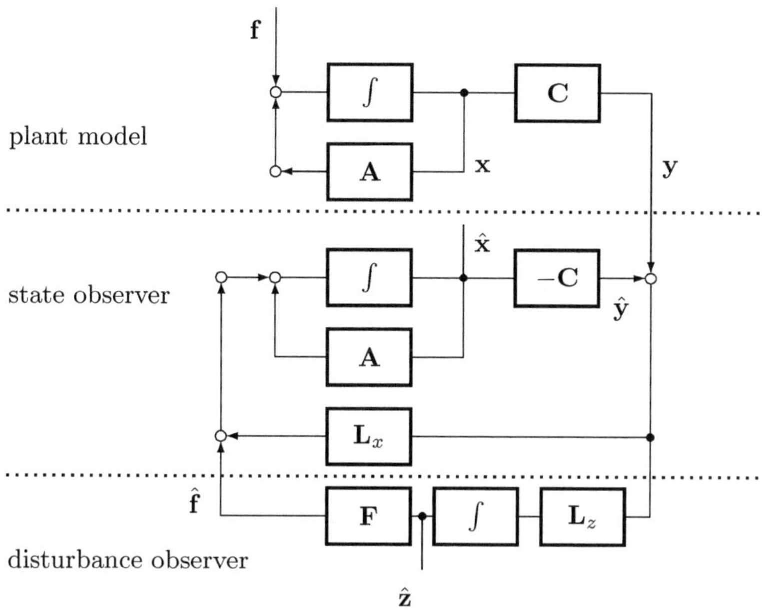

Figure 32. Observer scheme

A well known statement is that for sufficient high gains $\mathbf{L}_x$ (no disturbance reconstruction yet, $\mathbf{L}_z \equiv 0$) the estimated state suppresses the disturbance influence ("high gain observer"). This will be demonstrated with a single joint with friction in the gravity field, Figure 33.

The time histories in Figure 34 correspond to the initial conditions $(\mathbf{q}_o^T, \dot{\mathbf{q}}_o^T) = (0, 0)$ where $\mathbf{q}^T = (\gamma_M, \gamma_A)$. As (a) shows, the friction torque with its zero crossings at about $t = 1.8$ and $t = 3.4$, resp., puts hard conditions on the system. Nevertheless, the estimation (b) is perfect (notice the zero crossings for $\dot{\gamma}_M$). Of course, the gains $\mathbf{L}_x = (l_{x1}, l_{x2})$ need high values which are restricted due to measurement noise. The values used here conform with those which have succesfully been implemented for biped walking machines [21], [44]. The hardware realization is almost simple:

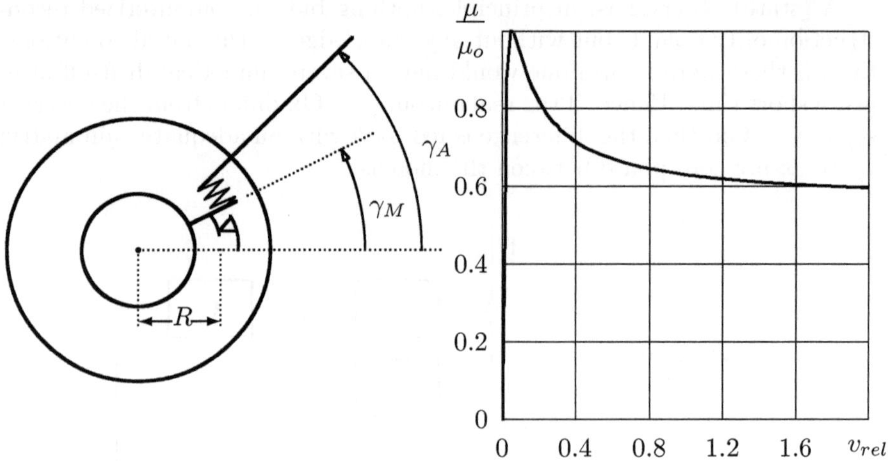

Figure 33. (a) single joint with friction, (b) Stribeck curve

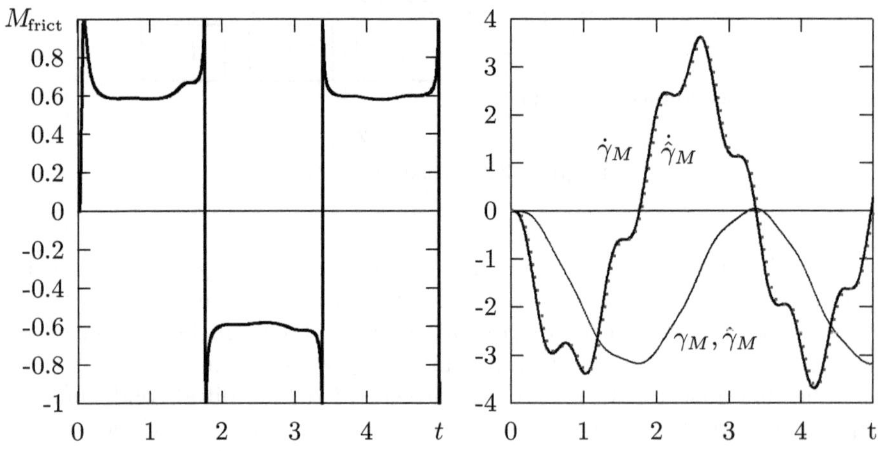

Figure 34. (a) M_{frict} normalized, (b) variables (solid), estimation (dotted)

from Figure 32 one has $\dot{\hat{\mathbf{x}}} = \mathbf{A}\hat{\mathbf{x}} - \mathbf{L}_x(\hat{\mathbf{y}} - \mathbf{y})$ which for $\mathbf{A} = \begin{bmatrix} 0 & 1 \\ 0 & 0 \end{bmatrix}$ yields Figure 35.

$$\frac{d}{dt}\begin{pmatrix} \hat{\gamma}_M \\ \dot{\hat{\gamma}}_M \end{pmatrix} = \begin{bmatrix} 0 & 1 \\ 0 & 0 \end{bmatrix}\begin{pmatrix} \hat{\gamma}_M \\ \dot{\hat{\gamma}}_M \end{pmatrix} - \begin{pmatrix} l_{x1} \\ l_{x2} \end{pmatrix}(\hat{\gamma}_M - \gamma_M)$$

electrical board (black box) — input (measurement)

Figure 35. Circuit scheme (state estimation)

The disturbance remains unknown. It contains the friction torque as well as the reactions from gear twisting $\Delta\gamma = (\gamma_M - \gamma_A)$. Modeling the disturbance via basic functions $\dot{\mathbf{z}} = 0$ ($\mathbf{z}$: step functions) such that $\mathbf{f} = \mathbf{Fz}$ leads to $\dot{\hat{\mathbf{x}}} = \mathbf{A}\hat{\mathbf{x}} + \mathbf{F}\hat{\mathbf{z}} - \mathbf{L}_x(\hat{\mathbf{y}} - \mathbf{y})$ and $\dot{\hat{\mathbf{z}}} = -\mathbf{L}_z(\hat{\mathbf{y}} - \mathbf{y})$ (lower part of Figure 32). Along with $\mathbf{A} = \begin{bmatrix} 0 & 1 \\ 0 & 0 \end{bmatrix}$, $\mathbf{F} = \begin{bmatrix} 0 \\ 1 \end{bmatrix}$ and $\mathbf{L}_z = l_z$ one obtains the hardware realization according to Figure 36.

$$\frac{d}{dt}\begin{pmatrix} \hat{\gamma}_M \\ \dot{\hat{\gamma}}_M \\ \hat{z} \end{pmatrix} = \left[\begin{array}{cc|c} 0 & 1 & 0 \\ 0 & 0 & 1 \\ \hline 0 & 0 & 0 \end{array}\right]\begin{pmatrix} \hat{\gamma}_M \\ \dot{\hat{\gamma}}_M \\ \hat{z} \end{pmatrix} - \begin{pmatrix} l_{x1} \\ l_{x2} \\ l_z \end{pmatrix}(\hat{\gamma}_M - \gamma_M)$$

electrical board (black box) — input (measurement)

Figure 36. Circuit scheme (disturbance estimation)

The estimated disturbance once more contains friction as well as the interacting forces. Since $\mathbf{f} \in I\!R^n$ refers to the state equation, we call that part wich enters the motion equations $\overline{\mathbf{f}} \in I\!R^f$. According to Figure 32, these are outer forces. The total outer and inner (reaction) forces are therefore denoted $\overline{\overline{\mathbf{f}}}$. Its estimation is depicted in Figure 37.

6.6 Decentralized Control

Disposing of the (estimated) state variables and the (estimated) disturbances, these may be used for control and for disturbance compensation.

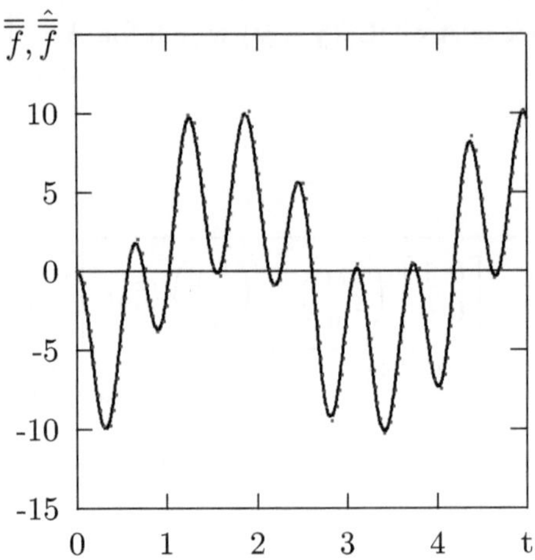

Figure 37. Disturbance (solid) and its estimation (dotted)

The motion equations

$$\mathbf{M}\ddot{\mathbf{q}} + \mathbf{G}\dot{\mathbf{q}} - \mathbf{Q} = \overline{\mathbf{f}}, \tag{180}$$

can be decomposed in the form

$$diag\{M_{oi}\}\ddot{\mathbf{q}} + \Delta\mathbf{M}\ddot{\mathbf{q}} + \mathbf{G}\dot{\mathbf{q}} - \mathbf{Q} = \overline{\mathbf{f}} \tag{181}$$

where $M_{oi} = const$ may be chosen such that it represents the desired (or nominal) value [39] . The ith component reads

$$M_{oi}\ddot{q}_i + \sum_{j=1}^{f} \Delta M_{ij}\ddot{q}_j + (\mathbf{G}\dot{\mathbf{s}} - \mathbf{Q})_i = \overline{f}_i \tag{182}$$

where now in

$$\ddot{q}_i = \frac{1}{M_{oi}}\left[-\sum_{j=1}^{f} \Delta M_{ij}\ddot{q}_j - (\mathbf{G}\dot{\mathbf{q}} - \mathbf{Q})_i + \overline{f}_i\right] := \overline{\overline{f}}_i, \tag{183}$$

$\overline{\overline{f}}_i$ contains all interaction terms (including, if need be, the parameter deviation of M_i from its desired (nominal) value) as well as all disturbances. One has thus

$$\frac{d}{dt}\begin{pmatrix} q \\ \dot{q} \end{pmatrix}_i = \begin{bmatrix} 0 & 1 \\ 0 & 0 \end{bmatrix}\begin{pmatrix} q \\ \dot{q} \end{pmatrix}_i + \begin{pmatrix} 0 \\ 1 \end{pmatrix}\overline{\overline{f}}_i. \tag{184}$$

Hence, the simplified linear plant used above has by no means been a singular example but represents somehow generality: we have come to decentralized joint control. In Figure 38, the joint model is used for test. It represents a 90° maneuver in a vertical plane (acting against gravity) including joint friction and gear twisting.

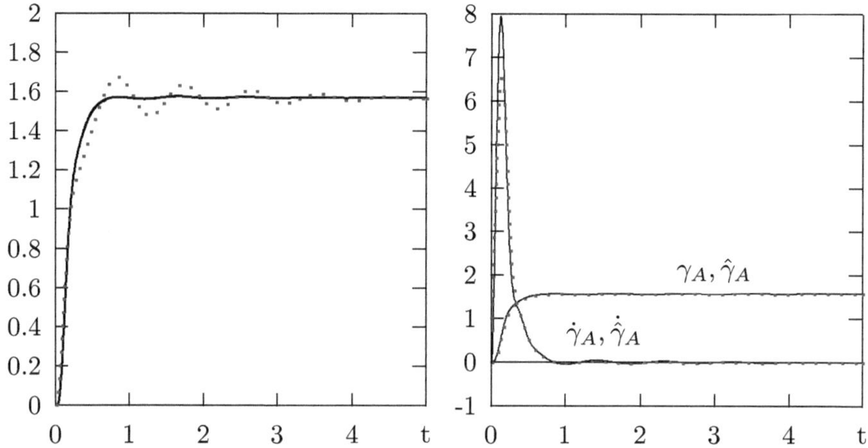

Figure 38. (a) slewing maneuver γ_A (dotted: γ_M), (b) "real values" (from simulation) and estimation (dotted)

Along with the closed loop control $u_{CL,i}$, the "exact linearization" from eq.(168) would require to eliminate the disturbances exactly,

$$\ddot{q}_i = u_{CL,i} + \underbrace{\left(\overline{\overline{f}}_i - \overline{\overline{f}}_i\right)}_{\equiv 0}, \tag{185}$$

which is hard to achieve. It needs exact knowlegde of parameter values and will fail in case of friction. The main difference now is, that the disturbances and the interactions are cancelled asymptotically,

$$\ddot{q}_i = u_{CL,i} + \underbrace{\left(\overline{\overline{f}}_i - \hat{\overline{\overline{f}}}_i\right)}_{\text{asympt.} \to 0} . \tag{186}$$

It gets along nearly without any parameter knowledge. This requires, of course, careful investigations primarily w.r.t. the gains $\mathbf{L}_x, \mathbf{L}_z$. Here, powerful computers come into play.

6.7 Big Data

Along with the fact that one does not need a model for the disturbances and interactions, and disposing of powerful computers, one may dispense of modeling at all and look at the corresponding hardware realization in the sense of a black box, which, for the practical engineer, is out of interest (buy, use, and don't care. The computer does the job).

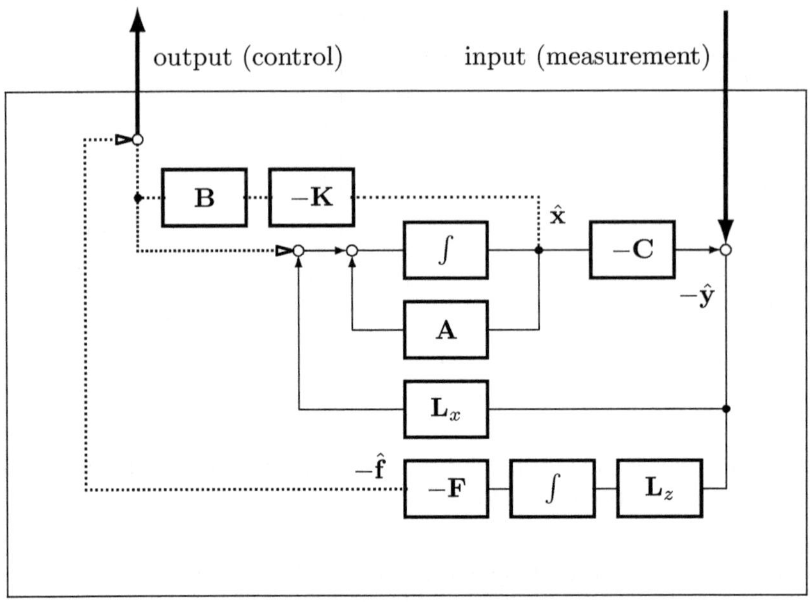

Figure 39. The black box

Data collection and statistics is indeed (economically) successful in a wide range. Statistics makes modeling obsolete. In 2009, two US scientists extracted the Hamilton function of a double pendulum just from measurement data [19]. However, does it belong two a rotational double pendulum or to a translational one (two masses with springs) which (for sufficient small amplitudes) have the same behaviour – or to a combination, and what about mass distribution and lengths? In mechanical engineering, attempts to do without modeling have a long tradition. To reconstruct a mechanical system just from measurement data has thereby definitely failed already fifty years ago. A simulation, and thereafter a control, cannot be done without a model.

7 Conclusion

Clearly, data and data collection may be helpful to find answers. But this is not the end of the story, just the beginning. An example for an overwhelming succes in modeling (may be the best model ever): until around 1600, Tycho Brahe collected obervation data. 1619, Johannes Kepler postulated a physical model, 1686 Isaac Newton and fifty years later, Leonhard Euler formulated a mathematical model (gravity and momentum theorem, resp.). On this basis, the first satellite was launched in 1957, the moon landing followed in 1969, the first space station Saljut 1 was engineered 1971 and the international ISS from 1998 till today – and, on the same basis, what about internet, smart phone etc.? Thereby, step by step, more insight could be obtained.
Karl Popper: He who decides one day that scientific statements ⋯ can be regarded as finally verified, retires from the game (F. Pfeiffer [43]).
Hence, in natural and engineering sciences,

- modeling remains indispensable,
- the computer is helpful, but nothing than a dummy slave.

Even more: everything what we see, what we believe, how we behave, in all the areas of life, is based on an abstraction process of thinking – what else are we dealing with than models? Models which describe reality, and which may perhaps converge to truth which, however, remains out of reach. From belief to truth with own capabilities: Münchhausen could not pull himself up by his own hair, this was a lie.

Bibliography

[1] d'Alembert, J. (1757), *Musicalische Setzkunst, nach den Lehrsätzen des Herrn Rameau.* Leipzig: Breitkopf.

[2] d'Alembert, J. (1758), *traité de dynamique.* 2nd ed., David, Paris.

[3] Brandl, H., Johanni, R., Otter, M. (1986), A Very Efficient Algorithm for the Simulation of Robots and Similar Multibody Systems..., *Proc. IFAC Symp. Vienna/Austria*, 365–370.

[4] Bremer, H. (1988), Über eine Zentralgleichung in der Dynamik, *Zeitschr. Ang. Math. u. Mech. (ZAMM)*, Vol. 68, 307-311.

[5] Bremer, H., (1988), *Dynamik und Regelung mechanischer Systeme.* Teubner, Stuttgart.

[6] Bremer, H., Pfeiffer, F. (1992), *Elastische Mehrkörpersysteme.* Teubner, Stuttgart.

[7] Bremer, H. (1999), On the dynamics of elastic multibody systems. *Appl. Mech. Rev.*, 52 (9), 275–303.

[8] Bremer, H. (2008), *Elastic Multibody Systems - A Direct Ritz Approach.* Springer.

[9] Bremer, H. (2014), Multiple beam systems – How to obtain the PDEs/BCs? How to solve?, *Acta Mechanica 225*, 1883 - 1900.

[10] Bresse, J.A.C. (1859), *Cours de Méchanique Appliquée*, Paris: Mallet Bachelier.

[11] Bublath, J. (1990), *Abenteuer Forschung*, ZDF (2nd German Television).

[12] Chetaev, N.G. (1961), *The Stability of Motion*, Pergamon Press.

[13] Euler, L. (1736), *Mechanica sive motus scientia analytice exposita*, German Translation by J. Ph. Wolfers (2 Volumes: 1848, 1850), Greifswald: C.A. Koch.

[14] Euler, L. (1744), *Methodus inveniendi lineas curvas maximi minimive gaudentes sive solutio problematis isoperimetrici latissimo sensu accepti*, 2nd appendix. Lausanne & Genf: Marcus Michaelis Bousquet.

[15] Euler, L. (1750), Découverte d'un nouveau principe de la mécanique, *Mém. Acad. Sci. Berlin* 6, 185-217, printed 1752.

[16] Euler, L. (1758), Du mouvement de rotations des corps solides autour d'un axe variable, *Mém. Acad. Sci. Berlin* 14, 154-193, printed 1765.

[17] Euler, L. (1775), Nova methodus motum corporum rigidorum determinandi, *Mém. Acad. Sci. Petropol.* 20, 208-238, printed 1776.

[18] Föppl, A. (1923), *Vorlesungen über Technische Mechanik*, Vol. IV, 7th ed., p. 273. Berlin, Leipzig: Teubner. First edition 1899.

[19] *http://www.heise.de/newsticker/meldung/Wissenschafts-KI-zum-Download-877627.html*

[20] Galerkin, B.G. (1915), Series solution of some problems in elastic equilibrium of rods and plates, *Vestn. Inzh. Tech.*, 897 - 908.

[21] Gattringer, H. (2006), *Realisierung, Modellbildung und Regelung einer zweibeinigen Laufmaschine*, Johannes Kepler Univ., PhD Thesis

[22] Gauss, C.F. (1829), Über ein neues allgemeines Grundgesetz der Mechanik, *J. f. Reine u. Ang. Mathematik*, Vol. 4: 232 - 235.

[23] Hamel, G. (1904), Die Lagrange-Eulerschen Gleichungen in der Mechanik, *Zeitschr. Ang. Math. u. Phys.*, Vol. 50: 1 - 57.

[24] Heun, K. (1901), Die Bedeutung des d'Alembertschen Prinzips für starre Systeme und Gelenkmechanismen, *Math. Annalen*, III, Nr. II.

[25] Hubinger, S. (2012), *Ein Beitrag zur semi- analytischen Berechnung von Kontinuumsschwingungen* , Johannes Kepler Univ., PhD Thesis.

[26] Kappus, R. (1939), Zur Elastizität endlicher Verschiebungen, *Z. Ang. Math. u. Mech. (ZAMM)*, 19, Nr.5: 271-285 and Nr.6: 344-361.

[27] Jourdain, P.E.B. (1909), Note on an Analogue of Gauss' Principle of Least Constraints, *The Quarterly Journal of Pure and Applied Mathematics*, XL: 153 - 157.

[28] Kilian, F. (2013), *Schwingungsdämpfungsmethoden elastischer Linearroboter*, Johannes Kepler Univ., PhD Thesis.

[29] Kleemann, U. (1989), *Regelung elastischer Roboter*, Düsseldorf: Fortschr.-Ber. VDI-Z., 8, Nr. 191.

[30] Klein, F. (1926), *Vorlesungen über die Entwicklung der Mathemathik im 19. Jahrhundert*, Berlin: Springer (Reprint 1979).

[31] Lagrange, J.L. de (1762), Essai sur une nouvelle méthode por déterminer les maxima et les minima des formules intégrales indéfinies, *Miscellanea Taurinensia*, 173-195, German translation by P. Stäckel (ed.): *Variationsrechnung*. Darmstadt: Wiss. Buchges.

[32] Lagrange, J.L. de (1764), Récherches sur la libration de la lune, *Acad. Royale des Sciences de Paris*, Reprint by Gauthier-Villars 1873.

[33] Lagrange, J.L. de (1780), Théorie de la libration de la lune, *Academie Royale des Sciences de Berlin*, Reprint by Gauthier-Villars 1870.

[34] Lagrange, J.L. de (1813-1815), *Mécanique analytique*, 2nd ed. (posthumous), German transl. by H. Servus. Berlin: Springer 1887.

[35] Likins, P.W., Bouvier, K.H. (1971), Attitude Control of Nonrigid Spacecraft, *Astronautics & Aeronautics*, Vol. 9: 67.

[36] Mach, E. (1883), *Die Mechanik in ihrer Entwickelung*, Berlin: Springer, [6th edition 1908].

[37] Maggi, G.A., (1903), *Principii di Stereodinamica*, Milano: Editore Libraio della Real Casa.

[38] Müller, P.C. (1971), Asymptotische Stabilität von linearen mechanischen Systemen mit positiv semidefiniter Dämpfungsmatrix, *Zeitschr. Ang. Math. u. Mech. (ZAMM)*, Vol. 51, T197-T198.

[39] Müller, P.C. (1995), Non-Linear Robot Control: Method of Exact Linearization and Decoupling by State Feedback and Alternative Control Methods, *J. Appl. Math. and Comp. Sci.*, Vol. 5, No 2, 359-371.

[40] Nelson, F.C. (2003), A Brief History of Early Rotor Dynamics, *Sound & Vibration*, 37: 9-11.

[41] Oberhuber, B. (2012), *Ein Beitrag zur modularen Modellierung und Regelung redundanter Robotersysteme*, Joh. Kepler Univ., PhD Thesis.

[42] *Die Pamir - Untergang eines Gross-Seglers (Doku)*, YouTube.

[43] Pfeiffer, F. (2005), *Mechanical System Dynamics*, Berlin-Heidelberg: Springer, 503.

[44] Pfeiffer, F. (2007), The TUM Walking Machines, *Phil.Trans.R.Soc.*, A, Vol. 1850, 109-131.

[45] Poinsot, L. (1837), *Elémens de statique*, Paris: Bachellier, 476.

[46] Prandtl, L. (1899), *Kipperscheinungen* (PhD thesis), Nürnberg: R. Stich.

[47] Radetsky, P. (1986), The Man who Mastered Motion, *Sience*, May 86, 52 - 60.

[48] Rayleigh, Lord J.W. Strutt (1877), *The Theory of Sound*, London: McMillan.

[49] Ritz, W. (1909), Theorie der Transversalschwingungen einer quadratischen Platte mit freien Rändern, *Ann. d. Phys.*, IV: 737 - 786.

[50] Schwertassek, R., Wallrapp, O. (1999), *Dynamik flexibler Mehrkörpersysteme.* Vieweg & Sohn, Braunschweig, Wiesbaden.

[51] Sommerfeld, A. (1902), Beiträge über den dynamischen Ausbau der Festigkeitslehre, *Physikalische Zeitschrift*, Vol. 3: 266 - 271.

[52] Szabó, I (1979), *Geschichte der mechanischen Prinzipien*, 2nd. ed., Stuttgart: Birkhäuser.

[53] Szabó, I. (1979), Bemerkungen zur Literatur über die Geschichte der Mechanik, *Humanismus und Technik* 22.

[54] Timoshenko, S.P. (1921), On the correction for shear of the differential equation for the transverse vibrations of prismatic bars, *Philos. Mag.*, series 6, 41: 744-746.

[55] Truesdell, C. (1984), *An Idiot's Fugitive Essays on Sience*, Berlin-Heidelberg-New York-Tokyo: Springer.

[56] Truckenbrodt, A. (1980), *Bewegungsverhalten und Regelung hybrider Mehrkörpersysteme mit Anwendung auf Industrieroboter*, VDI-Fortschr.-Ber. 8, Nr. 33.

[57] Varignon, P. (1725) *Nouvelle Mécanique ou Statique 2*, Paris: Claude Jombert.

[58] Vereshchagin, A.F. (1974), Computer Simulation of Complicated Mechanisms of Robot Manimulators, *Engineering and Cybernetics*, 6, 65–70.

[59] Weidemann, H.-J. (1993), *Dynamik und Regelung von sechsbeinigen Robotern und natürlichen Hexapoden.* VDI-Fortschr.-Ber. 8, 362.

List of Symbols

$\mathbf{A}, A$	transformation, state matrix, area, inertia moment
$\mathbf{a}, a$	state function, min. coefficient
$\mathbf{B}, B$	input matrix, inertia moment
$\mathbf{C}, C$	output matrix, inertia moment
$\mathbf{c}, c$	coefficient vector, coefficient, mass center distance
$\mathbf{D}, D$	damping matrix, gain (diff.)
$\mathbf{E}, E$	identity matrix, Young's modulus, excess-function
$\mathbf{e}, e$	unit vector, error
$\mathbf{F}, \overline{\mathbf{F}}$	functional matrix (or Jacobian)
$\mathbf{f}, f$	force, disturbance, frequency, function (general)
$\mathbf{G}, G$	gyroscopic matrix, torsional elasticity, Gibbs-function
$\mathbf{g}, g$	approximation functions, gravitation vector, grav. constant
H	Hamilton-function
$\mathbf{I}, I$	area moments of inertia
$\mathbf{J}, J$	mass moments of inertia, Jacobian, optimization criterion
$\mathbf{K}$	(conservative) restoring matrix, curvature feedback, state feedback matrix
$\mathbf{L}, L$	(angular) momentum, Lagrange-function, length
$\mathbf{M}, M$	mass matrix, moment
m	mass, time derivation grade
$\mathbf{N}$	(nonconserv.) restoring matrix
$\mathbf{P}, P$	gain (proportional), power
$\mathbf{p}$	(linear) momentum
$\mathbf{Q}$	(gen.) force, weighting matrix
$\mathbf{q}$	minimal (or generalized or Lagrangian) coordinates
$\mathbf{R}, R$	weighting matrix, Rayleigh-quot.
$\mathbf{r}$	(inertial) position vector
$\mathbf{s}$	quasi coordinates (minimal)
T, t	kinetic energy, time
$\mathbf{u}$	control vector
V	potential energy
$\mathbf{v}, v$	vector of shape functions, bending function
W	work
$\mathbf{w}, w$	vector of shape functions, bending function
x, y, z	spatial variables
$\mathbf{x}, \mathbf{y}, \mathbf{z}$	state vector, discribing variables, disturbance functions
α, β, γ	Cardan (or Tait-Bryan) angles
Δ	difference, Laplace-operator
δ	variation, Dirac-distribution ($\overline{\delta}$)
ϵ	Cauchy strain, variational parameter
$\boldsymbol{\vartheta}, \vartheta$	vector of shape functions, torsional function
$\boldsymbol{\kappa}$	(vector of) curvatures
$\boldsymbol{\lambda}, \lambda$	Lagrange-parameters, eigenvalue
ν	Poisson number, frequency
$\boldsymbol{\pi}$	quasi coordinates (rotation)
$\boldsymbol{\rho}, \rho$	quasi coordinates (translation), mass density
σ	Cauchy-stress
τ	tangential plane
$\boldsymbol{\Phi}, \varphi$	constraint, bending angles
$\boldsymbol{\Psi}$	matrix of shape functions and derivatives
Ω	angular velocity
$\boldsymbol{\omega}, \omega$	angular velocity, frequency

Modeling of Flexible Bodies

Ahmed A. Shabana
University of Illinois at Chicago, Chicago, IL, U.S.A.

This chapter will discuss some of the widely used techniques for modeling flexible multibody systems (MBS). Among the topics discussed in this chapter are the *floating frame of reference* (FFR) formulation and the *absolute nodal coordinate formulation* (ANCF). Discussion on the use of the CAD-based methods such as B-spline and NURBS (Non-uniform rational B-spline) will be also provided. The FFR formulation is used to study small deformation problems, while ANCF finite elements are used in the large deformation analysis.

1. Floating Frame of Reference (FFR) Formulation

It is not recommended to use a full nodal-based finite element (FE) representation to study *small deformation problems* since such a representation is not the most efficient approach for solving for the small deformations. The geometry of the small deformation takes simple shapes that can be captured using a lower dimension model. Furthermore, in the analysis of small deformations, with a proper selection of the coordinate systems, one can use finite elements which have smaller number of nodal coordinates. Conventional non-isoparametric beam, plate and shell finite elements can still be used with the FFR formulation which leads to a local linear elasticity problem that allows for employing linear modes to further reduce the number of the model degrees of freedom and eliminate high frequency modes of vibration. The FE/FFR approach is the most widely used method to solve the small deformation, large rotation problem in MBS applications. The FE/FFR was the basis for developing new computational algorithms that led in the early eighties to introducing new generation of codes that became known as *flexible multibody computer codes*. These computer codes are widely used in industry, universities and research institutions.

It is important to point out that the FFR formulation was introduced long before the FE method was introduced. However, the use of the FFR formulation has been limited to problems with very simple geometry such as beams and plates. For this reason, before the eighties, there was no successful integration of MBS and FE small deformation algorithms. The concept of the FE *intermediate element coordinate system*, introduced in the early eighties, was necessary in order to be able

F. Pfeiffer and H. Bremer (eds.), *The Art of Modeling Mechanical Systems*,
CISM International Centre for Mechanical Sciences 570,
DOI 10.1007/978-3-319-40256-7_3

to solve MBS problems in which components such as chassis of a vehicle can have complex geometry characterized by discontinuities. Therefore, understanding the concept of the FE intermediate element coordinate system is necessary in order to understand the FE/FFR formulation that is widely used in MBS computer codes.

1.1 Rotation and Angular Velocity

In this section, the general definition of the absolute angular velocity vector is presented. The components of the angular velocity vector are function of a selected set of orientation parameters and their time derivatives. These angular velocity vector components can be written as linear functions of the derivatives of the orientation coordinates. Recall that if $\mathbf{A}$ is an orthogonal matrix, one has $\mathbf{A}^{\mathrm{T}}\mathbf{A} = \mathbf{A}\mathbf{A}^{\mathrm{T}} = \mathbf{I}$, Differentiating $\mathbf{A}^{\mathrm{T}}\mathbf{A}$ with respect to time and using the orthogonality of the rotation matrix $\mathbf{A}^{\mathrm{T}}\mathbf{A} = \mathbf{I}$, one obtains $\dot{\mathbf{A}}^{\mathrm{T}}\mathbf{A} + \mathbf{A}^{\mathrm{T}}\dot{\mathbf{A}} = \mathbf{0}$, This equation shows that

$$\mathbf{A}^{\mathrm{T}}\dot{\mathbf{A}} = -\left(\dot{\mathbf{A}}^{\mathrm{T}}\mathbf{A}\right) = -\left(\mathbf{A}^{\mathrm{T}}\dot{\mathbf{A}}\right)^{\mathrm{T}} \tag{1}$$

Therefore, $\mathbf{A}^{\mathrm{T}}\dot{\mathbf{A}}$ is always a skew symmetric matrix denoted as $\tilde{\bar{\boldsymbol{\omega}}}$, that is

$$\mathbf{A}^{\mathrm{T}}\dot{\mathbf{A}} = \tilde{\bar{\boldsymbol{\omega}}} = \begin{bmatrix} 0 & -\bar{\omega}_3 & \bar{\omega}_2 \\ \bar{\omega}_3 & 0 & -\bar{\omega}_1 \\ -\bar{\omega}_2 & \bar{\omega}_1 & 0 \end{bmatrix} \tag{2}$$

The bar over a vector or a matrix is used in these lecture notes to indicate a vector or a matrix whose components are defined in a body or local coordinate system. Similarly, one can differentiate $\mathbf{A}\mathbf{A}^{\mathrm{T}} = \mathbf{I}$ with respect to time to show that $\dot{\mathbf{A}}\mathbf{A}^{\mathrm{T}}$ is a skew symmetric matrix, that is

$$\dot{\mathbf{A}}\mathbf{A}^{\mathrm{T}} = \tilde{\boldsymbol{\omega}} = \begin{bmatrix} 0 & -\omega_3 & \omega_2 \\ \omega_3 & 0 & -\omega_1 \\ -\omega_2 & \omega_1 & 0 \end{bmatrix} \tag{3}$$

Rigid Body Kinematic Description. In the case of an arbitrary rigid body motion, the position vector of a point on the body can be defined in the global coordinate system as

$$\mathbf{r} = \mathbf{r}_O + \mathbf{A}\bar{\mathbf{u}} \tag{4}$$

where $\mathbf{r}_O$ is the global position vector of the origin of the body coordinate system, $\mathbf{A}$ is the orthogonal transformation matrix that defines the orientation of the body coordinate system, and $\bar{\mathbf{u}}$ is the local position vector of the arbitrary point with respect to the body coordinate system. Differentiating the preceding equation with respect to time, the absolute velocity vector of the arbitrary point on the body can be written as $\dot{\mathbf{r}} = \dot{\mathbf{r}}_O + \dot{\mathbf{A}}\bar{\mathbf{u}}$. Since the rotation matrix $\mathbf{A}$ is an orthogonal matrix, the absolute velocity vector equation can be written in the following two equivalent forms:

$$\left.\begin{aligned} \dot{\mathbf{r}} &= \dot{\mathbf{r}}_O + \mathbf{A}\left(\bar{\boldsymbol{\omega}} \times \bar{\mathbf{u}}\right) \\ \dot{\mathbf{r}} &= \dot{\mathbf{r}}_O + \boldsymbol{\omega} \times \mathbf{u} \end{aligned}\right\} \tag{5}$$

where $\bar{\boldsymbol{\omega}}$ and $\boldsymbol{\omega}$ are called the angular velocity vectors defined, respectively, in the body coordinate system and the global coordinate system, and $\mathbf{u} = \mathbf{A}\bar{\mathbf{u}}$. The angular velocity vector defined in the body coordinate system and the global coordinate system can be written, respectively, as

$$\bar{\boldsymbol{\omega}} = \bar{\mathbf{G}}\dot{\boldsymbol{\theta}}, \qquad \boldsymbol{\omega} = \mathbf{G}\dot{\boldsymbol{\theta}} \tag{6}$$

where $\bar{\mathbf{G}}$ and $\mathbf{G}$ are two matrices that can be nonlinear functions of the orientation coordinates $\boldsymbol{\theta}$.

1.2 FFR Kinematic Description

In the FFR formulation, a body coordinate system $X_1^b X_2^b X_3^b$ that shares the large overall displacement of the body is introduced. The position vector of the origin of the body coordinate system is defined by the three-dimensional vector $\mathbf{r}_O$, while the orientation of the body-coordinate system is defined using the orthogonal transformation matrix **A**. The unconstrained motion of the coordinate system of the body can then be described using six independent coordinates; three translational coordinates $\mathbf{r}_O = \begin{bmatrix} \mathbf{r}_{O1} & \mathbf{r}_{O2} & \mathbf{r}_{O3} \end{bmatrix}^{\mathrm{T}}$, and three independent rotation parameters $\boldsymbol{\theta}$ that define the orthogonal matrix **A**. Using these coordinates, the global position vector of an arbitrary point on the finite element j of the body can be written as

$$\mathbf{r}^j = \mathbf{r}_O + \mathbf{A}\bar{\mathbf{u}}^j, \qquad j = 1, 2, \ldots, n_e \tag{7}$$

where n_e is the total number of finite elements used in the body discretization, and $\bar{\mathbf{u}}^j$ is the location of the arbitrary point on the finite element j with respect to the origin of the body coordinate system. In the preceding equation, $\mathbf{r}_O$ and $\mathbf{A}$ are the same for all finite elements, and therefore, the body coordinate system represents a common reference for all elements and serves as the basis for defining the connectivity between the finite elements of the body.

Intermediate Element Coordinate System. Modeling deformable bodies with complex geometries may require the use of finite elements which can have different orientations in the undeformed reference configuration. This has become known in the literature as *structural discontinuities* in order to distinguish this type of discontinuity from the discontinuities that result from the use of mechanical joints that allow large relative displacements. In order to be able to describe the kinematics of these elements, an *intermediate element coordinate system* $\bar{X}_1^j\bar{X}_2^j\bar{X}_3^j$ is introduced. The intermediate element coordinate system has an origin which is rigidly attached to the origin of the body coordinate system and has axes which are initially parallel to the axes of the element coordinate system $X_1^jX_2^jX_3^j$. Since it is assumed that the element shape function and the nodal coordinates can describe an arbitrary rigid body translation, the position vector of the arbitrary point on the finite element j with respect to the intermediate element coordinate system $\bar{X}_1^j\bar{X}_2^j\bar{X}_3^j$ can be written as $\bar{\mathbf{u}}^{ji} = \mathbf{S}^j\bar{\mathbf{e}}^{ji}$, $j = 1,2,\ldots n_e$. In this equation, $\mathbf{S}^j$ is the element shape function, and $\bar{\mathbf{e}}^{ji}$ is the element nodal coordinates defined in the intermediate element coordinate system. Using the assumption of small deformation and assuming that the elements are initially un-curved, one can write $\bar{\mathbf{e}}^{ji} = \bar{\mathbf{e}}_o^{ji} + \bar{\mathbf{e}}_f^{ji}$, where $\bar{\mathbf{e}}_o^{ji}$ is the vector of nodal coordinates in the reference configuration, and $\bar{\mathbf{e}}_f^{ji}$ is the vector of nodal coordinates that define the element small deformation with respect to the intermediate element coordinate system. The element nodal coordinates $\bar{\mathbf{e}}^{ji}$ defined in the intermediate element coordinate system $\bar{X}_1^j\bar{X}_2^j\bar{X}_3^j$ can be expressed in terms of the element coordinates $\mathbf{e}^j$ defined in the body coordinate system $X_1^bX_2^bX_3^b$ using the orthogonal transformation $\bar{\mathbf{T}}_n^{ji}$ as $\bar{\mathbf{e}}^{ji} = \bar{\mathbf{T}}_n^{ji}\mathbf{e}^j$. One can then write the vector $\bar{\mathbf{u}}^{ji}$ in terms of element nodal coordinates defined in the body coordinate system as $\bar{\mathbf{u}}^{ji} = \mathbf{S}^j\bar{\mathbf{T}}_n^{ji}\mathbf{e}^j$, $j = 1,2,\ldots n_e$. In this equation, $\bar{\mathbf{T}}_n^{ji}$ is a constant transformation since the intermediate element coordinate system $\bar{X}_1^j \bar{X}_2^j \bar{X}_3^j$ is assumed to have a constant orientation with respect to the body coordinate system $X_1^b X_2^b X_3^b$. Similarly, the components of the vector $\bar{\mathbf{u}}^{ji}$ can be defined in the body coordinate system as $\bar{\mathbf{u}}^j = \bar{\mathbf{T}}_u^{ji} \bar{\mathbf{u}}^{ji} = \bar{\mathbf{T}}_u^{ji} \mathbf{S}^j \bar{\mathbf{T}}_n^{ji} \mathbf{e}^j$, $j = 1, 2, \ldots n_e$, where $\bar{\mathbf{T}}_u^{ji}$ is also a constant transformation matrix that has dimension equal to the dimension of the vector $\bar{\mathbf{u}}^{ji}$. It follows that

$$\bar{\mathbf{u}}^j = \mathbf{S}_b^j \mathbf{e}^j , \qquad j = 1, 2, \ldots n_e \tag{8}$$

where $\mathbf{S}_b^j = \bar{\mathbf{T}}_u^{ji} \mathbf{S}^j \bar{\mathbf{T}}_n^{ji}$ is the element shape function defined in the body coordinate system.

The use of the concept of the finite element intermediate coordinate system is necessary in order to be able to model structural continuities that characterize many mechanical, structural, and aerospace components. Introducing this concept allowed for using the FFR formulation for complex structural components. Before introducing the finite element intermediate element coordinate system, the use of the FFR formulation was limited to very simple geometry. The birth of what is known as flexible MBS computer codes that perform coupled nonlinear analysis is attributed to introducing the concept of the FE intermediate element coordinate system. Using this concept, the FE/FFR formulation was systematically developed and its use was demonstrated in the early eighties.

1.3 Connectivity and Reference Conditions

In order to develop the FE/FFR equations of motion, the *connectivity* and the *reference conditions* must be defined first. The connectivity conditions lead to the element assembly, while the reference conditions eliminate the rigid body modes of the element shape function, and define a unique displacement field.

Connectivity Conditions. Using the Boolean matrix approach, the element nodal coordinates can be written in terms of the body nodal coordinates as $\mathbf{e}^j = \mathbf{B}_c^j \mathbf{e}_b$, $j = 1, 2, \ldots n_e$, where $\mathbf{e}_b$ is the vector of the body nodal coordinates, and $\mathbf{B}_c^j$ is the constant Boolean matrix that defines the connectivity conditions for the finite element j. Therefore, the position vector of the material points can be defined as $\bar{\mathbf{u}}^j = \mathbf{S}_b^j \mathbf{B}_c^j \mathbf{e}_b$, $j = 1, 2, \ldots n_e$.

Reference Conditions. As in the case of the individual elements, one can write the vector of body nodal coordinates as the sum of two vectors as $\mathbf{e}_b = \mathbf{e}_{bo} + \mathbf{e}_{bf}$, where $\mathbf{e}_{bo}$ is the vector of nodal coordinates in the initial un-deformed configuration, and $\mathbf{e}_{bf}$ is the vector of deformation nodal coordinates. By imposing the reference conditions, one can write $\mathbf{e}_{bf}$ in terms of a new reduced set of body nodal coordinates $\mathbf{e}_f$ as $\mathbf{e}_{bf} = \mathbf{B}_r \mathbf{e}_f$, where $\mathbf{B}_r$ is the matrix of reference conditions that eliminates dependent nodal coordinates and defines how the deformation is measured with respect to the body coordinate system. The number of reference conditions should not be less than the number of the rigid body modes of the finite element shape function. The position vector $\bar{\mathbf{u}}^j$ of the material point on the finite element can then be defined in the body coordinate system as

$$\bar{\mathbf{u}}^j = \mathbf{S}_b^j \mathbf{B}_c^j \left(\mathbf{e}_{bo} + \mathbf{B}_r \mathbf{e}_f \right), \qquad j = 1, 2, \ldots n_e \tag{9}$$

This position vector can be written as the sum of the position vector in the undeformed state plus the vector of deformation as

$$\bar{\mathbf{u}}^j = \bar{\mathbf{u}}_o^j + \bar{\mathbf{u}}_f^j, \qquad j = 1, 2, \ldots n_e \tag{10}$$

where

$$\bar{\mathbf{u}}_o^j = \mathbf{S}_b^j \mathbf{B}_c^j \mathbf{e}_{bo}, \qquad \bar{\mathbf{u}}_f^j = \mathbf{S}_b^j \mathbf{B}_c^j \mathbf{B}_r \mathbf{e}_f \tag{11}$$

This equation shows that $\bar{\mathbf{u}}_o^j$ does not depend on time, while $\bar{\mathbf{u}}_f^j$ is time dependent.

An efficient implementation of the FFR formulation should not restrict the type of normal modes used to define the body deformation. Some commercial MBS computer codes restrict the mode shapes to the free-free modes and provide the option for making corrections to these modes. The FFR formulation presented in these lecture notes does not put any restriction on the type of modes used. In fact, as demonstrated in the literature, different types of modes that are obtained using different types of boundary conditions can lead to the same solution. Therefore, the use of what is called cut-off frequencies by some commercial MBS codes can be misleading since the frequency of the linear problem has no effect on the solution of the nonlinear MBS problem.

1.4 Virtual Displacement, Position, Velocity, and Acceleration

Using the analysis and definitions presented in the preceding section, the global position vector of an arbitrary material point on element j can be written as

$$\mathbf{r}^j = \mathbf{r}_O + \mathbf{A}\left(\bar{\mathbf{u}}_o^j + \bar{\mathbf{u}}_f^j\right), \qquad j = 1, 2, \ldots, n_e \tag{12}$$

A virtual change in the position vector $\mathbf{r}^j$ is given by

$$\delta\mathbf{r}^j = \delta\mathbf{r}_O - \mathbf{A}\tilde{\bar{\mathbf{u}}}^j\bar{\mathbf{G}}\delta\boldsymbol{\theta} + \mathbf{A}\mathbf{S}_b^j\mathbf{B}_c^j\mathbf{B}_r\delta\mathbf{e}_f \tag{13}$$

In this equation, $\tilde{\bar{\mathbf{u}}}^j$ is the skew symmetric matrix associated with the vector $\bar{\mathbf{u}}^j$, and $\bar{\mathbf{G}}$ is the matrix that relates the angular velocity vector to the time derivatives of the orientation parameters. The preceding equation can be written for the finite element j as

$$\delta\mathbf{r}^j = \begin{bmatrix} \mathbf{I} & -\mathbf{A}\tilde{\bar{\mathbf{u}}}^j\bar{\mathbf{G}} & \mathbf{A}\mathbf{S}_b^j\mathbf{B}_c^j\mathbf{B}_r \end{bmatrix} \begin{bmatrix} \delta\mathbf{r}_O \\ \delta\boldsymbol{\theta} \\ \delta\mathbf{e}_f \end{bmatrix} \tag{14}$$

which can also be written as

$$\delta\mathbf{r}^j = \mathbf{S}_f^j\delta\mathbf{q} \tag{15}$$

where

$$\mathbf{S}_f^j = \begin{bmatrix} \mathbf{I} & -\mathbf{A}\tilde{\bar{\mathbf{u}}}^j\bar{\mathbf{G}} & \mathbf{A}\mathbf{S}_b^j\mathbf{B}_c^j\mathbf{B}_r \end{bmatrix}, \qquad \mathbf{q} = \begin{bmatrix} \mathbf{r}_O \\ \boldsymbol{\theta} \\ \mathbf{e}_f \end{bmatrix} \tag{16}$$

The vector $\mathbf{q}$ which contains the reference coordinates and the elastic nodal coordinates of the body can be written as

$$\mathbf{q} = \begin{bmatrix} \mathbf{q}_r \\ \mathbf{q}_f \end{bmatrix} \tag{17}$$

where

$$\mathbf{q}_r = \left[\mathbf{r}_O^{\mathrm{T}} \quad \boldsymbol{\theta}^{\mathrm{T}}\right]^{\mathrm{T}}, \qquad \mathbf{q}_f = \mathbf{e}_f \tag{18}$$

In this equation, $\mathbf{q}_r$ is the vector of *reference coordinates*, and $\mathbf{q}_f$ is the vector of elastic coordinates. It follows that the absolute velocity vector can also be written as

$$\dot{\mathbf{r}}^j = \mathbf{S}_f^j \dot{\mathbf{q}} \tag{19}$$

and the acceleration vector is

$$\ddot{\mathbf{r}}^j = \mathbf{S}_f^j \ddot{\mathbf{q}} + \dot{\mathbf{S}}_f^j \dot{\mathbf{q}} \tag{20}$$

The position, velocity, and acceleration equations can be used with the principle of virtual work to develop the equations of motion of the deformable body.

1.5 FFR Inertia Forces

The virtual work of the inertia forces of the finite element j can be written as $\delta W_i^j = \int_{V^j} \rho^j \ddot{\mathbf{r}}^{j\mathrm{T}} \delta \mathbf{r}^j dV^j$. In this equation, ρ^j is the mass density of the material points of the element, and V^j is the element volume. For the most part, the FFR formulation is used for small deformation problems. As such one can assume that the determinant of the matrix of position vector gradients $\mathbf{J}$ is equal to one. That is, $J = |\mathbf{J}| \approx 1$, and as a consequence, the density and volume are assumed to remain unchanged. Using the definitions of $\delta \mathbf{r}^j$ and $\ddot{\mathbf{r}}^j$, the virtual work of the inertia forces of the finite element can be written as

$$\delta W_i^j = \int_{V^j} \rho^j \left(\mathbf{S}_f^j \ddot{\mathbf{q}} + \dot{\mathbf{S}}_f^j \dot{\mathbf{q}}\right)^{\mathrm{T}} \mathbf{S}_f^j \delta \mathbf{q} dV^j \tag{21}$$

In this equation, $\mathbf{q}$ is the vector of generalized coordinates of the body. The preceding equation can be written as the sum of two terms; one linear in the accelerations and the other is quadratic in the velocities, that is

$$\delta W_i^j = \left\{ \ddot{\mathbf{q}}^{\mathrm{T}} \left(\int_{V^j} \rho^j \mathbf{S}_f^{j\mathrm{T}} \mathbf{S}_f^j dV^j \right) + \dot{\mathbf{q}}^{\mathrm{T}} \left(\int_{V^j} \rho^j \dot{\mathbf{S}}_f^{j\mathrm{T}} \mathbf{S}_f^j dV^j \right) \right\} \delta \mathbf{q} \tag{22}$$

or

$$\delta W_i^j = \left\{ \ddot{\mathbf{q}}^{\mathrm{T}} \mathbf{M}^j - \mathbf{Q}_v^{j\mathrm{T}} \right\} \delta \mathbf{q} \tag{23}$$

where $\mathbf{M}^j$ is the symmetric mass matrix of the finite element j, and $\mathbf{Q}_v^j$ is the vector of Coriolis and centrifugal forces. The mass matrix $\mathbf{M}^j$ and the vector $\mathbf{Q}_v^j$ are defined as

$$\mathbf{M}^j = \int_{V^j} \rho^j \mathbf{S}_f^{j\mathrm{T}} \mathbf{S}_f^j dV^j, \qquad \mathbf{Q}_v^j = -\left(\int_{V^j} \rho^j \mathbf{S}_f^{j\mathrm{T}} \dot{\mathbf{S}}_f^j dV^j \right) \dot{\mathbf{q}} \tag{24}$$

The definition of the matrix $\mathbf{S}_f^j$ shows that the element mass matrix and the vector of Coriolis and centrifugal forces are highly nonlinear functions of the reference and elastic coordinates.

Inertia Shape Integrals. While the mass matrix and the vector of Coriolis and centrifugal forces are highly nonlinear, one can show that the nonlinear mass matrix and the centrifugal and Coriolis forces can be expressed in terms of a unique set of finite element *inertia shape integrals*. Assuming that the element shape function can describe arbitrary translational motion in all direction, one can show that in the case of three-dimensional analysis, the inertia forces of the finite element can be written in terms of the following unique seven inertia shape integrals:

$$\left.\begin{aligned} \bar{\mathbf{S}}^j &= \int_{V^j} \rho^j \mathbf{S}_b^j dV^j \\ \mathbf{S}_{kl}^j &= \int_{V^j} \rho^j \mathbf{S}_{bk}^{j\ \mathrm{T}} \mathbf{S}_{bk}^j dV^j, \quad k,l = 1,23 \end{aligned}\right\} \tag{25}$$

where $\mathbf{S}_{bk}^j$ is the kth row of the element shape function matrix $\mathbf{S}_b^j$. Note that $\bar{\mathbf{S}}^j$ has the same dimension as the element shape function matrix, while $\mathbf{S}_{kl}^j$ are square matrices which have dimensions equal to the number of nodal coordinates of the finite element. In the case of the two dimensional analysis, one can show that the inertia shape integrals reduce to three integrals only defined as

$$\bar{\mathbf{S}}^j = \int_{V^j} \rho^j \mathbf{S}_b^j dV^j, \quad \tilde{\mathbf{S}}^j = \int_{V^j} \rho^j \mathbf{S}_b^{j\mathrm{T}} \tilde{\mathbf{I}} \mathbf{S}_b^j dV^j, \quad \mathbf{S}_{ff}^j = \int_{V^j} \rho^j \mathbf{S}_b^{j\mathrm{T}} \mathbf{S}_b^j dV^j \tag{26}$$

where $\tilde{\mathbf{I}}$ is a skew symmetric matrix defined as

$$\tilde{\mathbf{I}} = \begin{bmatrix} 0 & 1 \\ -1 & 0 \end{bmatrix} \tag{27}$$

The first inertia shape integral in Eq. 26 has dimension equal to the dimension of the element shape function matrix, while the last two inertia shape integrals are square matrices that have dimensions equal to the number of element nodal coordinates. The matrix $\tilde{\mathbf{S}}^j$ is a skew symmetric matrix, while $\mathbf{S}_{ff}^j$ is a symmetric matrix.

Body Inertia Forces. The virtual work of the inertia forces of the body can be obtained by summing up the expressions of the virtual work of the inertia forces of its finite element, that is

$$\delta W_i = \sum_{j=1}^{n_e} \delta W_i^j = \left(\sum_{j=1}^{n_e} \left\{ \ddot{\mathbf{q}}^{\mathrm{T}} \mathbf{M}^j - \mathbf{Q}_v^{j\mathrm{T}} \right\} \right) \delta \mathbf{q} \tag{28}$$

which can also be written as

$$\delta W_i = \left(\ddot{\mathbf{q}}^{\mathrm{T}} \sum_{j=1}^{n_e} \mathbf{M}^j - \sum_{j=1}^{n_e} \mathbf{Q}_v^{j\mathrm{T}} \right) \delta \mathbf{q} = \left(\ddot{\mathbf{q}}^{\mathrm{T}} \mathbf{M} - \mathbf{Q}_v^{\mathrm{T}} \right) \delta \mathbf{q} \tag{29}$$

where $\mathbf{M}$ is the body mass matrix, and $\mathbf{Q}_v$ is the vector of body centrifugal and Coriolis forces. The symmetric body matrix $\mathbf{M}$ and the vector $\mathbf{Q}_v$ are defined as $\mathbf{M} = \sum_{j=1}^{n_e} \mathbf{M}^j$, $\mathbf{Q}_v = \sum_{j=1}^{n_e} \mathbf{Q}_v^j$. As in the case of a single finite element, one can show that the inertia forces of the body can be expressed in terms of a unique set of inertia shape integrals obtained by summing up the inertia shape integrals of its elements. In the case of the three-dimensional analysis, the inertia shape integrals of the body are obtained as

$$\left. \begin{aligned} \bar{\mathbf{S}} &= \left(\sum_{j=1}^{n_e} \bar{\mathbf{S}}^j \mathbf{B}_c^j \right) \mathbf{B}_r \\ \mathbf{S}_{kl} &= \mathbf{B}_r^{\mathrm{T}} \left(\sum_{j=1}^{n_e} \mathbf{B}_c^{j\mathrm{T}} \mathbf{S}_{kl}^j \mathbf{B}_c^j \right) \mathbf{B}_r , \quad k,l = 1,23 \end{aligned} \right\} \tag{30}$$

In the two-dimensional analysis, one has the following body inertia shape integrals:

$$\overline{\mathbf{S}} = \left(\sum_{j=1}^{n_e} \overline{\mathbf{S}}^j \mathbf{B}_c^j \right) \mathbf{B}_r, \quad \tilde{\mathbf{S}} = \mathbf{B}_r^{\mathrm{T}} \left(\sum_{j=1}^{n_e} \mathbf{B}_c^{j\mathrm{T}} \tilde{\mathbf{S}}^j \mathbf{B}_c^j \right) \mathbf{B}_r, \quad \mathbf{S}_{ff} = \mathbf{B}_r^{\mathrm{T}} \left(\sum_{j=1}^{n_e} \mathbf{B}_c^{j\mathrm{T}} \mathbf{S}_{ff}^j \mathbf{B}_c^j \right) \mathbf{B}_r \tag{31}$$

Understanding the role of the inertia shape integrals allows one to develop an automated scheme for coupling existing commercial general purpose FE computer codes and general purpose flexible MBS computer codes.

1.6 FFR Elastic Forces

The elastic forces can be formulated using the virtual work which can be written for the finite element j as $\delta W_s^j = -\int_{V^j} \boldsymbol{\sigma}_{P2}^j : \delta \boldsymbol{\varepsilon}^j dV^j$, where $\boldsymbol{\sigma}_{P2}^j$ is the second Piola-Kirchhoff stress tensor, and $\boldsymbol{\varepsilon}^j$ is the Green-Lagrange strain tensor. In the case of small deformation, no distinction needs to be made between the second Piola-Kirchhoff stress tensor and Cauchy stress tensor. Using the constitutive equations, one can write $\delta W_s^j = -\int_{V^j} \left(\mathbf{E}^j : \boldsymbol{\varepsilon}^j \right) : \delta \boldsymbol{\varepsilon}^j dV^j$, where $\mathbf{E}^j$ is the fourth order tensor of elastic coefficients. One can show by using the strain-displacement relationship that the virtual work of the elastic forces can be written as $\delta W_s^j = \mathbf{Q}_s^{j\mathrm{T}} \delta \mathbf{e}_f^j$, where $\mathbf{Q}_s^j$ is the vector of elastic forces associated with the element nodal coordinates. The vector of element nodal deformation coordinates $\mathbf{e}_f^j$ can be written in terms of the body deformation nodal coordinates as $\mathbf{e}_f^j = \mathbf{B}_c^j \mathbf{B}_r \mathbf{e}_f$, one can write the virtual work of the element elastic forces as $\delta W_s^j = \mathbf{Q}_s^{j\mathrm{T}} \mathbf{B}_c^j \mathbf{B}_r \delta \mathbf{e}_f$. The virtual work of the elastic forces of the body can be obtained as

$$\delta W_s = \sum_{j=1}^{n_e} \delta W_s^j = \left(\sum_{j=1}^{n_e} \mathbf{Q}_s^{j\mathrm{T}} \mathbf{B}_c^j \right) \mathbf{B}_r \delta \mathbf{e}_f \tag{32}$$

which can be written as $\delta W_s = \mathbf{Q}_{sf}^{\mathrm{T}} \delta \mathbf{e}_f$, where

$$\mathbf{Q}_{sf} = \mathbf{B}_r^{\mathrm{T}} \left(\sum_{j=1}^{n_e} \mathbf{B}_c^{j\mathrm{T}} \mathbf{Q}_s^j \right) \tag{33}$$

Because in the FFR formulation the rigid body motion leads to zero strains, the virtual work of the elastic forces can also be written as $\delta W_s = \mathbf{Q}_s^{\mathrm{T}} \delta \mathbf{q}$, where

$$\mathbf{Q}_s = \begin{bmatrix} \mathbf{0} \\ (\mathbf{Q}_s)_f \end{bmatrix}, \qquad \mathbf{q} = \begin{bmatrix} \mathbf{q}_r \\ \mathbf{q}_f \end{bmatrix} = \begin{bmatrix} \mathbf{r}_O \\ \theta \\ \mathbf{e}_f \end{bmatrix} \tag{34}$$

The vector of the elastic forces $(\mathbf{Q}_s)_f$ can also be written as the product of a body stiffness matrix $\mathbf{K}$ multiplied by the vector of body nodal coordinates $\mathbf{q}$. The FFR formulation allows for adding nonlinear terms to the stiffness matrix in order to capture geometric stiffening effects. This formulation can also be used in the analysis of large deformation if the shape of deformation remains simple. In this case, nonlinear strain-displacement relationships must be used.

1.7 FFR Equations of Motion

The equations of motion of the flexible body can be obtained using the principle of virtual work in dynamics which can be written as $\delta W_i = \delta W_s + \delta W_e$, where δW_i is the virtual work of the inertia forces, δW_s is the virtual work of the elastic forces, and δW_e is the virtual work of the applied forces such as gravity, external forces, and magnetic forces. It can be shown that the virtual work of the applied forces can be written as $\delta W_e = \mathbf{Q}_e^{\mathrm{T}} \delta \mathbf{q}$. Using this equation and the expressions for the virtual work of the inertia and elastic forces previously obtained, the principle of virtual work in dynamics can be written as

$$\left\{ \ddot{\mathbf{q}}^{\mathrm{T}} \mathbf{M} - \mathbf{Q}_v^{\mathrm{T}} \right\} \delta \mathbf{q} = \mathbf{Q}_s^{\mathrm{T}} \delta \mathbf{q} + \mathbf{Q}_e^{\mathrm{T}} \delta \mathbf{q} \tag{35}$$

Using this equation in the case of unconstrained motion, one obtains

$$\mathbf{M}\ddot{\mathbf{q}} = \mathbf{Q}_s + \mathbf{Q}_e + \mathbf{Q}_v \tag{36}$$

Using the coordinate partitioning $\mathbf{q} = \begin{bmatrix} \mathbf{q}_r^{\mathrm{T}} & \mathbf{q}_f^{\mathrm{T}} \end{bmatrix}^{\mathrm{T}}$, the preceding equation can be written as

$$\begin{bmatrix} \mathbf{M}_{rr} & \mathbf{M}_{rf} \\ \mathbf{M}_{fr} & \mathbf{M}_{ff} \end{bmatrix} \begin{bmatrix} \ddot{\mathbf{q}}_r \\ \ddot{\mathbf{q}}_f \end{bmatrix} = \begin{bmatrix} \mathbf{0} \\ (\mathbf{Q}_s)_f \end{bmatrix} + \begin{bmatrix} (\mathbf{Q}_e)_r \\ (\mathbf{Q}_e)_f \end{bmatrix} + \begin{bmatrix} (\mathbf{Q}_v)_r \\ (\mathbf{Q}_v)_f \end{bmatrix} \tag{37}$$

In this equation, subscripts r and f refer, respectively, to reference and elastic coordinates. The dynamic coupling between the reference motion and the elastic deformation is represented by the two matrices $\mathbf{M}_{rf}$ and $\mathbf{M}_{fr}$.

1.8 Reduced Order Model

In the small deformation problems, the bodies assume simple deformation shapes that can be described using few simple functions. Using component mode reduction techniques, the vector of the body coordinates can be written in terms of another smaller set of coordinates, thereby significantly reducing the problem dimensionality and eliminating high frequency modes that do not have significant effect on the solution accuracy. In order to demonstrate the application of the component mode techniques, the vector of elastic forces $\mathbf{Q}_s$ can be written as

$$\mathbf{Q}_s = -\begin{bmatrix} \mathbf{0} & \mathbf{0} \\ \mathbf{0} & \mathbf{K}_{ff} \end{bmatrix} \begin{bmatrix} \mathbf{q}_r \\ \mathbf{q}_f \end{bmatrix} \tag{38}$$

where in this equation $\mathbf{K}_{ff}$ is the stiffness matrix associated with the elastic coordinates of the body. Substituting the preceding equation into the equations of motion, one obtains

$$\begin{bmatrix} \mathbf{M}_{rr} & \mathbf{M}_{rf} \\ \mathbf{M}_{fr} & \mathbf{M}_{ff} \end{bmatrix} \begin{bmatrix} \ddot{\mathbf{q}}_r \\ \ddot{\mathbf{q}}_f \end{bmatrix} + \begin{bmatrix} \mathbf{0} & \mathbf{0} \\ \mathbf{0} & \mathbf{K}_{ff} \end{bmatrix} \begin{bmatrix} \mathbf{q}_r \\ \mathbf{q}_f \end{bmatrix} = \begin{bmatrix} (\mathbf{Q}_e)_r \\ (\mathbf{Q}_e)_f \end{bmatrix} + \begin{bmatrix} (\mathbf{Q}_v)_r \\ (\mathbf{Q}_v)_f \end{bmatrix} \tag{39}$$

In order to use the component mode techniques, the case of free vibration with respect to the body reference is first considered. In this case, one has

$$\mathbf{M}_{ff}\ddot{\mathbf{q}}_f + \mathbf{K}_{ff}\mathbf{q}_f = \mathbf{0} \tag{40}$$

One can assume a solution of this equation in the form $\mathbf{q}_f = \mathbf{a}e^{i\beta t}$, where $i = \sqrt{-1}$, $\mathbf{a}$ is the vector of amplitude, t is time, and β is the frequency. Substituting this assumed solution in the preceding equation, one obtains the *generalized eigenvalue problem* $\left(\mathbf{K}_{ff} - \beta^2\mathbf{M}_{ff}\right)\mathbf{a} = \mathbf{0}$. This equation can be solved for the eigenvalues β_k^2 , $k = 1, 2, \ldots, n_f$ where n_f is the number of elastic nodal coordinates. The *eigenvectors* or *mode shapes* associated with the eigenvalues β_k^2 can also be determined. Using the eigenvectors, a constant coordinate transformation from the physical nodal coordinates $\mathbf{q}_f$ to the new modal elastic coordinates $\mathbf{p}_f$ can then be written as $\mathbf{q}_f = \mathbf{B}_m \mathbf{p}_f$,

where $\mathbf{B}_m$ is the *modal transformation matrix* whose columns are the low-frequency n_m mode shapes, and $\mathbf{p}_f$ is the reduced vector of modal coordinates. The total vector of the body coordinates can be written in terms of the new reduced set of coordinates as

$$\mathbf{q} = \begin{bmatrix} \mathbf{q}_r \\ \mathbf{q}_f \end{bmatrix} = \begin{bmatrix} \mathbf{I} & \mathbf{0} \\ \mathbf{0} & \mathbf{B}_m \end{bmatrix} \begin{bmatrix} \mathbf{p}_r \\ \mathbf{p}_f \end{bmatrix} \tag{41}$$

Substituting this transformation into the equations of motion and pre-multiplying by the transpose of the coefficient matrix in the coordinate transformation of the preceding equation leads to

$$\begin{bmatrix} \mathbf{M}_{rr} & \bar{\mathbf{M}}_{rf} \\ \bar{\mathbf{M}}_{fr} & \bar{\mathbf{M}}_{ff} \end{bmatrix} \begin{bmatrix} \ddot{\mathbf{p}}_r \\ \ddot{\mathbf{p}}_f \end{bmatrix} + \begin{bmatrix} \mathbf{0} & \mathbf{0} \\ \mathbf{0} & \bar{\mathbf{K}}_{ff} \end{bmatrix} \begin{bmatrix} \mathbf{p}_r \\ \mathbf{p}_f \end{bmatrix} = \begin{bmatrix} (\mathbf{Q}_e)_r \\ (\bar{\mathbf{Q}}_e)_f \end{bmatrix} + \begin{bmatrix} (\mathbf{Q}_v)_r \\ (\bar{\mathbf{Q}}_v)_f \end{bmatrix} \tag{42}$$

In this equation,

$$\left.\begin{array}{l} \bar{\mathbf{M}}_{rf} = \bar{\mathbf{M}}_{fr}^{\mathrm{T}} = \mathbf{M}_{rf}\mathbf{B}_m, \qquad \bar{\mathbf{M}}_{ff} = \mathbf{B}_m^{\mathrm{T}}\mathbf{M}_{ff}\mathbf{B}_m \\ \bar{\mathbf{K}}_{ff} = \mathbf{B}_m^{\mathrm{T}}\mathbf{K}_{ff}\mathbf{B}_m, \quad (\bar{\mathbf{Q}}_e)_f = \mathbf{B}_m^{\mathrm{T}}(\mathbf{Q}_e)_f, \quad (\bar{\mathbf{Q}}_v)_f = \mathbf{B}_m^{\mathrm{T}}(\mathbf{Q}_v)_f \end{array}\right\} \tag{43}$$

The modal transformation can be constructed at a preprocessing stage in order to reduce the array size.

1.9 Small Deformation FE/MBS Code Interface

The inertia shape integrals play a fundamental role in developing an efficient FE/MBS interface for the small deformation analysis. One can show by using equations previously presented that the modal form of the inertia shape integrals in the three-dimensional analysis is given by

$$\left.\begin{aligned}(\bar{\mathbf{S}})_m &= \bar{\mathbf{S}}\mathbf{B}_m = \left(\sum_{j=1}^{n_e}\bar{\mathbf{S}}^j\mathbf{B}_c^j\right)\mathbf{B}_r\mathbf{B}_m \\ (\mathbf{S}_{kl})_m &= \mathbf{B}_m^{\mathrm{T}}\mathbf{S}_{kl}\mathbf{B}_m = \mathbf{B}_m^{\mathrm{T}}\mathbf{B}_r^{\mathrm{T}}\left(\sum_{j=1}^{n_e}\mathbf{B}_c^{j\mathrm{T}}\mathbf{S}_{kl}^j\mathbf{B}_c^j\right)\mathbf{B}_r\mathbf{B}_m, \quad k,l=1,23\end{aligned}\right\} \tag{44}$$

In the two-dimensional analysis, the modal form of the body inertia shape integrals is given by

$$\left.\begin{aligned}(\bar{\mathbf{S}})_m &= \bar{\mathbf{S}}\mathbf{B}_m = \left(\sum_{j=1}^{n_e}\bar{\mathbf{S}}^j\mathbf{B}_c^j\right)\mathbf{B}_r\mathbf{B}_m \\ (\tilde{\mathbf{S}})_m &= \mathbf{B}_m^{\mathrm{T}}\tilde{\mathbf{S}}\mathbf{B}_m = \mathbf{B}_m^{\mathrm{T}}\mathbf{B}_r^{\mathrm{T}}\left(\sum_{j=1}^{n_e}\mathbf{B}_c^{j\mathrm{T}}\tilde{\mathbf{S}}^j\mathbf{B}_c^j\right)\mathbf{B}_r\mathbf{B}_m \\ (\mathbf{S}_{ff})_m &= \mathbf{B}_m^{\mathrm{T}}\mathbf{S}_{ff}\mathbf{B}_m = \mathbf{B}_m^{\mathrm{T}}\mathbf{B}_r^{\mathrm{T}}\left(\sum_{j=1}^{n_e}\mathbf{B}_c^{j\mathrm{T}}\mathbf{S}_{ff}^j\mathbf{B}_c^j\right)\mathbf{B}_r\mathbf{B}_m\end{aligned}\right\} \tag{45}$$

where m refers to modal form.

When dealing with metals which have high stiffness, it is recommended to scale the mode shapes to be orthonormal with respect to the stiffness matrix. This will reduce the magnitude differences between the reference and elastic coordinates. In order to explain this, one can write the body strain energy as $U=(1/2)\mathbf{q}_f\mathbf{K}_{ff}\mathbf{q}_f=(1/2)\mathbf{p}_f\bar{\mathbf{K}}_{ff}\mathbf{p}_f$. If the mode shapes are made orthonormal with respect to the stiffness matrix, the stiffness matrix associated with the modal coordinates becomes an identity matrix, that is, $\bar{\mathbf{K}}_{ff}=\mathbf{I}$. In the case of stiff metals, the original stiffness matrix $\mathbf{K}_{ff}$ has entries with very large magnitude. In order to have the same strain energy, which should not be affected by the coordinate transformation, the modal coordinates must assume larger magnitude, as compared to $\mathbf{q}_f$, in order for the strain energy $U=(1/2)\mathbf{p}_f\mathbf{p}_f$ to have the same value. This modal transformation provides a natural scaling of the modal coordinates and can lead to significant improvements in the efficiency and robustness of the FFR implementation.

2. ABSOLUTE NODAL COORDINATE FORMULATION (ANCF)

In the preceding section, the important problem of small deformation and large rotation of flexible bodies which is typical of many MBS applications is discussed. In the case of small deformations, more efficient formulations that require the use of smaller number of coordinates can be used. In this section, the more general large rotation and large deformation FE *absolute nodal coordinate formulation* (ANCF) is discussed. This formulation imposes no restrictions on the amount of rotation or deformation within the finite element, is consistent with the nonlinear theory of continuum mechanics, leads to a constant mass matrix and zero centrifugal and Coriolis forces, and it accounts for the dynamic coupling between the rigid body motion and the elastic deformation. The formulation also allows for using different methods for formulating the elastic forces. The generalized elastic forces can be formulated using a general continuum mechanics approach, classical beams and plate theories, or by introducing a local reference frame in order to define the deformation and strains in a local reference frame. Regardless of the method used to formulate the elastic forces, the resulting ANCF equations can be solved using a non-incremental solution procedure.

2.1 ANCF Assumed Displacement Field

In the FE method, assumed polynomials are used to describe the positions or displacements of the material points of the finite element. The polynomial coefficients are replaced by the element nodal coordinates which have physical meaning. In the case of ANCF finite elements, all the components of the position vector are interpolated using polynomials that have the same order. For a given domain defined by the spatial coordinates x_1, x_2 and x_3, the k th component of the position vector can be written as

$$r_k = a_{0k} + a_{1k}x_1 + a_{2k}x_2 + a_{3k}x_3 + a_{4k}x_1^2 + a_{5k}x_2^2 + \dots, \quad k = 1,2,3 \tag{46}$$

The coefficients $a_{ik}, i = 1,2,3,\dots$, in the case of dynamics, are assumed to depend only on time. Using the separation of variables, the assumed displacement field can be written as the product of functions that depend only on the spatial coordinates $\mathbf{x} = \begin{bmatrix} x_1 & x_2 & x_3 \end{bmatrix}^{\mathrm{T}}$ and a vector of time dependent coordinates. Using this standard procedure, one can write the position vector $\mathbf{r}$ as

$$\mathbf{r}(\mathbf{x},t)=\mathbf{S}(\mathbf{x})\mathbf{e}(t) \tag{47}$$

where $\mathbf{S}(\mathbf{x})$ is called the *shape function matrix*, and $\mathbf{e}$ is the vector of nodal coordinates. If the continuum is divided into a number of n_e finite elements, the displacement field of a finite element j can be written using the absolute nodal coordinates as

$$\mathbf{r}^j\left(\mathbf{x}^j,t\right)=\mathbf{S}^j\mathbf{e}^j, \qquad j=1,2,\ldots,n_e \tag{48}$$

where $\mathbf{r}^j$ is the global position vector of an arbitrary point on the finite element j, $\mathbf{S}^j=\mathbf{S}^j\left(\mathbf{x}^j\right)$ is a *shape function matrix* that depends on the element spatial coordinates $\mathbf{x}^j=\begin{bmatrix}x_1^j & x_2^j & x_3^j\end{bmatrix}^{\mathrm{T}}$ defined in the element reference configuration, and $\mathbf{e}^j=\mathbf{e}^j(t)$ is the vector of time dependent nodal coordinates that define the displacements and possibly their spatial derivatives at a set of nodal points. Following the simple and general continuum mechanics description, the vector of nodal coordinates can be selected in the three-dimensional analysis to consist of three translations and nine components of the position vector gradients. In this case, the vector of nodal coordinates $\mathbf{e}^j$ of the finite element at a node k can be written as

$$\mathbf{e}^{jk}=\begin{bmatrix}\mathbf{r}^{jk\mathrm{T}} & \mathbf{r}_{x_1}^{jk\mathrm{T}} & \mathbf{r}_{x_2}^{jk\mathrm{T}} & \mathbf{r}_{x_3}^{jk\mathrm{T}}\end{bmatrix}^{\mathrm{T}} \tag{49}$$

where $\mathbf{r}^{jk}$ is the absolute (global) position vector of the node k of the finite element j, and $\mathbf{r}_{x_l}^{jk}$ is the vector of position gradients obtained by differentiation with respect to the spatial coordinate x_l, $l=1,2,3$. It is clear that the last three vectors of the vector of the element coordinates is $\mathbf{r}_{\mathbf{x}}^{jk}=\mathbf{J}^{jk}$, which is the matrix of the position vector gradients at node k, where $\mathbf{x}^j=\begin{bmatrix}x_1^j & x_2^j & x_3^j\end{bmatrix}^{\mathrm{T}}$.

Example. A special case of the element formulation discussed in this section is the two-dimensional two-node ANCF beam element. Each node of this element has 6 coordinates; two position coordinates and four gradient components. The total number of element coordinates is then 12, and therefore, each component of the position vector field can be approximated with a polynomial which has 6 coefficients, that is,

$$\mathbf{r}=\begin{bmatrix} r_1 \\ r_2 \end{bmatrix}=\begin{bmatrix} a_0+a_1x_1^j+a_2x_2^j+a_3x_1^jx_2^j+a_4\left(x_1^j\right)^2+a_5\left(x_1^j\right)^3 \\ b_0+b_1x_1^j+b_2x_2^j+b_3x_1^jx_2^j+b_4\left(x_1^j\right)^2+b_5\left(x_1^j\right)^3 \end{bmatrix} \tag{50}$$

The polynomial coefficients must be replaced with nodal coordinates that have physical meaning. To this end, the following conditions are imposed at node 1:

$$\mathbf{r}^{j1}=\mathbf{r}^j(0,0)=\begin{bmatrix} e_1 \\ e_2 \end{bmatrix}, \quad \mathbf{r}_{x_1}^{j1}=\mathbf{r}_{x_1}^j(0,0)=\begin{bmatrix} e_3 \\ e_4 \end{bmatrix}, \quad \mathbf{r}_{x_2}^{j1}=\mathbf{r}_{x_2}^j(0,0)=\begin{bmatrix} e_5 \\ e_6 \end{bmatrix} \tag{51}$$

Similarly, the following conditions are imposed at node 2:

$$\mathbf{r}^{j2}=\mathbf{r}^j(l,0)=\begin{bmatrix} e_7 \\ e_8 \end{bmatrix}, \quad \mathbf{r}_{x_1}^{j2}=\mathbf{r}_{x_1}^j(l,0)=\begin{bmatrix} e_9 \\ e_{10} \end{bmatrix}, \quad \mathbf{r}_{x_2}^{j2}=\mathbf{r}_{x_2}^j(l,0)=\begin{bmatrix} e_{11} \\ e_{12} \end{bmatrix} \tag{52}$$

where l is the length of the finite element, and $\mathbf{r}_{x_i^j}^j=\partial\mathbf{r}^j/\partial x_i^j,\ i=1,2$. Using the conditions at the two nodes, the polynomial coefficients can be replaced by the nodal coordinates. One can show that the displacement field of the element can be written as $\mathbf{r}^j(\mathbf{x}^j,t)=\mathbf{S}^j(\mathbf{x}^j)\mathbf{e}^j(t)$, where $\mathbf{e}^j=\begin{bmatrix} e_1 & e_2 & \dots & e_{12}\end{bmatrix}^{\mathrm{T}}$ is the vector of nodal coordinates as defined previously in this example, and $\mathbf{S}^j$ is the element shape function which can be written as

$$\mathbf{S}^j=\begin{bmatrix} s_1\mathbf{I} & s_2\mathbf{I} & s_3\mathbf{I} & s_4\mathbf{I} & s_5\mathbf{I} & s_6\mathbf{I}\end{bmatrix} \tag{53}$$

where the shape functions $s_i,\ i=1,2,\dots,6$ are defined as

$$\left.\begin{aligned} s_1&=1-3\xi^2+2\xi^3, & s_2&=l\left(\xi-2\xi^2+\xi^3\right) \\ s_3&=l\eta(1-\xi), & s_4&=3\xi^2-2\xi^3 \\ s_5&=l\left(-\xi^2+\xi^3\right), & s_6&=l\xi\eta \end{aligned}\right\} \tag{54}$$

and $\xi=x_1^j/l,\quad \eta=x_2^j/l$.

Element Connectivity. In order to formulate the equations of motion of the continuum, the finite elements must be assembled. Let $\mathbf{e}_b=\begin{bmatrix}\mathbf{e}^{1\mathrm{T}} & \mathbf{e}^{2\mathrm{T}} & \dots & \mathbf{e}^{n_e\mathrm{T}}\end{bmatrix}^{\mathrm{T}}$ denote the vector of nodal coordinates of all elements before the assembly, where $\mathbf{e}^j$ is the vector of nodal coordinates of the finite element j, and n_e is the total number of finite elements. Let $\mathbf{e}$ be the vector of all nodal coordinates of the body after the assembly. The vector of the element j coordinates can be written in terms of the nodal coordinates of the body as $\mathbf{e}^j = \mathbf{B}^j\mathbf{e}$, where $\mathbf{B}^j$ is a Boolean matrix that includes zeros and ones and maps the element coordinates to the body coordinates. If the finite elements have different orientations, as in the case of slope discontinuities; a constant transformation can be defined by first defining the element coordinates by the vector $\overline{\mathbf{e}}^j$ and write this vector in terms of element nodal coordinates defined in the same coordinate system as the body nodal coordinates, that is, $\overline{\mathbf{e}}^j = \mathbf{T}^j\mathbf{e}^j$, where $\mathbf{T}^j$ is an element transformation matrix and $\mathbf{e}^j$ is a vector of nodal coordinates defined in the same coordinate system as the body nodal coordinates. The Boolean matrix $\mathbf{B}^j$ can then be used to write the vector of the element nodal coordinates in terms of the body nodal coordinates as $\overline{\mathbf{e}}^j = \mathbf{T}^j\mathbf{B}^j\mathbf{e}$.

2.2 ANCF Inertia and Elastic Forces

Unlike the FFR formulation, ANCF finite elements lead to a simple expression for the inertia forces and complex expression for the elastic forces. The formulation of the finite element inertia and elastic forces is discussed in this section.

ANCF Inertia Forces. In order to obtain the ANCF finite element inertia forces, one must obtain an expression for the acceleration vector. Differentiating the global position vector $\mathbf{r}^j = \mathbf{S}^j\mathbf{e}^j$ with respect to time, the absolute velocity vector of an arbitrary material point on the element j can be written as $\mathbf{v}^j = \dot{\mathbf{r}}^j = \mathbf{S}^j\dot{\mathbf{e}}^j$, $j = 1,2,\ldots,n_e$. Differentiating this equation with respect to time, the acceleration vector can be written as $\mathbf{a}^j = \ddot{\mathbf{r}}^j = \mathbf{S}^j\ddot{\mathbf{e}}^j$, $j = 1,2,\ldots,n_e$. Using the acceleration vector, the virtual work of the inertia forces of the finite element can be written as $\delta W_i^j = \int_{V^j} \rho^j \mathbf{a}^{j\mathrm{T}} \delta\mathbf{r}^j dV^j$, where ρ^j and V^j are, respectively, the mass density and volume of the finite element. The virtual change in the position vector of an arbitrary point can be written as $\delta\mathbf{r}^j = \mathbf{S}^j\delta\mathbf{e}^j$. Using these equations, one can show that the virtual work of the inertia forces can be written as

$$\delta W_i^j = \left\{ \ddot{\mathbf{e}}^{j\mathrm{T}} \int_{V^j} \rho^j \mathbf{S}^{j\mathrm{T}} \mathbf{S}^j dV^j \right\} \delta\mathbf{e}^j \tag{55}$$

This equation can be written as $\delta W_i^j = \left\{ \ddot{\mathbf{e}}^{j\mathrm{T}} \mathbf{M}^j \right\} \delta \mathbf{e}^j$, where $\mathbf{M}^j$ is the constant symmetric mass matrix of the finite element j defined as

$$\mathbf{M}^j = \int_{V^j} \rho^j \mathbf{S}^{j\mathrm{T}} \mathbf{S}^j dV^j \tag{56}$$

The fact that the ANCF mass matrix is constant can be used to significantly simplify the governing equations. In this case, the centrifugal and Coriolis forces are identically zero regardless of the amount of rotation or deformation of the ANCF finite element.

ANCF Elastic Forces. While ANCF finite elements lead to a constant mass matrix, the elastic forces are highly nonlinear functions of the element coordinates. An expression for the virtual work of the elastic forces can be obtained in terms of the Green-Lagrange strain tensor and the Second Piola-Kirchhoff stress tensor for the finite element j as $\delta W_s^j = -\int_{V^j} \boldsymbol{\sigma}_{P2}^j : \delta \boldsymbol{\varepsilon}^j dV^j$, where $\boldsymbol{\sigma}_{P2}^j$ is the second Piola-Kirchhof stress tensor, and $\boldsymbol{\varepsilon}^j$ is the Green-Lagrange strain tensor. The virtual strain can be expressed in terms of the virtual changes of the position vector gradients as

$$\delta \boldsymbol{\varepsilon}^j = \frac{1}{2} \left\{ \left(\delta \mathbf{J}^{j\mathrm{T}} \right) \mathbf{J}^j + \mathbf{J}^{j\mathrm{T}} \left(\delta \mathbf{J}^j \right) \right\} \tag{57}$$

The second Piola-Kirchhoff stresses are related to the Green-Lagrange strains using the constitutive equations $\boldsymbol{\sigma}_{P2}^j = \mathbf{E}^j : \boldsymbol{\varepsilon}^j$, where $\mathbf{E}^j$ is the fourth order tensor of elastic coefficients. Using these relationships and the expression of the virtual work, one can show that the virtual work of the elastic forces of the finite element j can be written as

$$\delta W_s^j = -\frac{1}{2} \int_{V^j} \left(\mathbf{E}^j : \boldsymbol{\varepsilon}^j \right) : \left\{ \left(\delta \mathbf{J}^{j\mathrm{T}} \right) \mathbf{J}^j + \mathbf{J}^{j\mathrm{T}} \left(\delta \mathbf{J}^j \right) \right\} dV^j = -\mathbf{Q}_s^{j\mathrm{T}} \delta \mathbf{e}^j \tag{58}$$

In this equation, $\mathbf{Q}_s^j$ is the vector of the elastic forces associated with the nodal coordinates of the finite element j.

While the procedure for developing the elastic forces using the general continuum mechanics approach is outlined in this section, ANCF allows for using different methods for formulating the elastic forces. In addition to using the general continuum mechanics approach, classical beams and plate theories or a local reference frame can be introduced in order to define the deformation and strains in a local reference frame. Regardless of the method used to formulate the elastic forces, the resulting ANCF equations can be solved non-incrementally as previously mentioned.

2.3 ANCF Dynamic Equations of Motion

In the case of unconstrained motion, the principle of virtual work in dynamics can be written as $\delta W_i = \delta W_s + \delta W_e$, where δW_i is the virtual work of the inertia forces, δW_s is the virtual work of the elastic forces due to the deformation, and δW_e is the virtual work of the applied forces such as gravity and magnetic forces. The virtual work of the applied forces acting on the body can be defined by summing up the virtual work of the forces acting on its finite elements, that is

$$\delta W_e = \sum_{j=1}^{n_e} \delta W_e^j = \sum_{j=1}^{n_e} \mathbf{Q}_e^{j\mathrm{T}} \delta \mathbf{e}^j \tag{59}$$

In this equation, $\mathbf{Q}_e^j$ is the vector of generalized applied nodal forces of element j. The virtual work of the inertia forces of the body can be obtained by summing up the virtual work of the inertia forces of its finite elements, that is,

$$\delta W_i = \sum_{j=1}^{n_e} \delta W_i^j = \sum_{j=1}^{n_e} \left(\mathbf{M}^j \ddot{\mathbf{e}}^j\right)^{\mathrm{T}} \delta \mathbf{e}^j \tag{60}$$

Similarly, the virtual work of the stress forces of the body can be written as

$$\delta W_s = \sum_{j=1}^{n_e} \delta W_s^j = -\sum_{j=1}^{n_e} \mathbf{Q}_s^{j\mathrm{T}} \delta \mathbf{e}^j \tag{61}$$

Substituting these virtual work expressions into the principle of virtual work in dynamics, one obtains $\sum_{j=1}^{n_e} \left(\mathbf{M}^j \ddot{\mathbf{e}}^j + \mathbf{Q}_s^j - \mathbf{Q}_e^j\right)^{\mathrm{T}} \delta \mathbf{e}^j = 0$, which can be written more explicitly as

$$\left\{\begin{bmatrix} \mathbf{M}^1 & \mathbf{0} & \dots & \mathbf{0} \\ \mathbf{0} & \mathbf{M}^2 & \dots & \mathbf{0} \\ \vdots & \vdots & \ddots & \mathbf{0} \\ \mathbf{0} & \mathbf{0} & \dots & \mathbf{M}^{n_e} \end{bmatrix}\begin{bmatrix} \ddot{\mathbf{e}}^1 \\ \ddot{\mathbf{e}}^2 \\ \vdots \\ \ddot{\mathbf{e}}^{n_e} \end{bmatrix} + \begin{bmatrix} \mathbf{Q}_s^1 \\ \mathbf{Q}_s^2 \\ \vdots \\ \mathbf{Q}_s^{n_e} \end{bmatrix} - \begin{bmatrix} \mathbf{Q}_e^1 \\ \mathbf{Q}_e^2 \\ \vdots \\ \mathbf{Q}_e^{n_e} \end{bmatrix}\right\}^{\mathrm{T}} \begin{bmatrix} \delta\mathbf{e}^1 \\ \delta\mathbf{e}^2 \\ \vdots \\ \delta\mathbf{e}^{n_e} \end{bmatrix} = 0 \tag{62}$$

Since $\delta\mathbf{e}^j = \mathbf{B}^j\delta\mathbf{e}$, and $\ddot{\mathbf{e}}^j = \mathbf{B}^j\ddot{\mathbf{e}}$, where $\mathbf{B}^j$ is a Boolean matrix that defines the element connectivity and $\mathbf{e}$ is the vector of the body nodal coordinates, the equation $\sum_{j=1}^{n_e}\left(\mathbf{M}^j\ddot{\mathbf{e}}^j + \mathbf{Q}_s^j - \mathbf{Q}_e^j\right)^{\mathrm{T}}\delta\mathbf{e}^j = 0$ can be written in terms of the nodal coordinates of the body as $\left\{\sum_{j=1}^{n_e}\left(\mathbf{M}^j\mathbf{B}^j\ddot{\mathbf{e}} + \mathbf{Q}_s^j - \mathbf{Q}_e^j\right)^{\mathrm{T}}\mathbf{B}^j\right\}\delta\mathbf{e} = 0$. In the case of unconstrained motion, the elements of the vector $\mathbf{e}$ are independent, and as a result, their coefficients in the preceding equation must be equal to zero, that is $\sum_{j=1}^{n_e}\left\{\mathbf{B}^{j\mathrm{T}}\mathbf{M}^j\mathbf{B}^j\ddot{\mathbf{e}} + \mathbf{B}^{j\mathrm{T}}\mathbf{Q}_s^j - \mathbf{B}^{j\mathrm{T}}\mathbf{Q}_e^j\right\} = \mathbf{0}$. By performing the summation in this equation, one can show that the finite element equations of motion of the body can be written as

$$\mathbf{M}\ddot{\mathbf{e}} + \mathbf{Q}_s - \mathbf{Q}_e = \mathbf{0} \tag{63}$$

where $\mathbf{M}$ is the body symmetric mass matrix, $\mathbf{Q}_s$ is the vector of body elastic forces, and $\mathbf{Q}_e$ is the vector of the body applied forces. These matrices and vectors are defined as

$$\mathbf{M} = \sum_{j=1}^{n_e}\mathbf{B}^{j\mathrm{T}}\mathbf{M}^j\mathbf{B}^j, \qquad \mathbf{Q}_s = \sum_{j=1}^{n_e}\mathbf{B}^{j\mathrm{T}}\mathbf{Q}_s^j, \qquad \mathbf{Q}_e = \sum_{j=1}^{n_e}\mathbf{B}^{j\mathrm{T}}\mathbf{Q}_e^j \tag{64}$$

Because the mass matrix is constant, a transformation based on *Cholesky coordinates* that leads to an identity mass matrix can be used, leading to an optimum sparse matrix structure in the MBS algorithms. Because the elastic forces are highly nonlinear in general, numerical integration methods must be used in the evaluation of these forces.

2.4 Geometry and ANCF Finite Elements

Some of ANCF finite elements employ a complete set of gradient vectors as nodal coordinates. These elements, called *fully parameterized elements*, allow, in a straight forward manner, for the use of a general continuum mechanics approach to formulate the elastic forces and allow also for using tensor transformation to describe discontinuities. Fully parameterized elements also allow for using more general constitutive relationships. Elements which do not employ a full parameterization are called *gradient deficient*. The use of a general continuum mechanics approach with these elements is not as straight forward as compared to the fully parameterized elements. However, gradient deficient elements that constrain some deformation modes can be more efficient in some applications.

Both fully parameterized and gradient deficient ANCF finite elements, which were introduced to study large deformation of very flexible components, perform well and can be very efficient in the case of very flexible bodies. Efficiency is achieved in the analysis of large deformations of very flexible bodies because a *non-incremental solution procedure* can be adopted when ANCF finite elements are used. As the structure stiffness decreases, ANCF finite elements become more efficient. Fully parameterized ANCF finite elements can be less efficient and can suffer from locking problems when these elements are used with the general continuum mechanics approach in the analysis of thin and stiff structures.

Curve Geometry. The geometry of deformation can be better understood using the concepts of differential geometry. The centerline of a beam element, for example, represents a space curve which can be uniquely defined in terms of one parameter only. If α is the parameter that defines the curve over the interval $a \le \alpha \le b$, the curve can then be defined using the following parametric form:

$$\mathbf{r}(\alpha) = \begin{bmatrix} r_1(\alpha) & r_2(\alpha) & r_3(\alpha) \end{bmatrix}^{\mathrm{T}} \tag{65}$$

The tangent vector to the curve at α is obtained by differentiating the preceding equation as

$$\mathbf{r}_\alpha = \frac{d\mathbf{r}}{d\alpha} = \begin{bmatrix} \dfrac{dr_1(\alpha)}{d\alpha} & \dfrac{dr_2(\alpha)}{d\alpha} & \dfrac{dr_3(\alpha)}{d\alpha} \end{bmatrix}^{\mathrm{T}} \tag{66}$$

If at a given point α, $|d\mathbf{r}(\alpha)/d\alpha| = 0$, one has a *singular point*. The parameter α can be selected to be the *arc length* s. If s is the arc length, one has $|d\mathbf{r}/ds| = |\mathbf{r}_s(s)| = 1$, that is, if the arc length is used as a parameter, the tangent vector $\mathbf{r}_s$ is a unit vector. The *curvature vector* is defined as

$$\mathbf{r}_{ss}(s)=\frac{d^2\mathbf{r}}{ds^2}=\frac{d\mathbf{r}_s}{ds} \tag{67}$$

The magnitude of the curvature vector $\kappa(s)=|\mathbf{r}_{ss}(s)|$ is called the *curvature* which can be used to measure the amount of bending of the curve. Because the tangent and curvature vectors are orthogonal, a unit vector along the curvature vector defines the unit normal $\mathbf{n}$ to the curve. The normal vector is, therefore, given as

$$\mathbf{n}(s)=\frac{\mathbf{r}_{ss}(s)}{\kappa(s)} \tag{68}$$

The unit tangent and normal vectors form the *osculating plane*. The *radius of curvature* of the curve at s is $R=1/\kappa(s)$. A unit vector normal to the osculating plane, called the *binormal vector*, is given by

$$\mathbf{b}(s)=\mathbf{r}_s(s)\times\mathbf{n}(s) \tag{69}$$

The three orthogonal unit vectors $\mathbf{r}_s$, $\mathbf{n}$, and $\mathbf{b}$ form a triad called the *Frenet frame*. One can show that $\mathbf{b}_s(s)=-\tau(s)\mathbf{n}(s)$, where τ is called the *torsion.*

Surface Geometry. A surface can be completely described in terms of two parameters s_1 and s_2, and therefore, a surface can be defined in the following parametric form:

$$\mathbf{r}(s_1,s_2)=\begin{bmatrix} r_1(s_1,s_2) & r_2(s_1,s_2) & r_3(s_1,s_2)\end{bmatrix}^{\mathrm{T}} \tag{70}$$

The Jacobian matrix is given by

$$\mathbf{J}=\begin{bmatrix} \dfrac{\partial\mathbf{r}}{\partial s_1} & \dfrac{\partial\mathbf{r}}{\partial s_2}\end{bmatrix}=\begin{bmatrix} \dfrac{\partial r_1}{\partial s_1} & \dfrac{\partial r_1}{\partial s_2} \\ \dfrac{\partial r_2}{\partial s_1} & \dfrac{\partial r_2}{\partial s_2} \\ \dfrac{\partial r_3}{\partial s_1} & \dfrac{\partial r_3}{\partial s_2}\end{bmatrix} \tag{71}$$

This Jacobian matrix must have a rank equal to two, that is, $(\partial\mathbf{r}/\partial s_1)\times(\partial\mathbf{r}/\partial s_2)\neq 0$, which implies that the two tangent vectors $\mathbf{r}_{s_1}=\partial\mathbf{r}/\partial s_1$ and $\mathbf{r}_{s_2}=\partial\mathbf{r}/\partial s_2$ at the point of intersection of the coordinate lines s_1 and s_2 are linearly independent. The unit vector normal to the surface at this point is $\mathbf{n}=(\mathbf{r}_{s_1}\times\mathbf{r}_{s_2})/|\mathbf{r}_{s_1}\times\mathbf{r}_{s_2}|$.

First Fundamental Form. The *first fundamental form* of a surface is defined as

$$I = d\mathbf{r}\cdot d\mathbf{r} = d\mathbf{r}^{\mathrm{T}} d\mathbf{r} \tag{72}$$

The first fundamental form I can be used as a measure of distance or length. Because $d\mathbf{r}=\mathbf{r}_{s_1}ds_1+\mathbf{r}_{s_2}ds_2$, the first fundamental form can be written as

$$\begin{aligned} I &= (\mathbf{r}_{s_1}ds_1+\mathbf{r}_{s_2}ds_2)^{\mathrm{T}}(\mathbf{r}_{s_1}ds_1+\mathbf{r}_{s_2}ds_2) \\ &= E_I\,(ds_1)^2+2F_I\,ds_1ds_2+G_I\,(ds_2)^2 \end{aligned} \tag{73}$$

In this equation, the coefficients $E_I=\mathbf{r}_{s_1}^{\mathrm{T}}\mathbf{r}_{s_1}$, $F_I=\mathbf{r}_{s_1}^{\mathrm{T}}\mathbf{r}_{s_2}$, and $G_I=\mathbf{r}_{s_2}^{\mathrm{T}}\mathbf{r}_{s_2}$ are called the *coefficients of the first fundamental form.*

Second Fundamental Form. The *second fundamental form* of a surface is

$$\begin{aligned} II &= -d\mathbf{r}\cdot d\mathbf{n} = -\left(\mathbf{r}_{s_1}ds_1+\mathbf{r}_{s_2}ds_2\right)^{\mathrm{T}}\left(\mathbf{n}_{s_1}ds_1+\mathbf{n}_{s_2}ds_2\right) \\ &= L_{II}\,(ds_1)^2+2M_{II}\,ds_1ds_2+N_{II}\,(ds_2)^2 \end{aligned} \tag{74}$$

where $L_{II}=-\mathbf{r}_{s_1}^{\mathrm{T}}\mathbf{n}_{s_1}$, $M_{II}=-\frac{1}{2}\left(\mathbf{r}_{s_1}^{\mathrm{T}}\mathbf{n}_{s_2}+\mathbf{r}_{s_2}^{\mathrm{T}}\mathbf{n}_{s_1}\right)$, and $N_{II}=-\mathbf{r}_{s_2}^{\mathrm{T}}\mathbf{n}_{s_2}$ are the coefficients of the second fundamental form. Because $\mathbf{r}_{s_1}$ and $\mathbf{r}_{s_2}$ are perpendicular to the unit normal $\mathbf{n}$ for all values of the parameters s_1 and s_2, one has the following identities:

$$\left.\begin{aligned} \mathbf{r}_{s_1s_1}^{\mathrm{T}}\mathbf{n} &= -\mathbf{r}_{s_1}^{\mathrm{T}}\mathbf{n}_{s_1}, \quad \mathbf{r}_{s_1s_2}^{\mathrm{T}}\mathbf{n} = -\mathbf{r}_{s_1}^{\mathrm{T}}\mathbf{n}_{s_2}, \\ \mathbf{r}_{s_2s_1}^{\mathrm{T}}\mathbf{n} &= -\mathbf{r}_{s_2}^{\mathrm{T}}\mathbf{n}_{s_1}, \quad \mathbf{r}_{s_2s_2}^{\mathrm{T}}\mathbf{n} = -\mathbf{r}_{s_2}^{\mathrm{T}}\mathbf{n}_{s_2} \end{aligned}\right\} \tag{75}$$

Using these identities, alternate forms of the coefficients of the second fundamental form can be provided as $L_{II} = \mathbf{r}_{s_1 s_1}^{\mathrm{T}} \mathbf{n}$, $M_{II} = \mathbf{r}_{s_1 s_2}^{\mathrm{T}} \mathbf{n}$, and $N_{II} = \mathbf{r}_{s_2 s_2}^{\mathrm{T}} \mathbf{n}$, where $\mathbf{r}_{s_i s_j} = \left(\partial^2 \mathbf{r} / \partial s_i \partial s_j\right)$. Using these definitions and the fact that $d^2\mathbf{r} = \mathbf{r}_{s_1 s_1}(ds_1)^2 + 2\mathbf{r}_{s_1 s_2} ds_1 ds_2 + \mathbf{r}_{s_2 s_2}(ds_2)^2$, one can show that that the second fundamental form can be written as

$$II = d^2\mathbf{r} \cdot \mathbf{n} \tag{76}$$

The surface is elliptic if $L_{II} N_{II} - M_{II}^2 > 0$, hyperbolic if $L_{II} N_{II} - M_{II}^2 < 0$, parabolic if $L_{II} N_{II} - M_{II}^2 = 0$, and planar if $L_{II} = M_{II} = N_{II} = 0$.

Let $\mathbf{c} = \mathbf{c}\left(s_1(\alpha), s_2(\alpha)\right)$ be a regular curve defined on the surface $\mathbf{r} = \mathbf{r}(s_1, s_2)$, where α is a parameter. The *normal curvature vector* to the curve $\mathbf{c}$, which has an arc length s, at point P denoted by $\mathbf{K}_n$ is defined as the projection of the curvature vector $\mathbf{c}_{ss}$ of the curve onto the normal $\mathbf{n}$ to the surface at point P and is given by $\mathbf{K}_n = (\mathbf{c}_{ss} \cdot \mathbf{n})\mathbf{n}$. The norm of the normal curvature vector is called the *normal curvature* and is defined as $k_n = \mathbf{c}_{ss} \cdot \mathbf{n}$. From the definition of the curvature vector of a curve, the curve curvature vector at a point P on the surface $\mathbf{r}$ is

$$\mathbf{c}_{ss} = \frac{d\mathbf{r}_s}{ds} = \frac{d\mathbf{r}_s}{d\alpha}\left(\frac{d\alpha}{ds}\right) = \frac{1}{\left|d\mathbf{r}/d\alpha\right|}\frac{d\mathbf{r}_s}{d\alpha} \tag{77}$$

where $\mathbf{r}_s$ is the tangent vector to the curve at P. In deriving Eq. 77, the fact that $\left|d\mathbf{r}/d\alpha\right| = \left|\mathbf{r}_s\right|\left(ds/d\alpha\right) = \left(ds/d\alpha\right)$, which is the consequence of the fact that $\left|\mathbf{r}_s\right| = 1$, is used. Because $\mathbf{r}_s$ and $\mathbf{n}$ are two orthogonal vectors, one has $\frac{d}{d\alpha}\left(\mathbf{r}_s^{\mathrm{T}}\mathbf{n}\right) = 0$ which leads to $\left(d\mathbf{r}_s/d\alpha\right)^{\mathrm{T}}\mathbf{n} = -\mathbf{r}_s^{\mathrm{T}}\left(d\mathbf{n}/d\alpha\right)$. Using the relationship developed in this section for the surface, one can show that

$$k_n = \frac{L_{II}(ds_1)^2 + 2M_{II} ds_1 ds_2 + N_{II}(ds_2)^2}{E_I(ds_1)^2 + 2F_I ds_1 ds_2 + G_I(ds_2)^2} = \frac{II}{I}, \qquad (ds_1)^2 + (ds_2)^2 \neq 0 \tag{78}$$

The directions which define the maximum or minimum values of the normal curvature can be obtained using the conditions $\partial k_n / \partial(ds_1) = \partial k_n / \partial(ds_2) = 0$. These conditions lead to

$$\begin{bmatrix} L_{II} - k_n E_I & M_{II} - k_n F_I \\ M_{II} - k_n F_I & N_{II} - k_n G_I \end{bmatrix} \begin{bmatrix} ds_1 \\ ds_2 \end{bmatrix} = \begin{bmatrix} 0 \\ 0 \end{bmatrix} \tag{79}$$

For a nontrivial solution, one has the following characteristic equation:

$$(E_I G_I - F_I^2)(k_n)^2 - (E_I N_{II} + G_I L_{II} - 2F_I M_{II})k_n + L_{II} N_{II} - M_{II}^2 = 0 \tag{80}$$

The solution of this quadratic characteristic equation defines the roots k_1 and k_2, called the *principal curvatures*, which can be used to determine the *principal directions*. The *mean curvature* K_m and the *Gaussian curvature* K_G at a point P on the surface are defined using the principal curvatures as $K_m = \frac{1}{2}(k_1 + k_2)$, and $K_G = k_1 k_2$.

3. COMPUTATIONAL GEOMETRY METHODS

Computational geometry (CG) methods such as Bezier, B-spline, and NURBS (**N**on-**U**niform **R**ational **B-S**plines) are used in CAD systems to develop solid models of system components that can have complex geometry. The CAD solid models are converted to FE meshes in order to perform the analysis and obtain the solutions in response to loading conditions. This conversion can be very costly and time consuming since many of the existing finite elements are not based on correct geometric description, and therefore, there is no linear mapping between CG methods and the finite elements used to develop the analysis mesh. CG methods, however, can be converted to ANCF finite elements using linear mapping. In this section, the geometry methods used in CAD systems are discussed.

3.1 Bezier Curve

A Bezier curve with m-degree is defined as

$$\mathbf{r}(\xi) = \sum_{i=0}^{m} S_{i,m}(\xi)\mathbf{P}_i, \qquad 0 \le \xi \le 1 \tag{81}$$

where $S_{i,m}(x)$ are the *basis* or *blending functions*, and the coefficients $\mathbf{P}_i$ are the *control points*. The basis functions $S_{i,m}(x)$ are the *m*-degree *Bernstein polynomials* defined as

$$S_{i,m}(\xi)=\frac{m!}{i!(m-i)!}\xi^i(1-\xi)^{m-i} \tag{82}$$

In the special case of cubic curve ($m=3$), the basis functions $S_{0,3}, S_{1,3}, S_{2,3}$, and $S_{3,3}$ are defined as

$$\left.\begin{aligned} S_{0,3}&=(1-\xi)^3, & S_{1,3}&=3\xi(1-\xi)^2,\\ S_{2,3}&=3\xi^2(1-\xi), & S_{3,3}&=\xi^3 \end{aligned}\right\} \tag{83}$$

Therefore, the cubic Bezier curve is defined by

$$\mathbf{r}=(1-\xi)^3\mathbf{P}_0+3\xi(1-\xi)^2\mathbf{P}_1+3\xi^2(1-\xi)\mathbf{P}_2+\xi^3\mathbf{P}_3 \tag{84}$$

This equation is the same as the equation that can be obtained using the ANCF cable element if the gradient vectors are replaced by control points.

3.2 B-Spline Representation

A B-spline curve with *p*-degree is defined as

$$\mathbf{r}(u)=N_{0,p}(u)\mathbf{P}_0+N_{1,p}(u)\mathbf{P}_1+\cdots+N_{n,p}(u)\mathbf{P}_n=\sum_{i=0}^{n}N_{i,p}(u)\mathbf{P}_i \tag{85}$$

where $N_{i,p}(u)$ are B-spline basis functions of degree *p*, $\mathbf{P}_i$ are the control points, and n is the number of the control points. The B-spline basis functions $N_{i,p}(u)$ are defined as

$$\left.\begin{aligned} N_{i,0}(u)&=\begin{cases}1 & \text{if } u_i\le u<u_{i+1}\\ 0 & \text{otherwise}\end{cases}\\ N_{i,j}(u)&=\frac{u-u_i}{u_{i+j}-u_i}N_{i,j-1}(u)+\frac{u_{i+j+1}-u}{u_{i+j+1}-u_{i+1}}N_{i+1,j-1}(u) \end{aligned}\right\} \tag{86}$$

where u_i, $i = 0,1,2,\ldots,n+p+1$ are called the *knots*; which represent a non-decreasing sequence, that is $u_i \le u_{i+1}$. The vector $\mathbf{U} = \{u_0 \quad u_1 \quad \cdots \quad u_{n+p+1}\}$ is called the *knot vector*. The knots do not have to be distinct; distinct knots are called *breakpoints* and define segments with non-zero length. Each non-zero knot span corresponds to a segment of the B-spline curve. The number of the non-distinct knots at a point is referred to as the knot multiplicity.

3.3 ANCF/ B-Spline Linear Mapping

Because a B-spline curve can be represented as a series of connected Bezier curves, B-spline representation can be converted to ANCF representation. In the case of the ANCF cable elements, the global position vector $\mathbf{r}$ of an arbitrary point on the finite element centerline can be defined as $\mathbf{r}(x) = \mathbf{S}(x)\mathbf{e}(t)$, where $\mathbf{S}$ is the element shape function matrix expressed in terms of the element spatial coordinate x; $\mathbf{e} = \begin{bmatrix} \mathbf{r}^{A^T} & \mathbf{r}_x^{A^T} & \mathbf{r}^{B^T} & \mathbf{r}_x^{B^T} \end{bmatrix}^T$ is the vector of nodal coordinates that consist of absolute position and gradient coordinates of the first and second nodes of the element (denoted respectively as A and B); and t is time. For the three-dimensional cable elements, the shape function matrix can be written as $\mathbf{S} = \begin{bmatrix} s_1\mathbf{I} & s_2\mathbf{I} & s_3\mathbf{I} & s_4\mathbf{I} \end{bmatrix}$, where $\mathbf{I}$ is the identity matrix, and $s_i, i = 1,2,3,4$, are shape functions defined as

$$\left.\begin{aligned} s_1 &= 1 - 3\xi^2 + 2\xi^3, \quad & s_2 &= l\left(\xi - 2\xi^2 + \xi^3\right), \\ s_3 &= 3\xi^2 - 2\xi^3, & s_4 &= l\left(-\xi^2 + \xi^3\right) \end{aligned}\right\} \tag{87}$$

In this equation, l is the length of the finite element, and $\xi = x/l$. The cubic Bezier curve control points $\mathbf{P}_0$, $\mathbf{P}_1$, $\mathbf{P}_2$, and $\mathbf{P}_3$ can be written in terms of the ANCF nodal coordinates as

$$\begin{bmatrix} \mathbf{P}_0 \\ \mathbf{P}_1 \\ \mathbf{P}_2 \\ \mathbf{P}_3 \end{bmatrix} = \begin{bmatrix} \mathbf{I} & \mathbf{0} & \mathbf{0} & \mathbf{0} \\ \mathbf{I} & \frac{l}{3}\mathbf{I} & \mathbf{0} & \mathbf{0} \\ \mathbf{0} & \mathbf{0} & \mathbf{I} & -\frac{l}{3}\mathbf{I} \\ \mathbf{0} & \mathbf{0} & \mathbf{I} & \mathbf{0} \end{bmatrix} \begin{bmatrix} \mathbf{r}^A \\ \mathbf{r}_x^A \\ \mathbf{r}^B \\ \mathbf{r}_x^B \end{bmatrix} \tag{88}$$

Using this linear mapping, one can obtain ANCF representation that has identical geometry to the B-spline representation. One can also show that the relationship between the ANCF shape functions and the Bernstein polynomials $S_{0,3}$, $S_{1,3}$, $S_{2,3}$, and $S_{3,3}$ is defined as

$$\left.\begin{aligned} s_1 &= S_{0,3} + S_{1,3}, \qquad & s_2 &= \frac{l}{3} S_{1,3}, \\ s_3 &= S_{2,3} + S_{3,3}, \qquad & s_4 &= -\frac{l}{3} S_{2,3} \end{aligned}\right\} \tag{89}$$

Bibliography

Bonet, J., and Wood, R.D. (1997). *Nonlinear Continuum Mechanics for Finite Element Analysis*: Cambridge University Press.

Boresi, A.P., and Chong, K.P. (2000). *Elasticity in Engineering Mechanics*, Second Edition. John Wiley & Sons. Dierckx, P. (1993). *Curve and Surface Fitting with Splines.* New York: Oxford University Press.

Farin, G. (1999). *Curves and Surfaces for CAGD*, A Practical Guide. San Francisco: Morgan Kaufmann, Publishers, Fifth Edition.

Hu, W., Tian, Q., Hu, H. Y. (2014). Dynamics Simulation of the Liquid-Filled Flexible Multibody System via the Absolute Nodal Coordinate Formulation and SPH Method. *Nonlinear Dynamics*, 75:653-671.

Kreyszig, E. (1991). *Differential Geometry*: Dover Publications.

Liu, C., Tian, Q., Hu, H. Y. (2011). Dynamics of Large Scale Rigid-Flexible Multibody System Composed of Composite Laminated Plates. *Multibody System Dynamics*, 26:283-305.

Ogden, R.W. (1984) *Non-Linear Elastic Deformations*: Dover Publications.

Omar, M. A. and Shabana, A. A. (2001). A Two-Dimensional Shear Deformable Beam for Large Rotation and Deformation Problems. *J. Sound Vibration,* 243.3:565-576.

Piegl, L., Tiller, W. (1997). *The NURBS Book.* New York: Springer-Verlag, Second Edition.

Rogers, D.F. (2001). *An Introduction to NURBS with Historical Perspective*. San Diego, CA: Academic Press.

Shabana, A.A. (2012). *Computational Continuum Mechanics*: Cambridge University Press, Second Edition.

Shabana, A.A. (2014). *Dynamics of Multibody Systems*: Cambridge University Press, Fourth Edition.

Sanborn, G.G., and Shabana, A.A. (2009). On the Integration of Computer Aided Design and Analysis Using the Finite Element Absolute Nodal Coordinate Formulation. *Multibody System Dynamics*, 22:181-197.

Tian, Q., Chen, L.P., Zhang, Y.Q., Yang, J.Z. (2009). An Efficient Hybrid Method for Multibody Dynamics Simulation Based on Absolute Nodal Coordinate Formulation. *ASME Journal of Computational and Nonlinear Dynamics*, 4: 021009-1-021009-14.

Tian, Q., Sun, Y. L., Liu, C., Hu, H. Y., Paulo, F. (2013). Elasto-Hydro-Dynamic Lubricated Cylindrical Joints for Rigid-Flexible Multibody Dynamics. *Computers & Structures,* 114-115:106-120.

Tian, Q., Xiao, Q. F., Sun, Y. L., Hu, H. Y., Liu, H., Paulo, F. (2015). Coupling Dynamics of a Geared Multibody System Supported by Elasto-Hydro-Dynamic Lubricated Cylindrical Joints. *Multibody System Dynamics*, 33:259-284.

Art of Modeling in Contact Mechanics

Michel Raous
LMA, CNRS, Aix-Marseille Univ., Centrale Marseille, France
raous@lma.cnrs-mrs.fr *

Abstract In this chapter, we will first address general issues of the art and craft of modeling - contents, concepts, methodology. Then, we will focus on modeling in contact mechanics, which will give the opportunity to discuss these issues in connection with non-smooth problems. It will be shown that the non-smooth character of the contact laws raises difficulties and specificities at every step of the modeling process. A wide overview will be given on the art of modeling in contact mechanics under its various aspects: contact laws, their mechanical basics, various scales, underlying concepts, mathematical analysis, solvers, identification of the constitutive parameters and validation of the models. Every point will be illustrated by one or several examples.

1 Modeling: the bases

It would be ambitious to try to give a general definition of the concept either of a model itself or of model processing. Modeling relates to the general process of production of scientific knowledge and also to the scientific method itself. It could be deductive (from the general to the particular, as privileged by Aristotle) or inductive (making sense of a corpus of raw data). Descartes (38) saw in the scientific method an approach to be followed step by step to get to a truth. Modeling can be effectively regarded as a scientific method that proceeds step by step, but its objective is more modest: to give sense of an observation or an experiment, and above all to predict behaviors within the context of specific assumptions. This concept of "proceeding step by step" is fundamental in modeling.

In this first chapter, we will examine the notion of model in the general context of mechanical systems. The main features of these models are their

*The author is grateful to Prof. Marius Cocou for his very constructive comments on the mathematical aspects and Prof. Alfredo Soldati for our very interesting discussion on the general topic.

F. Pfeiffer and H. Bremer (eds.), *The Art of Modeling Mechanical Systems*,
CISM International Centre for Mechanical Sciences 570,
DOI 10.1007/978-3-319-40256-7_4

contents, their purposes, the way to build them, the way to use them, the expectations and challenges.

Some of the reflections presented in this chapter find inspiration in the book entitled Mathematical Modeling Techniques written by Aris Rutherford, from University of Caltech, in the 80s (108).

1.1 The objectives

The objective of modeling is twofold: explaining and understanding, giving sense to experimental observations on the one hand, predicting behaviors on the other one.

- Understanding. The models we are talking about all start from experimental observations to which we want to give sense (deductive approach) and that we wish to explain, understand and describe using the basic laws of mechanics and adding "properly" chosen theoretical ingredients. The term "properly" is fundamental and could be very constraining.

With the goal of understanding, the approach consists in gathering the inherent elements based on concepts necessary to build the systems of equations to be solved. Behind that, there are some questions that should be answered: do the solutions exist, are these solutions unique, are they stable, are there methods (existing or to be developed) for solving them? In other words, a model is part of a theoretical mechanical and mathematical framework, and it will be essential to make sure that this framework has been properly chosen, that there is consistency between the equations and the framework, and that the possible limitations can be identified.

The main ingredients when building models are based on the fundamental laws of mechanics and thermodynamics, as well as on those of physics, electricity and chemistry, and they need either the use or the development of mathematical tools. In the following we will present how, in some cases, modeling requires new mathematical developments, this being particularly true for contact problems. The problems studied today are increasingly complex, due to interactions between phenomena of different natures, and multiphysic models can be considered.

Using or developing ingredients that have a meaning with respect to some concepts to construct a scientifically coherent model is an essential feature of the art of modeling. This refers to the objectives of "understanding" and "giving sense" in modeling.

Before going any further into the analysis of the modeling processes, let us make mention of other types of approaches. The notion of modeling that we present in the following is actually restrictive, and there are branches of the modeling that differ markedly in their concept. Let us present two of them.

Note 1. When we associate modeling to "equations", we cut off other approaches which are directly based on experiments without theoretical link or at least without a formal theoretical link. Let us give an example of experience-based modeling by referring to the work of the Catalan architect Gaudi, known, among others things, for the basilica church "La Sagrada Familia" in Barcelona. For the design of its arches, Gaudi developed the reversed chain model. He created mockups, using chains simulating the arches weighted with small bags to simulate the bearing forces applied by the supporting pillar on the arch. Then, a mirror placed below gave the reversed image of the structure and showed the final shape of the real arch design (see Fig. 1). He also used small hanging bags (full of salt or flour) for the representation of the shapes and the volumes of the domes, the cupolas, the arches and the pillars. The modeling process uses directly measurements of the mechanical behavior of the mockup without setting a theoretical framework as we will do in the following. However Gaudi's architectural pieces of art brilliantly illustrate the efficiency of the reversed hanging chain model and of this kind of modeling approach.

Figure 1. Reversed chain model and cathedral church La Sagrada Familia

Note 2. Another class of modeling that we will overlook in this presentation is the one based on statistical considerations. In that case, from a (if possible large) number of experiments or random draws, we build a predictive behavior or determine a solution to the problem using sophisticated statistical tools. The Monte Carlo methods have been the basis for this kind of approach. The Monte Carlo method was invented by Metropolis-Ulam (73). This class of methods is based on random drawings chosen according to suitable probability distributions. Let us give an example. The acoustic field radiated by a source in a volume of any shape may be determined without solving the wave equation but by using a ray theory: the acoustic field is then evaluated as the superposition of the trajectories of sound shots fired from the source in random directions, the absorption by the walls being taken into account in the reflection rule. This approach works quite well as long as the edge diffraction phenomena are not too important. Of course, the higher the number of shots is, the better the approximation is.

Examples of various methods employing statistical and probability tools in the modeling process itself include variance reduction, polynomial chaos, kriging (used in geostatistics, meteorology, environmental sciences and electromagnetism), as well as fractal analysis (see Cherepanov-Balakin (18)). The latter, more recent, is widely used in many fields. Among them can be cited characterization and analysis of rough surfaces, dynamic cracking and elasticity of polymers or rubbers (18), analysis or monitoring of industrial processes, fault detection in machines, etc.

- **Predicting.** Beyond providing the satisfaction of a better understanding, the main purpose of modeling is to predict, that is to say, replace experimentation by simulation to predict the behavior of a mechanical system submitted to new loads, i.e., different from those used for building and validating the model. Replacing the experiment by a simulation via computation or analytical analysis is a crucial challenge, for two reasons at least. In the first place, replacing a time-consuming and costly experiment by some calculations will be comfortable and economical (although it can also be very expensive at times). Secondly, calculations are indispensable when experiments are not feasible or very difficult to conduct, as in cases such as accidental events or natural disasters (a tsunami, the collapse of a dam or a pipe in an irradiated environment, a nuclear explosion, the movement of a cyclone, etc.). Weather forecasting is a daily example of modeling. In some cases, experiments are performed using mockups but the scale effects remain a problem which should be considered with great caution.

It is to be understood that, for a reliable behavior prediction, the challenges of modeling are the validation of the model, the study of its reliability and the identification of its domains of validity.

The objective "understanding" is the basis for the objective "predicting", and it has an even more fundamental purpose. On the other hand, the economic stakes involved in "predicting" are enormous: a validated model will drastically reduce the development period and the development cost for a new product (in the broad sense of the term).

Modeling is the alternative to a trial-error experimental process, which consists in the construction of experimental protocols conducted by testing the influence of the parameters considered as significant. This approach will be possible for systems exhibiting a linear behavior (which of course will have to be verified) but will soon become intractable in case of nonlinear behaviors, due to the required number of experiments and given the time for and the cost of making each specimen and conducting each experiment.

Developing a good model will represent an important challenge.

A validated model will permit the optimization of a mechanical system. It will be an efficient tool in the hands of the engineer who will be able to test the influence of every parameter rapidly. A model is not a universal tool that would replace the scientist. It is only a sophisticated tool that will help the designer or the researcher during the process of creating or optimizing a new product. Their skills in developing, selecting and using the models are indeed of central importance in the art of modeling.

Let us now get an overview on the various steps of the construction of a model:

- the founding step is the choice of the equations chosen to describe the phenomenon which has to be modeled: some assumptions have to be made, the main characteristics to take into account have to be identified, the scale of the analysis has to be chosen, etc.,
- then comes the formulation step, with the choice of the formulation framework,
- the mathematical analysis of the problem should give information about the existence, uniqueness and stability of solutions, and more generally information about the conditioning of the problem to be solved,
- then comes the step of solving the problem which is obviously a key step which leads to numerous developments and may give rise to very difficult problems; this should be completed on the one hand by the numerical analysis of the approximate problem, of the discrete problem and of the algorithms and on the other hand by the validation of the numerical method,
- once the numerical tool is ready, the basic constitutive parameter have to be identified using the model to simulate a test experiment: the sensitivity of the solutions to some changes of the parameter values should also be tested,
- and finally the last step is the validation of the model by simulating various experiments and evaluating the error between the theoretical and the experimental results.

Some of these steps can be more or less important and more or less difficult depending on the cases, but it will be important and useful to keep all of them in mind.

1.2 Construction of a model

The art of the modeler will be about highlighting the key points and identifying the nature of the significant phenomena and their contributions in various experiments. The basic step in building a model is the choice of the ingredients to use and of the assumptions to make. It will be essential to bear in mind that a model is composed of "equations" and "assumptions." These underlying assumptions will be the bases of the validity domain of the model and they have to be clearly defined and remembered. This may seem obvious but transgressions are common: for example, using numerical integration methods based on limited developments for computing solutions that are not differentiable would be incorrect.

Concerning the assumptions, there are also some subtleties, such as the difference between simplification and neglect. The former permits to solve the problem but is not necessarily legitimate (and that should be studied) while the latter consists in neglecting a secondary aspect of the problem (and that is a justified assumption).

We can try to list and classify by nature some of these choices and assumptions that will be made during the construction of the model and of the systems of equations to solve, in order to correctly simulate the experiment.

- The physics that should be taken into account. The first step is to identify the nature of the significant phenomena which have to be considered: mechanical, chemical, thermal, electrical effects or other effects. This defines the degree of complexity of the model. As was said before, multiphysic modeling is increasingly used today. This step is very important because neglecting a significant effect (incomplete model) would have as harmful effects as taking into account a side effect (unnecessarily complicated model).
- The "equations" and their framework. The "equations" will be written in the general framework of the fundamental principles of mechanics and thermodynamics. Their implementation gives rise to differences in points of view, equations of motion or energy principles, partial differential equations or minimum problems, etc. The degree of generality or abstraction of the framework will define the degree of generality of the model. We will speak of classes of models when this degree of generality is great. This will be illustrated in Section 2.5 by a work where we introduce a unified model for adhesive interfaces based on considerations such as the concept of generalized standard materials extended to interfaces.
- The role of time. Time plays a special role in the development of a model. Several questions arise and have to be answered. The first question is whether a dynamic study should be conducted, if the effects of inertia are to be taken into account, or a quasi-static approach is sufficient, if masses are small enough or phenomena slow enough.

  Another question concerns the possible influence of the velocities on the nature of the behaviors and if viscous dissipation should be considered.

A third question is whether the effects of variations in time of the parameters should be taken into account, either directly (effect of aging for example) or indirectly, these parameters depending on some of the variables such that displacements or velocities.

When the phenomenon depends on the loading path (plasticity, damage, etc.), then we are in a rather special case where time is replaced by a dimensionless variable characterizing the loading path.

- Choice of the scale. Another specific issue when building a model is the choice of the scale to conduct the analysis. It could be only the choice of a given scale or that of the combination of several scales (multiscale model). With the phenomenal increase in computer power, multi-scale approaches are experiencing spectacular growth.

1.3 Choosing a formulation

Choosing a formulation relates several levels. First, we can choose a discrete or a continuous formulation. Secondly, it has to be noted that the choice of the variables is directly connected to the form of the formulation (primal, dual or mixed formulation).

A discrete formulation consisting in assembling masses, springs and various elements in general has as objective to "understand". A discrete formulation can be very sophisticated, including nonlinear or specific items. Such example is presented in Section 1.9.The target to "simulate" and to "predict" the behavior of complex structures rather goes through continuous formulations associated with finite element methods which may work for any geometry. But many variants are possible and the combination of discrete and continuous formulations is often useful for complex assemblies.

The mathematical framework is closely relates to the formulation. To take some examples in continuum mechanics, the general framework for smooth problems will be that of functional analysis and Sobolev spaces. In some cases, such as contact problems discussed later the convenient framework will be the ones of convex and non-convex analysis. As for the variational formulations, we will distinguish among primal (displacement or velocity formulation), dual (stress formulations) or mixed formulations, each having its own advantages and disadvantages depending on the situation. In the case of non-smooth problems with respect to time, a convenient framework will be that of distributions and differential measures.

For non-smooth problems, the formulations and their mathematical framework may be drastically simplified by using regularizations. However, these processes may create a problem markedly different from the initial problem, which can have an impact on the validity of the model. A special section in the following will be devoted to that important issue.

1.4 The framework and the mathematical analysis

Although often considered as unnecessary or irrelevant, or else as an exercise for mathematicians, mathematical analysis has a leading role in the art of modeling. For smooth problems (with respect to time as well as space variables) the mathematical framework is very classical and does not need special care. However, it can be noted that only the writing of boundary conditions for partial differential equations uses the concept of trace in Sobolev spaces.

Note that whole areas of mathematics have been developed from problems in mechanics: some aspects of functional analysis, convex analysis, the notion of Γ-convergence, etc. (see Dautray-Lions (33)). Mechanical problems, especially non-smooth problems, have promoted very high level mathematical developments which are still ongoing. Unfortunately, despite giving an adequate framework, they sometimes lead to results that cannot be directly used by engineers. However, they always help in the understanding of both the model and the computational behaviors. The theoretical framework and the mathematical analysis are very constructive in setting the problems properly, understanding the main theoretical properties and developing convenient numerical methods for the resolution. We want to stress that using the right mathematical context adapted to the ingredients (equations) put into the model is a key point in the art of modeling, the only way to find and to characterize the right solutions, including sometimes some unexpected ones.

We will illustrate that in the following. Contact mechanics will be a good context to discuss modeling in these various aspects. Although the laws seem very simple (Coulomb's law dates back to the 18th century), everything is very complicated because they relate to non-smooth mechanics: threshold laws, the graphs of the contact laws are multivalued mappings and not functions, etc.

1.5 Resolution

It is obviously a key step in the art of modeling. The resolution could be analytical: it is then limited to elementary geometries, but it simplifies parametric studies and helps to identify the essential phenomena more easily. More generally, analytical approaches are useful to "understand" and hence they must not be neglected in modeling.

But today, the methods of resolution are mostly numerical and the great increase in computer power has paved the way not only for new methods but also for new ways of thinking about modeling approaches. There is now in the art of modeling a close link between resources and computing power and design patterns.

We would like to point out a downside risk associated with this increasing computational power. It is the risk of replacing "thinking" by "large number of degrees of freedom". This is undoubtedly one of the main common pitfalls of modeling. A lot of computations using millions of Degrees Of Freedom (DOF) will very often provide less on the knowledge and the understanding of a problem than a well thought out model using a small number of DOFs or a discrete mechanical model.

The range of existing methods is very wide and in constant development. Without trying to be exhaustive, one may mention the finite element methods (FEM) and their variants ($(FEM)^2$, subdomains, multigrid methods, diffuse elements, etc.), finite volume methods, integral methods, spectral methods, etc.

It is worth stressing several points about numerical methods and computations.

First, it will be important to keep in mind that a numerical method must be consistent with the assumptions and with the mathematical framework.

Another extremely important point is the validation of the numerical method itself (see Section 4). This must be done using standard solutions that can be analytical or qualified by other methods (benchmarks). This is not trivial matter and it is sometimes possible to find numerical methods validated with an experiment - which is absurd. What is more, this step of validation of the numerical method will generally be an opportunity to evaluate and control the numerical errors (distinguishing between discretization errors and calculation errors), check convergence conditions, properly select calculation parameters, etc.

The numerical analysis of the approached problems and of the discrete problems is often overlooked because it is often difficult. However, it will provide important information about the stability of the algorithms (and possibly provide the stability conditions to be verified), on convergence, error estimates and other properties that will be valuable during the numerical implementation.

1.6 Identification of the parameters

The parameter identification step should be distinguished from the model validation step - which we will be discussed in the following. This essential step comes after the model construction step, and consists in finding a solution to the optimization problem: finding the parameter values which minimize the gap (the error) between one experiment and the results given by the model simulation.

Generally, the identification of constitutive parameters is conducted with experiments that are either independent of the problem addressed or closely linked to it. For example, when modeling the behavior of a structure, the mechanical properties of the materials are usually determined with experiments on specimens, thus independently of the system studied.

On the contrary, in other cases, these characteristics depend so much on the environment of the system in real situation that it becomes necessary to use other processes. This will generally be the case for interface models, as we will see later.

1.7 Validation of the model

This is of course the key step in modeling since it will allow us to give reliable behavior predictions when using the model. Validation is part of the most important steps in modeling; it will define the quality of the model, specifies the validity domain and confirm its ability to be used by the designer. It is essential and requires special care (see Sectionchapter4).

Validation is carried out on one or more experiments by evaluating the error between the results of the experiments and those of the simulations, using the values of the constitutive parameters obtained by the identification process. Let us note in passing that the norm selected to evaluate this error is of great importance since a norm L^2 (squared error), which inherently carries an energy sense, will smooth the gaps, while a norm L^∞ (sup norm) will measure the maximum deviation which can in some cases be the decisive dimensioning factor.

Note that the experimental conditions should verify the assumptions made initially for the construction of the model (load amplitudes, speed, etc.).

At this level, adjustments to the model will be performed by iterations between the model prediction and the experimental results. Two approaches may be followed: either improving the model if the gap between theory and experiment is not small enough or checking whether it is possible to simplify the model while keeping an appropriate gap value. The second approach is often overlooked.

1.8 To conclude this introductory chapter

In this chapter, we have addressed many aspects of modeling. As a complement, we would like to stress two key points in the art of modeling.

This first key point is: "Make as simple as possible and as complex as necessary, no more, no less!".

We will illustrate this with the example of the famous work of Pablo Picasso, "Bulls" (Picasso, 1945) presented on Fig. 2.'Bull' is a suite of eleven lithographs that have become a master class in how to develop an artwork from the academic to the abstract. In this series of images, Picasso visually dissects the image of a bull to discover its essential presence through a progressive analysis of its form. Each plate is a successive stage in an investigation to find the absolute "spirit" of the beast. In this work, Picasso progressively simplifies the shape of the bull and keeps the quintessence of the draw. In the final print of the series, Picasso reduces the bull to a simple outline which is so carefully considered through the progressive development of each image that it captures the absolute essence of the creature in as concise an image as possible. A critic made that comment: "He has ended up where he should have started! He had gone in successive stages through all the other bulls. When you look at that line you cannot imagine how much work it involved. He had in mind to retrieve the bull's constituent parts, his dream bull - bred of pure lines - an elemental, disembodied, quintessential bullishness".

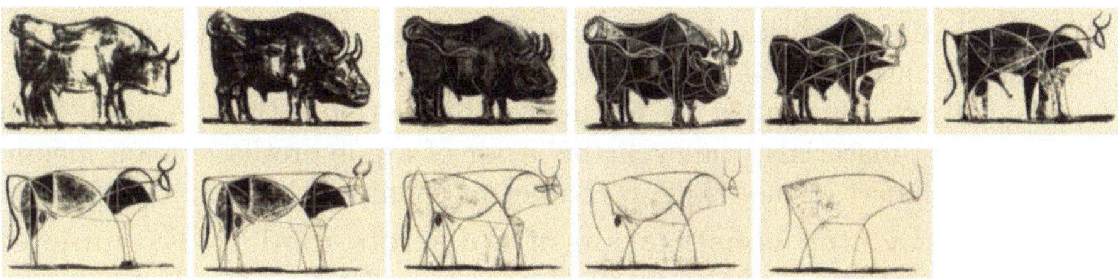

Figure 2. The process followed by Picasso in the eleven lithographs "Bull, 1945"

The result is the Platonic idea of a bull. The reverse process to get back to the initial complete image of the bull is the gain of knowledge. This development of knowledge to obtain the ideal model from a reality is the basis of the art of modeling. The process of Picasso is in fact exactly our only hope to gain knowledge.

In the same state of mind, we may also cite Fernando Pessoa: "Thought must start from the irreducible".

The art of modeling consists precisely in building the simplest model containing the key characteristics of the phenomena and permitting its suitable and validated simulation. Making the model as simple as possible and as complex as necessary is really fundamental in the art of modeling: making it too simple would be insufficient and making it too complex would be useless and costly.

A second key point in the art of modeling is: you should not fall into the trap of habit. One of the pitfalls the modeler must avoid when faced with a new problem is the temptation to do "what he does best "rather than "what he should do", i.e., fitting the problem within the framework of his own mechanical engineering culture, of his usual tools (even following necessary new developments) instead of using a totally different approach, which should be more convenient, but not familiar to him. It is not easy to resist a natural temptation and it is thus often difficult to avoid the trap of habit, as it requires a sound knowledge of models and modeling, when models are becoming increasingly sophisticated, and it is difficult to be an all-rounder. It will be more constructive to work within multiple skill groups or to liaise with a network of specialists, while having developed a sufficiently broad knowledge, from experimentation to numerical methods and mathematics, to be able to dialogue with those specialists. This variety of knowledge in modeling is well known to industrialists, who sometimes present the same problem to several academic research laboratories that differ in their approaches.

To conclude on this chapter, we can say that, besides the broad spectrum of his scientific qualities, his in-depth knowledge, the thoroughness of his analysis, intuition will be a key asset for the modeler.

We will now illustrate some features of this section with an example from mechanics of materials, the cyclic behavior of a polyurethane foam, before getting to the art of modeling in contact mechanics in the next sections.

1.9 Step by step construction of a discrete model - Example of the modeling of the cyclic behavior of a polymeric foam

This example will be used to illustrate three points discussed in this section.

- "Discrete model and analogical models" (springs, dashpots, etc.) help to "understand" and to identify the driving phenomena. We use a discrete model composed by assembling non-linear springs, dashpots and other convenient elements. The goal is not to simulate the behavior of a whole structure but to identify and to model the main phenomena occurring in this kind of material during cyclic loadings.
- "From simplicity to complexity": the model is developed step by step. This second point relates to the idea to put into a model what is necessary, no more, no less. We start with simple elements and we progressively add other necessary elements.
- "Identification of the constitutive parameters" and "validation of the model" which have to be clearly distinguished

This is a discrete model used to "understand" and which is "as simple as possible and as complicated as necessary". All details can be found in Del Piero-Pampolini-Raous (82) (81) (36).

Main characteristics of the behavior: the experiments. When a sample of an open-cell polymeric foam is submitted to quasi static compression, localization of the deformation can be observed with the occurrence of bands orthogonal to the loading direction (see Fig. 3).

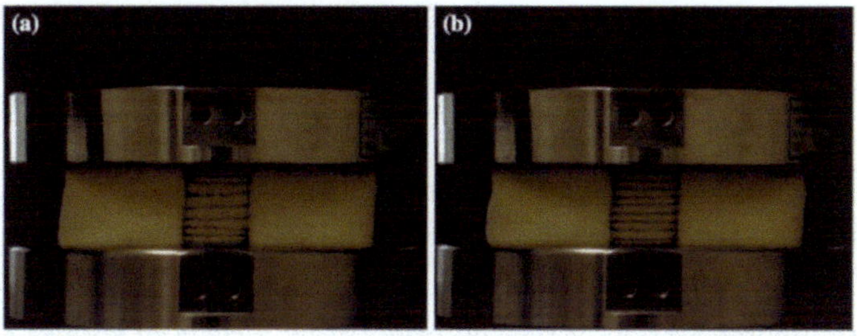

Figure 3. Localization of deformation during compression

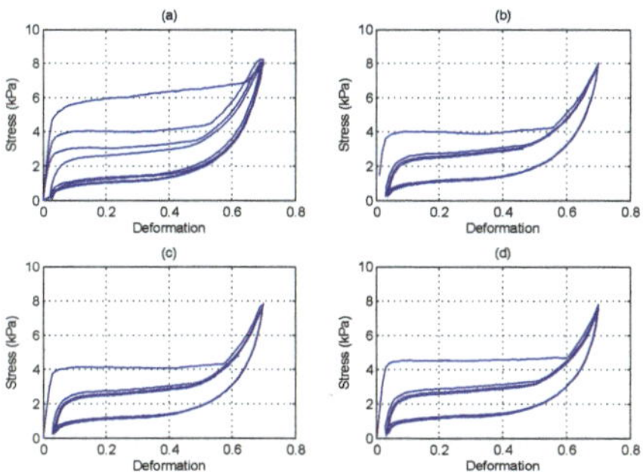

Figure 4. Recovering (a) Virgin material, Resting period : (b) 16 hours ; (c) 52 hours ; (d) 33 days.

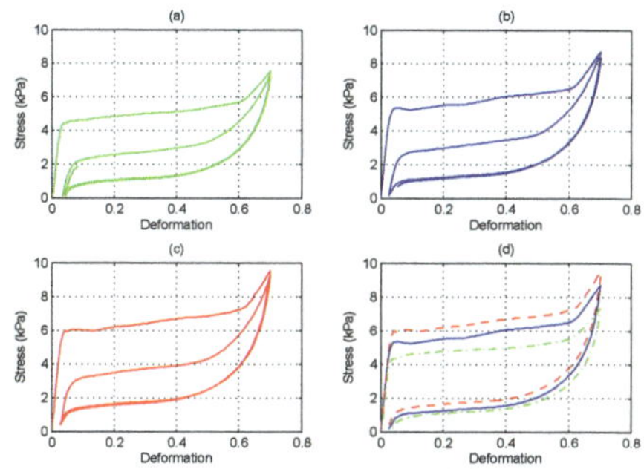

Figure 5. Loading velocity influence: (a) 0.1 mm/min ; (b) 5 mm/min ; (c) 100 mm/min ; (d) Comparison of the 3 results.

When a cyclic compression is applied, the following characteristics of the stress/strain cycles are noted (see Fig. 4 and Fig. 5):

- specific shape of the cycles,
- first cycle different from the next ones,
- influence of the loading velocity,
- influence of the resting periods on the behavior,
- analogy with Mullins effect for elastomers (strain softening).

Discrete problem and nonlinear elasticity. A discrete model, an assembly of springs, dashpots and other convenient elements, is considered. The specimen will be considered as discretized into a finite number of layers (Fig. 6). The art of modeling will consist in choosing the good characteristics of the various constitutive elements of the chains. We will first address the localization phenomenon observed during compression. For that, we consider that the springs exhibit a nonlinear elasticity based on the non-convex strain energy density w presented in Fig. 6 (details can be found in (81) (36)). Details on the analysis and on the computation of the response

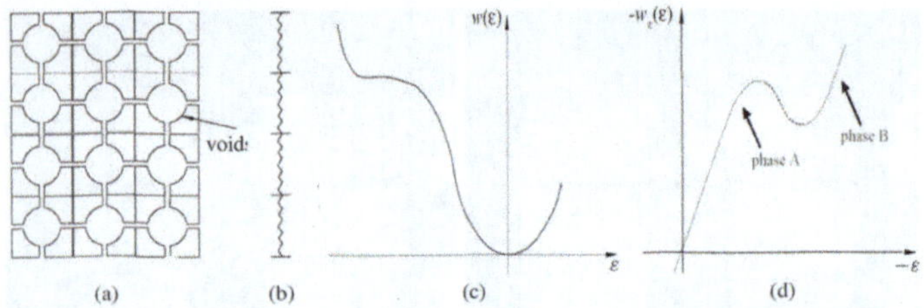

Figure 6. Representation of the cell lyers (a) by a chain of springs (b) with a non convex strain energy (c) which gives the behavior law (d) for each spring.

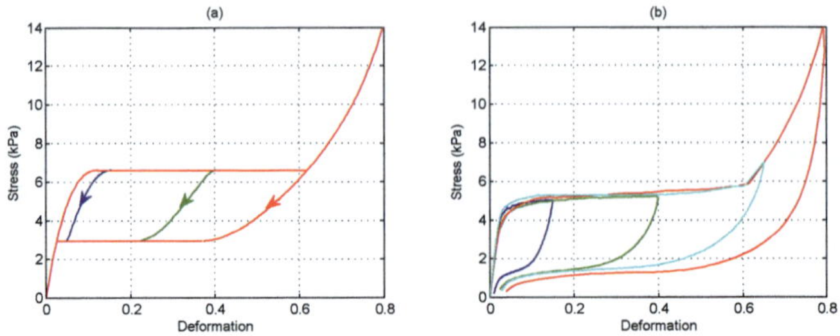

Figure 7. Comparison between the nonlinear elasticity model (a) and experiment (b) for different loading amplitudes.

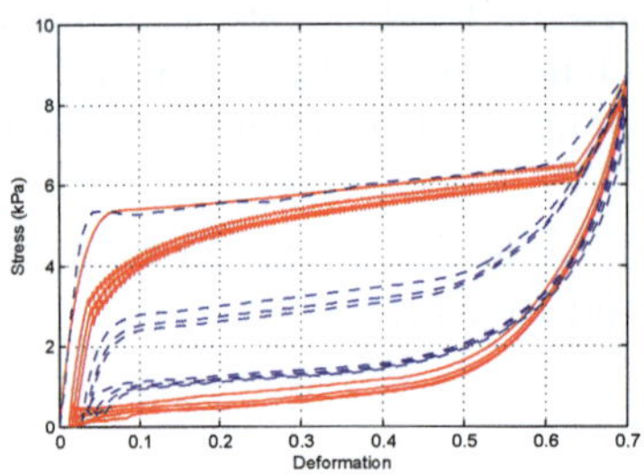

Figure 8. Theory/experiment comparison when using the nonlinear elasticity model with viscosity

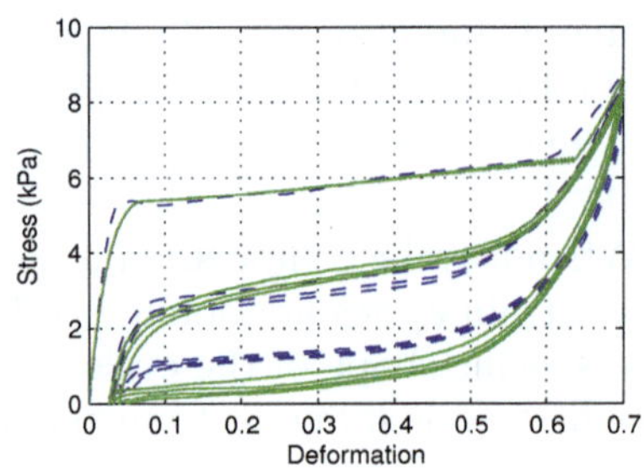

Figure 9. Theory/experiment comparison when using nonlinear elasticity, viscosity and damage.

can be found in (82). The existence of two phases is characterized: one phase corresponding to the virgin cells and the other corresponding to the totally squeezed cells.

A good agreement is observed between the model and the experiment for the first compression step, but, either when unloading or when various loading amplitudes are considered, there is a big gap between theory and experiment results (even qualitatively), as shown in Fig. 7.

Viscosity effects are added. The model has to be completed by taking into account other phenomena. We will now introduce the viscosity effects. The idea is to incorporate what is necessary gradually into the model to describe the experimental behavior. We introduce a Zener model with Maxwell viscosity (see ref). This model may encounter both the loading rate dependence (classical viscosity effect during the cycles) and a long period dependence (recovery of the initial properties after resting periods which is a relaxation phenomenon). For that, we introduce several relaxation times with different time scales in the viscosity model. In Fig. 49, it can be observed that a good agreement between theory (full line) and experiment (dotted line) is now obtained regarding the shape of the cycles. Nevertheless, even if the first cycle is very well described, it remains that the gap between the first cycle and the next ones is not described.

Damage is added. Once again, a new effect has to be taken into account and we introduce damage in the model. Now, as shown in Fig. 50 a very good agreement is obtained between the model simulation (full line) and the experimental results (dotted line). The complete model has been constructed step by step, progressively adding new concepts that are necessary to get an accurate description of the behaviors observed during the experiments.

1.10 A model to "understand"

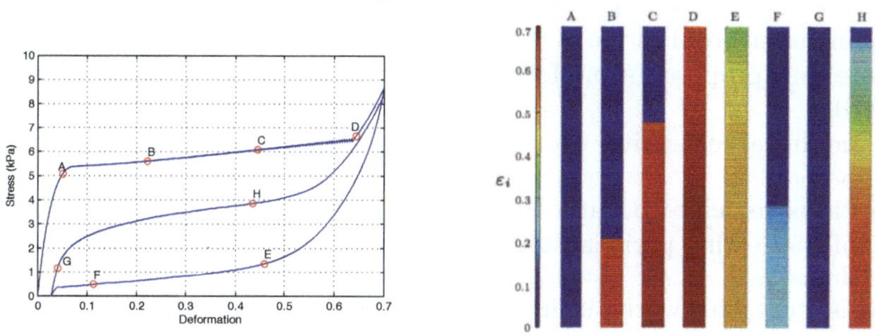

Figure 10. Evolution of the deformation condition (phase changes) during a cycle (the color scale for the deformations is given on the far left)

Let us note that this discrete model is very helpful to understand the behavior of polymeric foam. "Understanding" was presented as the first goal in the art of modeling in Section 1. We explain that this cyclic behavior is due to nonlinear elasticity, viscosity and damage. The model shows clearly the phase-change mechanism occurring in the polymeric foam during the loading, which gives the shape of the stress/strain curve. With a simulation conducted with a chain of 120 elements, the panel of colors (Fig. 10) gives the deformation levels in the column (this is a one dimensional model) corresponding to the different points on the stress/strain curve: A, D, E and G are single phase configurations, while B, C, F and H are two phase configurations (in B, C and F the two phases are block separate). The points C, E and H show situations with different stresses for the same deformation. This is due to the dependence of the solution on the loading path. A good understanding of the phenomenon is achieved.

How to validate the model? The good theory-experiment agreement shown in Fig. 50 is the result of the process of identification of the constitutive parameters. The identification was a quite difficult process and details of the methodology will be presented in Section 4.4 and in Pampolini-Raous (82). At this point (construction of the model and identification of the parameters), we only showed that there is at least one set of parameter values that permits to fit the model with one experiment. The model is not yet validated and not ready to be used for "prediction", the other goal of the art of modeling. In order to validate the model, we will use it, with the parameters evaluated during the identification process, to simulate other kinds of loadings. This time, a good agreement between theoretical predictions and experimental results will confirm the validation of the model. This will be shown at Section 4.4.

1.11 Conclusions about the art of modeling for this first example

This discrete model has helped to understand what the main mechanical phenomena involved in the cyclic behavior of polymeric foam are and, what is more, has proposed some forms for the various ingredients that have to be used:

- nonlinear elasticity with non-convex strain energy (a form is proposed),
- viscoelasticity with at least two time scales (cycle period and duration of the resting periods); a Zener-Maxwell model could be convenient,
- damage.

What is more, this work on this discrete model gives an order of magnitude of the various constitutive parameters for the mechanical model.

Let us note that the model has been constructed step by step in order to make "as simple as possible but as complex as necessary". The way to conduct "identification and validation" will be presented at Section 4.4.

The next step would be now to develop a continuous formulation of the problem and to use finite element methods to simulate the behavior of a polymeric foam structure in engineering systems, such as energy absorbers, seat cushions, packaging materials and lightweight composite sandwich structures, etc. That is not simple but the present model helps to understand what the main ingredients are that should be put into such a model.

2 Building models in contact mechanics

We will now address the art of modeling in contact mechanics. As already mentioned, contact mechanics is a field of interest for discussing the art of modeling, since many of the issues addressed in the previous section will raise difficult questions.

In this section, we will focus on the step "building the model" - which is the most mechanical part. We give a panorama of various contact laws (unilateral contact, friction, adhesion, wear, etc.) and present the diversity of the applications in various fields. The different scales that can be used are presented: we give some elements and references on multiscale analysis which is expanding strongly. We show how a model is built on the basis of concepts. We stress the advantages versus the dangers of regularization.

This is illustrated with two examples:

- a law coupling unilateral contact, friction and adhesion (the RCCM model) to illustrate the importance of concepts as the basis of a new contact model (thermodynamic analysis) and the mathematical complications; this is applied to an engineering problem: how to improve the resistance of a composite material to the crack propagation by controlling and optimizing the behavior of the fiber/matrix interface,
- a unified model coupling contact, friction and adhesion to illustrate the inductive/ deductive process when one elaborates a model on the basis of fundamental mechanics (generalized standard materials).

2.1 Basic interface models

Let us present a panorama of the main interface laws corresponding to the main usual behaviors that we want to describe: non penetration into the obstacle, friction, adhesion, healing adhesion, wear, etc. Details can be found in Raous (95) (103) (97) and Sauer (109).

- Unilateral conditions. The basis is to write that the solid cannot penetrate into the obstacle supposed to be rigid. It can be generalized to the contact between two deformable solids. Let us consider a rigid obstacle, the solid occupying a domain Ω, let Γ_C the part of the boundary Γ initially in contact with the obstacle, u_c the displacement (trace on Γ_C of the displacement u defined on Ω), R the contact force and n the outward normal vector to Γ_C. Using the following partition between normal and tangential components, the non-penetration condition is described by the following Signorini conditions which constitute a complementarity formulation. The graph of this behavior law is given in Fig.11. This is not a function but a multivalued mapping. It is here that difficulties begin!

$$u_c = u_n n + u_t \qquad R = R_n n + R_t$$

$$\begin{aligned} u_n &\leq 0 \\ R_n &\leq 0 \\ u_n R_n &= 0 \end{aligned} \tag{1}$$

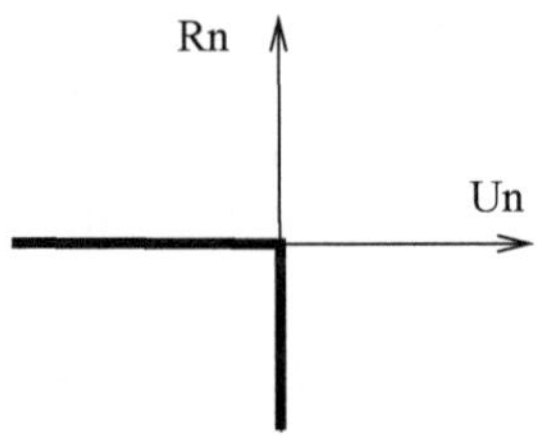

Figure 11. Unilateral conditions - Signorini conditions

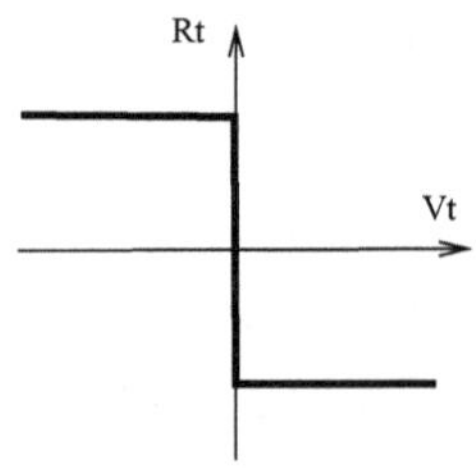

Figure 12. Friction law: the Coulomb law

- Friction: Coulomb - Amontons friction law (32) The basic law of friction is the Coulomb law (1821) which was developed on the basis of experimental observations with moderate amplitude forces. First, it is a law with a threshold, no movement occurs unless the tangential force reaches a certain threshold, secondly, this threshold is directly proportional to the normal force and thirdly, the movement occurs in the direction opposite to that of the applied tangential force. This is expressed by the following law:

$$\|R_t\| \leq \mu|R_n| \tag{2}$$

$$\text{if} \quad \|R_t\| < \mu|R_n| \quad \text{then} \quad \dot{u}_t = 0 \tag{3}$$

$$\text{if} \quad \|R_t\| < \mu|R_n| \quad \text{then} \quad \dot{u}_t = -\lambda R_t \quad \text{with } \lambda \geq 0 \tag{4}$$

This law seems to be very simple; only one parameter, the friction coefficient μ, is involved. However, at the same time, it is very complicated because it is not a function but a multivalued mapping (see Fig. 12), as for the unilateral conditions. It will drastically increase the difficulties when writing the mathematical formulations and developing the numerical methods. We should stress that this non smooth character is precisely the fundamental richness of this law; it will make it possible to model specific behaviors such as occurrence of instabilities (squeal, stick-slip, etc.), existence of multiple solutions, etc. Therefore, it will be important to conveniently deal with this non smooth character; we are in the context of non-smooth mechanics.

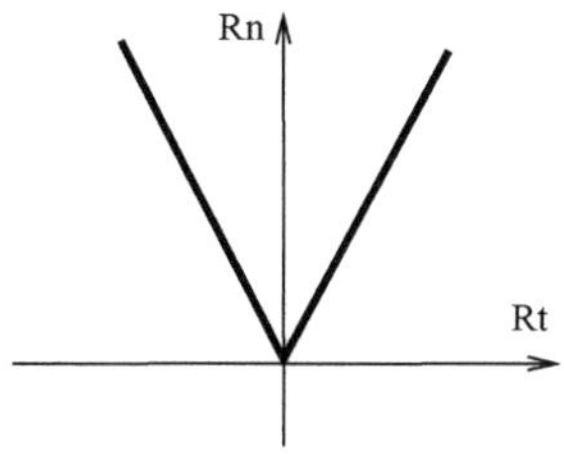

Figure 13. The Coulomb cone

The configuration space is given by the Coulomb cone in the force space (see Fig. 13). When the representative point is inside the cone, no movement occurs, when it reaches the boundary, sliding occurs with the direction of the tangential force (and not orthogonally to the boundary of the domain, as it is the case for example in classical plasticity). Thus, we should note that there is no normality rule and this will lead to extra difficulties for the formulation (no minimum principle, as we will see in Section 3.3). As announced in Section 1.3, various formulations can be given and modeling consists in selecting one among them. Other formulations of the Coulomb law can be given. Let us here briefly give some of them (details can be found in Moreau (76)):

- maximum dissipation principle

$$R_t \in C \qquad \forall S_t \in C \qquad (S_t - R_t)\dot{u}_t \geq 0 \tag{5}$$

  where, in 2D $\qquad C = [-\mu R_n, +\mu R_n]$
- subdifferential formulation

$$-\dot{u}_t \in \partial I_{C_t(R_n)}(R_t) \tag{6}$$

  where $\partial I_{C_t(R_n)}(R_t)$ is the subdifferential of the indicator function of $C_t(R_n)$ with $\qquad C_t(R_n) = \{P \text{ such that } |P| \leq -\mu R_n\}$
- dual subdifferential formulation

$$R_t \in \partial \phi_{R_n}(-\dot{u}_t) \tag{7}$$

  with $\qquad \phi_{R_n}(-\dot{u}_t) = \mu R_n \|\dot{u}_t\|$

- **Possible regularizations.** The comments presented in this section are of great importance for the art of modeling. Regularization is a natural tendency when we have to deal with a non-smooth problem. It allows us to replace the initial problem by a much simpler one, which is also simpler to solve. We want to stress a very important point: a regularized problem is different from the initial problem and could even model a very different mechanical phenomenon. The modeler should be very aware of that. Contact problems may illustrate that point very well.

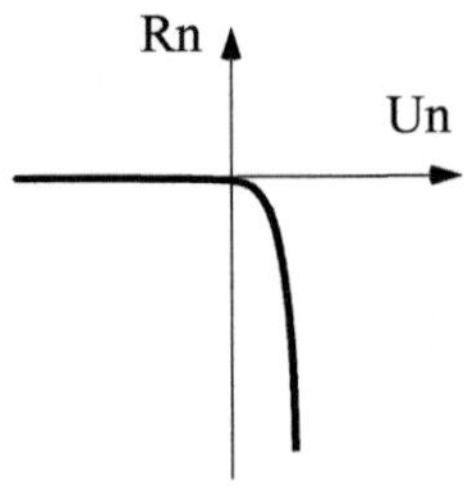

Figure 14. Compliance or penalization of the Signorini problem

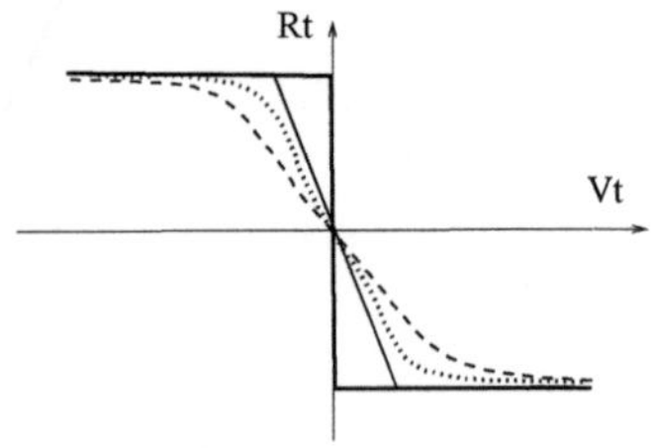

Figure 15. Regularizations for the Coulomb friction Hyperbolic tangent (dotted line), square root (dashed line), polynomial (solid line)

For contact problems, regularization will consist in replacing the multi-valued mappings by functions in order to obtain a smooth (but non-linear) problem.

- Regularization of the unilateral conditions: compliance or penalization of the contact laws. As shown in Fig. 14, the contact behavior is now characterized by a (non-linear) function. When penetration into the obstacle occurs, a strong force is introduced so as to push the deformable solid out of the obstacle. A classical choice is the following one, where $(u_n)_+$ denotes the positive part of the normal displacement and C_n and m_n are two prescribed regularization parameters:

$$-R_n = C_n(u_n)_+^{m_n} \tag{8}$$

This is very comfortable for the analysis and the resolution but let us look at the mechanical behavior. It has to be noted that we always have penetration into the obstacle with two consequences: one is a bad determination of the contact forces (which now depend on the compliance parameters) and the other one is a penetration into the obstacle that could be unrealistic.

The term compliance is used when dealing with the mechanical model while the term penalization is used when solvers are involved. Sometimes, in order to give a mechanical interpretation of this regularization, it is said that the penalization of the unilateral contact (or the compliance laws) is associated with the squeeze of the asperity. Actually, when the squeeze of an asperity is computed (in large plastic deformations, see Raous-Sage (94)), it turns out that the coefficients C_n and m_n are huge and do not correspond necessarily to the suitable values usually chosen to make computations easy to do. Therefore, it could be recommended either not to use regularization for the nonpenetration conditions or - especially when a standard computational code is used - to check the penetration in order to control if its magnitude is admissible for the problem under consideration.

- Regularization of the Coulomb friction. In Fig 15, various functions used to regularize the Coulomb friction law are presented. Again, the Coulomb multivalued mapping is replaced by functions (which are non-linear) and this will make the problem a lot simpler.

  But it should be stressed that the new mechanical problem is totally different from the initial one: sliding always occurs, except if the applied tangential force is equal to zero! This means that with a very small force, you can get the refrigerator sliding across the kitchen if you wait for a sufficiently long period of time! Obviously, this is wrong!

Therefore, special attention has to be paid to the choice of the regularization parameters relative to the time scale. Very often, when the values of these parameters are chosen so as to give a good approximation of the Coulomb law, the regularized problem remains a stiff problem, ill-conditioned from a numerical point of view. When the values of these parameters are chosen for computational convenience, the quality of the solution has to be checked with respect to the Coulomb conditions, to be sure that this quality is sufficient for the problem.

Otherwise, the choice of the regularization parameters has a great influence on the tangential forces. Thus, let us emphasize that such regularization is not suitable for a subtle analysis, such as the study of instabilities or squeal, among others, because the results of the analysis will depend on the values of the regularization parameters (and even stability cannot be achieved if the regularization is strong).

Therefore, in conclusion - and this is very general in the art of modeling - regularization makes the mathematics, the formulations and the computations simpler but it transforms the problem in another one and we must be very careful when using it (choice of the parameters, verifications a posteriori on the solutions, etc.). Regularization is not always convenient.

Of course, regularization will be licit and used for problems where accurate determination of the contact condition is not needed.

2.2 Panorama of interface models

Beyond these basic laws which seem elementary but contain the essential characteristics for describing contact behaviors, many other laws have been developed and new ones are still under research. We will discuss a brief overview of these laws, without going into the details which can be found in the literature (references are in (103)). This paragraph intends to show the variety of interface models.

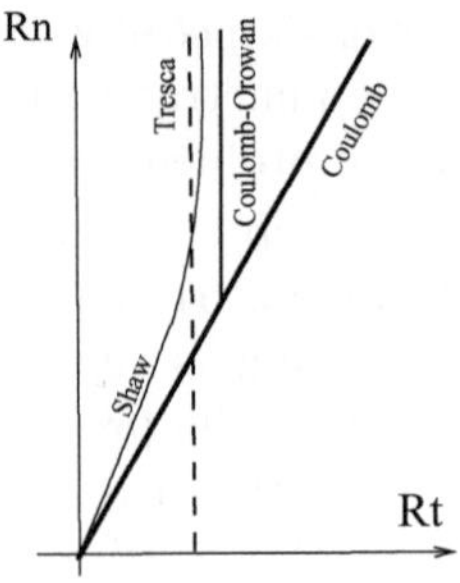

Figure 16. Friction cone for variants of the Coulomb law

- **Other friction laws.** In Fig. 16, configuration spaces for variants of the Coulomb law are given. These laws express that the friction no longer depends on the normal force when the latter becomes large. It is a kind of saturation of the friction threshold, which is often considered in metal forming. The Coulomb cone then becomes either a truncated cone (Coulomb-Orowan, Shaw) or a cylinder (Tresca).

- **Adhesion.** We now consider cases where the interface resists (with a resistive force) the separation of the solid from the obstacle. Therefore, the contact normal force can be positive. When the normal force increases, the adhesion progressively decreases as far as we finally turn back to the usual unilateral problem including friction. The intensity of the adhesion can be considered as a damage variable. These laws are used to describe interfaces as well as ductile cracks (Cohesive Zone Models CZM). More details can be found in the references given in Raous (103) and Sauer (109).

- Normal adhesion (resistance to traction). For the compressive force, we still have the strict unilateral conditions (Signorini problem). This is not regularization. In Fig. 17, some models are given.
- Tangential adhesion. These models are still based on the notion of intensity of adhesion i.e., of interface damage. When the adhesion totally collapses, one goes back to the usual Coulomb law. Some of these models are not coupled with friction. See Fig. 18.
- Recoverable adhesion. This is the case where, after total separation (and so total collapse of the adhesion), some adhesion is recovered when the solid is put again in contact with the obstacle (see Fig. 19). This could be a Van Der Walls force or some healing phenomenon. A model was proposed in Raous-Schryve-Cocou (100) (30). An example would be the case of an adhesive tape which can be used several times but which is a bit less efficient each time.

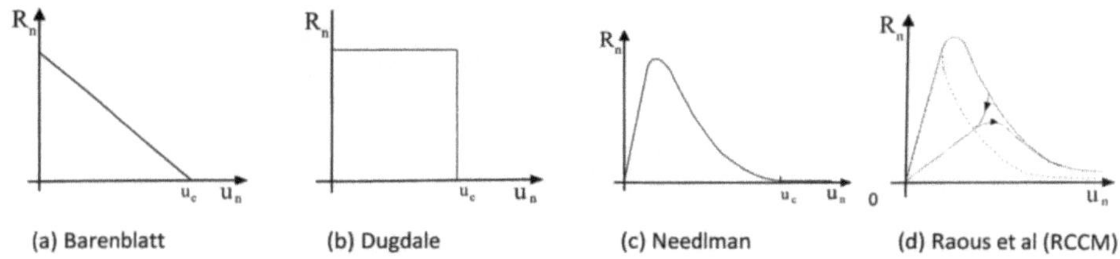

Figure 17. Adhesion laws for normal resistance

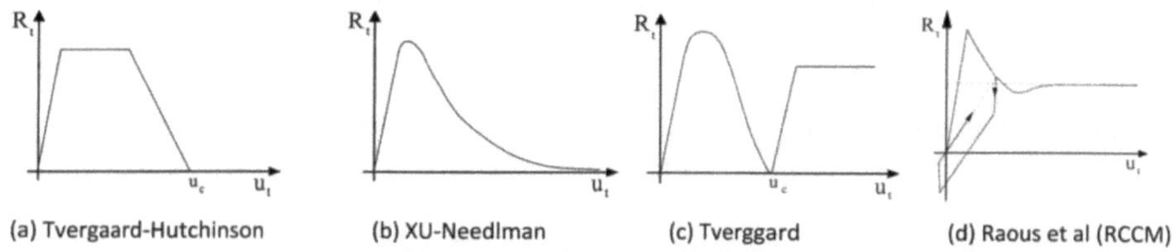

Figure 18. Tangential adhesion laws

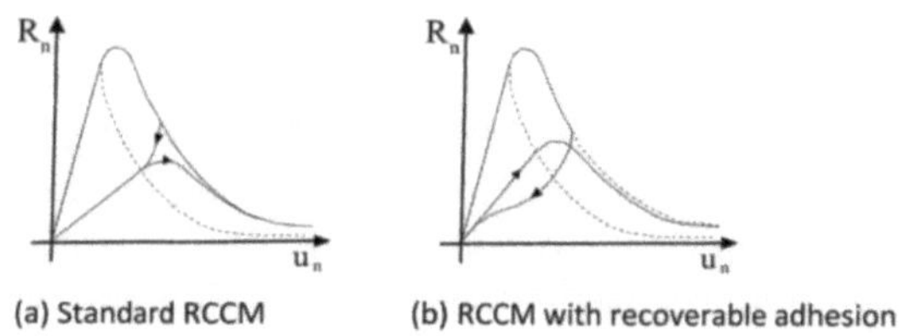

Figure 19. Recoverable adhesion

- **Wear, abrasion, grinding, polishing.** A number of laws regarding these phenomena can be found in the literature. The best known is the Archards law (7) for wear and abrasion modeling: the volume of the removed debris due to wear is proportional to the work done by friction forces.

- **Friction laws using variable friction coefficients.** A dependence of the friction coefficient on the velocity, on the pressure, or other quantities is sometimes introduced. A classical law is the Stribecks law used for metal forming with a dependence of the friction coefficient on the velocity V_t, the normal pressure p and the viscosity of the lubricant b (see Fig. 20).

The dependence of the friction coefficient on the sliding velocity is also a classical feature (see Fig. 21). However, great care is needed when such dependence (especially velocity dependence) is used in contact modeling. It is obviously legitimate when the velocity dependence can be clearly

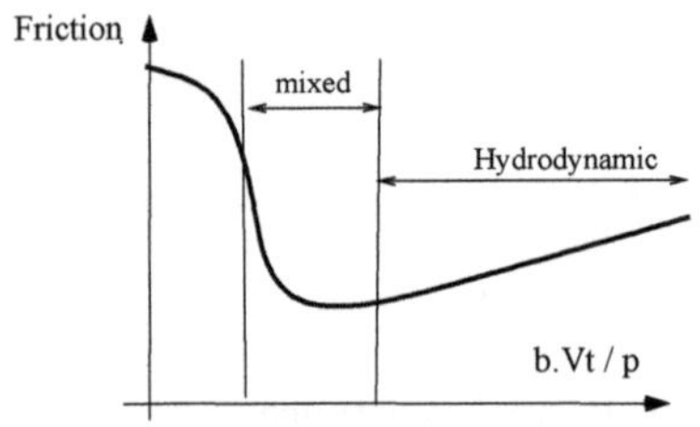

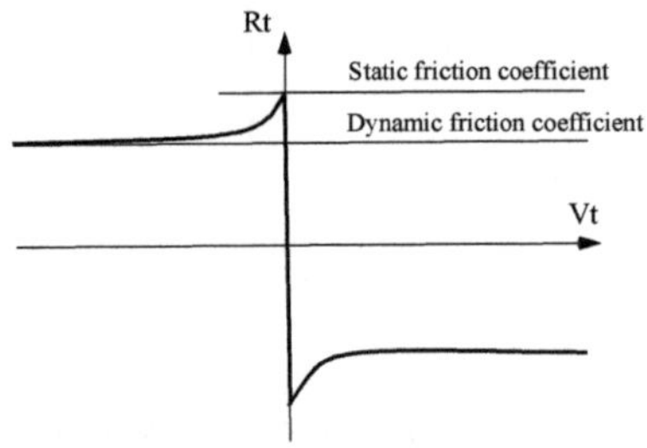

Figure 20. Stribeck law for metal forming

Figure 21. Variable friction coefficient

measured in some experiments. But it should be noted that dependence on the velocity is often introduced in the models too soon. For example, for a long time the reduction of the friction coefficient when velocity increases (or just considering static and dynamic friction coefficients) has been presented as a necessary condition for stick-slip to occur, which is not true: it is not a necessary condition. This will be shown in the example of Section 3.11.

- Interface models for faults in geophysics. Earthquake rupture is classically modeled as a friction-dominated process, where friction arises from the roughness of the contact between the two sides of a fault. Earthquake initiation is strongly assumed to be triggered by friction instabilities. A lot of work has been done to characterize the conditions of instability and the very beginning of earthquakes (nucleation duration and length). Choosing a realistic friction law for this interface is a key point for modeling nucleation of an earthquake and generation of the accompanying waves. Among many possible choices for the friction laws, it can be noted:

- slip-weakening friction (49) (120),
- rate-and-state friction (107) (17) (14).

The high number of parameters used in some of these laws is the sign of the complexity of the phenomenon to be modeled. In general, laws with many constitutive parameters should not be recommended. Despite the fact that giving convenient values (identification step) to these parameters will not be easy, choosing a model with many parameters could sometimes indicate that the dominant characteristics have not been conveniently identified. That relates to the previous fundamental remark to make it as simple as possible and as complex as necessary. Research is ongoing on the use of the RCCM model (adhesion and friction coupling; 4 parameters) to describe the behavior of fault interfaces in geophysics (102) (121).

Conclusion In this section we have given an overview on the most usual interface laws in order to show the complexity of frictional contact modeling. We have pointed out some of the main issues regarding the art of modeling.

- A key point in contact analysis is the non-smooth character of the behaviors (multivalued mapping). Using regularization to eliminate this non-smooth character of the behaviors is not recommended.
- Make it as simple as possible and as complicated as necessary. That means keeping the number of variables and of parameters accounting for the dominant phenomena as small as possible (for example, it is better not to use a variable friction coefficient when it is not necessary).
- Constitutive assumptions are part of the model. It is important to keep in mind that a model is built for a certain purpose (no universality) and may provide an adequate description of a physical system under certain conditions (magnitude of the forces, velocities, etc .)

2.3 Choice of the scale in contact mechanics: various scales and multiscale analysis

Friction and adhesion are phenomena that can occur at very different scales, from the nanometer (thin films) to the kilometer (geophysics) and, of course, their physical and mechanical natures can be very different. This means that the physical ingredients to put in the model could be very diverse depending on the scale of the phenomenon.

For a given phenomenon, another point is the interactions between the global and the local behavior, for example between the scale of the rocks and that of the fault in geophysics, between the scale of the asperities and that of the plate in metal forming, between the scale of the molecule and that of a thin protective film layer.

For most models, the law is set directly at the global scale. Today, new approaches combine local and global analysis in order to put more physics into the global model. This is made possible by the advances made in tribology measurements and by the extraordinary increase in computer power allowing millions of degrees of freedom to be treated for numerical simulations.

- Experimental knowledge at the local scale: micro tribology and asperity measurement. Tribology is the science dedicated to the study and the experimental investigations of friction, wear, lubrication. Experiments and accurate measurements at the local scales (millimeter, micrometer and nanometer) have permitted a better understanding of the effects of normal pressure, speed, temperature and other environmental conditions on the tribological behaviors, and also a better understanding of breakout

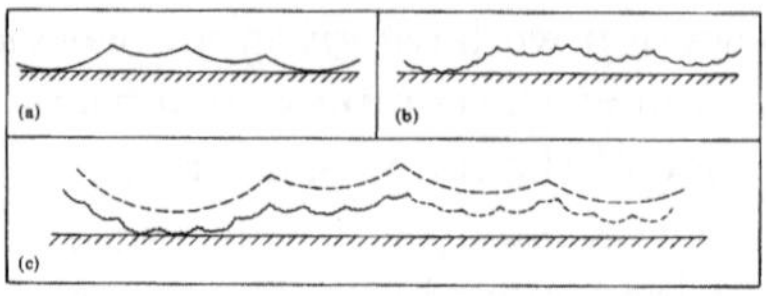

Figure 22. Archard modelization on the basis of sphere discretization and Hertz model

and movement of debris in an interface. Very interesting movies on this interface life and a theory of the third body in the interface have been proposed by Yves Berthier (13) at the LaMCoS in Lyon. This brings a strong contribution to the understanding of the global phenomenon, but the way to use and to put all these experimental measurements in the model is not easy. Measuring surface asperities has been done for quite a long time to characterize the plate surface in metal forming and stamping. They are described by their statistical characteristics. The hope would be to deduce the value of the friction coefficient from these statistical data. Introducing the local data into a global model is a big challenge.

- **Analytical approach.** First totally empirical, the Coulomb law got some justifications either from physical considerations by F.P. Bowden, D. Tabor (118) or local analysis by J.F. Archard (8) (9) or J.A. Greenwood(47). The Archards model was a first multiscale model. Archard approached the curve contact surface by a series of spheres and each sphere by another series of spheres of smaller radius (see Fig. 22) and so on. Then he used the analytical solution of the Hertzs problem on each small sphere and got a global interface law. The Hertz solution is the analytical solution of the contact between a rigid sphere and an elastic half plane (K.L. Johnson (55)).

Another example is the Greenwood-Williamson approach (47), based on a Gaussian asperity height distribution (see also B.N.J. Persson (85)).

- **Mathematical approach.** Considering a third body with its own behavior, contact laws can be obtained by evaluating the limit of the solution when the thickness goes to zero. This has been conducted with simple (either linear or non-linear) behavior laws for the third body and has permitted to justify various contact laws. These mathematical tasks are complex because they deal with the notion of limit in a non-smooth context and they depend on the complexity of the chosen behavior law (Licht-Lichaille (67), Bouchitte et al (15), Dumont et al (39), Rizzoni et al (104), Serpilli (111)).

- **Numerical approaches - Multiscale simulations.** Over the years, some numerical methods have been developed for determining accurately the solutions in a localized zone in a structure where peculiar phenomena have to be taken into account and where refinement of the meshes is lo-

Figure 23. : Applications of multilevel approaches for contact problems (Courtesy of P. Wriggers)

cally needed. Then iterative methods combining computations with several mesh levels have been developed: substructuration, FAQ methods, $(\mathrm{FEM})^2$, etc.). These methods avoid using a global refined mesh and so reduce the number of necessary DOF for determining an accurate solution to a problem including localized effects. The huge increase in computer power made computations with an extremely large possible number of DOF and also made it possible to mesh different zones with an extremely large refinement factor. These methods are of great interest for contact problems and have been used and developed in recent years. Let us cite the work of the research team led by Peter Wriggers at the ICM in Hannover, and that led by Jean-Franois Molinari at the EPFL in Lausanne.

To study the contact between a tire and the road and taking into account the asperities, Peter Wriggers and co-authors used multilevel approaches combined with a homogenization technique at the intermediate level. The problem is separated into macro and micro-scale problems. Thus local effects can be taken into account while keeping the computational time in a reasonable range, because the intermediate homogenization drastically reduces the computational time (80%). Details, applications and many references can be found in Temizer-Wriggers (119), Wriggers-Reinelt (130), De Laurenzis-Wriggers (34), Drosopouilos et al (42), Wagner et al (124), etc.

This multilevel methods make it possible to take into account local interactions, as the ones between the tire and road asperities, or the interaction with particles when a third body is modeled in the contact (Fig. 23).

Jean-François Molinari, Guillaume Anciaux and co-workers at the EPFL have developed multi-level models from atomistic to macroscopic levels. It can be observed that many surfaces tend to be nearly self-affine fractals (a self-affine fractal surface maintains its statistical properties when magnified). In many numerical studies, from the atomic scale up to the geologic scale, a spectral generator of surfaces allows to employ representative surfaces and ultimately infer surface deformation mechanisms by connecting the contact properties with the statistical/fractal characteristics of the

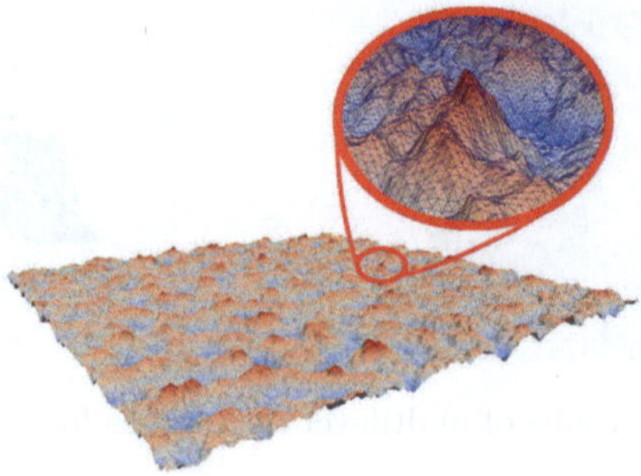

Figure 24. Discretization of surface asperities (courtesy of G. Anciaux)

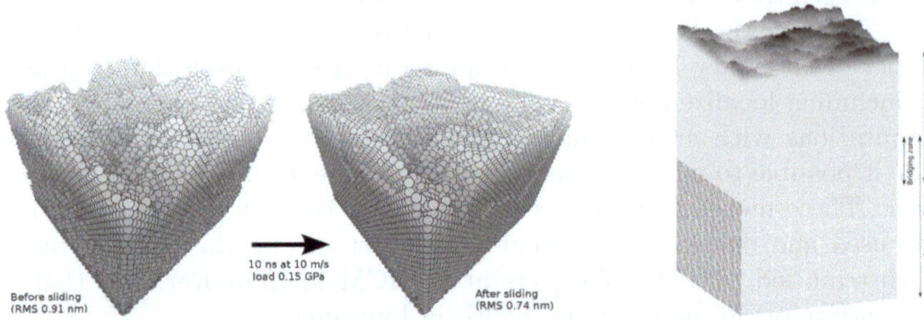

Figure 25. Flattening of nano-scale asperities during sliding and coupled MD-FE model for rough surface (2) (courtesy of G. Anciaux)

Figure 26. Simulation of a scratching indenter moving at constant speed. (courtesy of G. Anciaux)

original surface Yastrebov et al. (131) (see Fig. 24). The importance of the choice of the modeling scale was stressed in Section 1.2. Here the scale plays a paramount role. By using such profiles at the molecular level, the impact of the surface topology on all sorts of permanent deformations can be demonstrated. For instance, roughness flattening at the nano-scale leads to decreasing wear rates and to reduced friction coefficients (Spijker et al. (115) (116), see Fig. 25). In order to limit the cost of these simulations, routinely containing millions of atoms, Molecular Dynamics (MD) coupled to Continuum Schemes are employed - Anciaux et al (3) - (see Fig. 25).

With this multi-level model, two features have been outlined: the importance of the heat fluxes (see Ramisetti et al. (89) and the role of plastic deformations which takes the form of a collective motion of defects through crystalline materials (dislocations). In Fig. 26, this latter effect is well illustrated with a scratching indenter which deforms a bulk substrate and where some dislocations get trapped (Junge-Molinari (56)). Recent advances in a three-way coupling between molecular dynamics, finite elements and discrete dislocations permit a powerful multi-level modeling and make it possible to simulate with greater accuracy sliding contact with consistent surface topologies which would reveal mechanisms often impossible to observe through experiments (see Cho et al. (19)).

Conclusion about the choice of the scale in the art of modeling. In modeling and especially in contact modeling, great progress has been made on the experimental measurements, on the mathematical analysis and a lot on the computational capacities. Very interesting new opportunities have been opened towards a better understanding and a better modeling of contact behaviors by using local/global approaches:

- getting a better understanding and quantification of what friction or wear is, and what local behaviors are involved;
- modeling local effects of which some of them are
  - strong interaction between the solid and the asperities (same order of magnitude of the local deformations and the asperity size),
  - in tire-to-road contact, thin water film evolution and water flowing around the tire treads when it rains (various regimes), ice on the road, melting ice and cold temperature (coupling with thermal effects)
  - in metal forming, oil flows on Lasertex plates due to residual lubrication,
- connecting the global friction (friction coefficient or new global friction laws) and local surface characteristics (geometries, asperities, coupling, etc.)

2.4 Construction of a model coupling adhesion, friction and unilateral contact: thermomechanical and energetic basis.

In what follows, we want to illustrate with two examples the assertion made previously that models are based on concepts. The first example presents the construction of the RCCM model coupling unilateral contact, friction and adhesion. The second example is the generalization of this kind of models and the construction of a unified model including most of the models developed on this subject.

Models are based on general laws and conservation principles, complemented by constitutive relations which characterize the medium and should comply with some basic principles (causality, etc.). The RCCM (Raous-Cangémi-Cocou-Monerie) model is a model coupling adhesion, unilateral contact and friction. It is based on the concept of intensity of adhesion introduced by M. Frémond (45). Details on this model can be found in Raous (95), Raous et al. (96), Raous-Monnerie (99) and Cocou et al (25). We will stress the thermomechanical basis of the model and the difficulties due to the non-smooth character of the law. The application of the model to an industrial problem is presented: how to reduce crack propagation in a composite material by controlling the characteristics of the fiber/matrix interface.

- **The RCCM model: thermomechanical basis.** The problem is written here for the contact between two deformable solids and u_n denotes the gap and u_t the relative tangential displacement. Because of the chosen convention, the unilateral conditions are inverse to the ones given by (1) (see complementarity problem (9)). The variables in the interface are:

- u_n, u_t the normal and tangential displacements,
- R_n, R_t the normal and tangential forces ($R_t^r = C_t\, u_t\beta^2$ is the reversible part of R_t),
- β the adhesion intensity (interface damage).

Unilateral conditions with adhesion

$$-R_n + C_n\, u_n\beta^2 \geq 0\,, \quad u_n \geq 0\,, \quad \left(-R_n + C_n\, u_n\beta^2\right) u_n = 0 \tag{9}$$

Coulomb friction with adhesion

$$\|R_t - R_t^r\| \leq \mu(1-\beta)\left|R_n - C_n\, u_n\beta^2\right| \quad \text{with}$$

$$\begin{array}{l} \|R_t - R_t^r\| < \mu(1-\beta)\left|R_n - C_n\, u_n\beta^2\right| \Rightarrow \dot{u}_t = 0 \\ \|R_t - R_t^r\| = \mu(1-\beta)\left|R_n - C_n\, u_n\beta^2\right| \Rightarrow \exists\lambda \geq 0\,,\ \dot{u}_t = \lambda(R_t - R_t^r) \end{array} \tag{10}$$

Evolution of adhesion intensity

$$\dot{\beta} = -(1/b)\left(w - (C_n\, u_n^2 + C_t\,\|u_t\|^2)\beta\right)^- . \tag{11}$$

The parameters of the model are:

- μ the friction coefficient
- C_n, C_t the initial stiffnesses of the interface
- ω the adhesion energy (the Dupré energy)
- b the viscosity of the interface

The graphs presenting the interface behavior of the RCCM model are given in Fig. 27 to 30. The energies involved in the loading/unloading process are presented in Fig. 31. As was said before, the analysis of the model in terms of energy is fundamental. The thermomechanical bases are the following:

- a hypothesis of material boundary is set for the contact zone: that means that we can associate to this part of the boundary a surfacic energy E and a specific entropy S and so the free energy of Helmoltz $\psi = E - ST$,
- the variables are u_n, u_t, β,
- the thermodynamical forces are R_n, R_t, G_β .

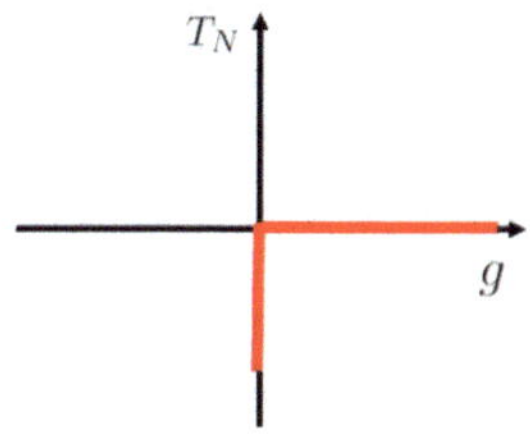

Figure 27. Signorini graph.

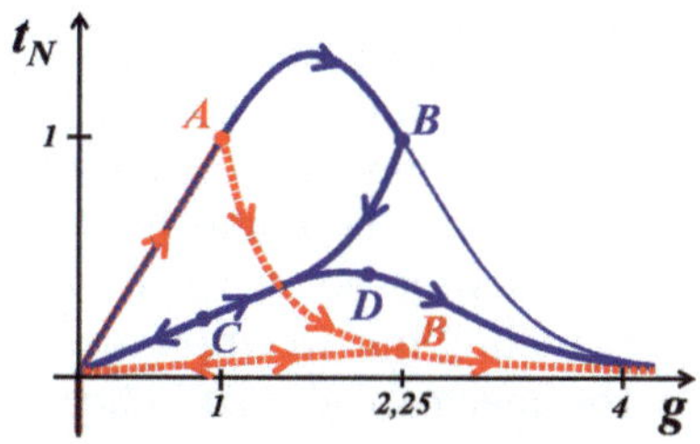

Figure 28. Normal behavior of RCCM model.

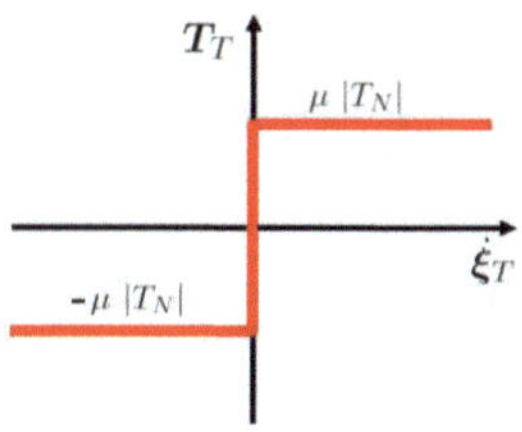

Figure 29. Coulomb graph.

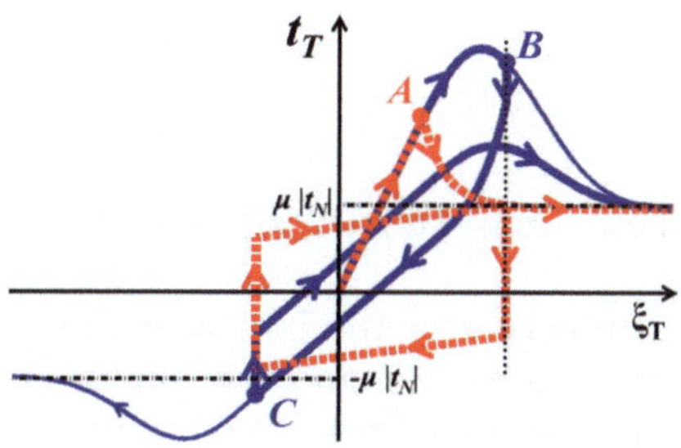

Figure 30. Tangentiel behavior of RCCM model.

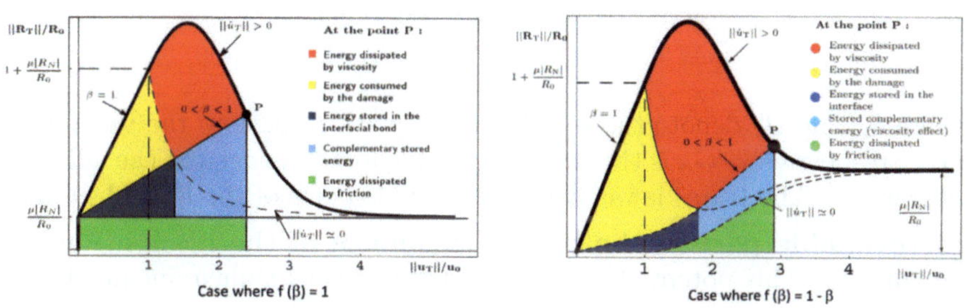

Figure 31. The energies involved.

The next steps will be:

- making an appropriate choice of the free energy ψ. It can be noted that it is non convex relative to (u_n, β) and non-differentiable. Then we apply the state laws in the sense of partial sub-differential and differential inclusions and we obtain the reversible part of the behavior.
- making an appropriate choice for the potential of dissipation ϕ compatible with the Clausius Duhem inequality. It can be noted that it is non differentiable. Thus the complementary laws are written in terms of differentiable inclusions. We obtain the non-reversible part of the behavior (dissipation). The main steps are given below and details can be found in Raous (95) and Raous et al (96). The main objective is not to go into all the details in this course on the art of modeling but to emphasize the theoretical basis of the model both on the mechanical part (choice of the energies) and on the mathematical formulation (sub-differentials because of the non-convexity and differential inclusions because of the non-differentiability).

- Reversible part of the behavior: choice of the free energy Ψ.

$$\Psi(u_n, u_t, \beta) = \frac{C_n}{2}\, u_n^2 \beta^2 + \frac{C_t}{2}\, \|u_t\|^2 \beta^2 - w\, h(\beta) + I_{\widetilde{K}}(u_n) + I_P(\beta) \qquad (12)$$

where $\widetilde{K} = \{v \,/\, v \geq 0\}$ and $P = \{\gamma \,/\, 0 \leq \gamma \leq 1\}$.

Introducting of the indicator functions $I_{\widetilde{K}}$ and I_P imposes the unilateral condition $u_n \geq 0$ and the condition $\beta \in [0, 1]$. To write the state laws, the two difficulties (lack of convexity and lack of differentiability) are overcome by using local or partial subdifferentiation. The state laws can then be written as follows :

$$R_n^r \in \partial_{u_n} \Psi(u_n, u_t, \beta) \qquad (13)$$

$$R_t^r \in \partial_{u_t} \Psi(u_n, u_t, \beta) \qquad (14)$$

$$-G_\beta \in \partial_\beta \Psi(u_n, u_t, \beta) \qquad (15)$$

where ∂_u and ∂_β denote the subdifferential with respect to the variables u and β respectively. G_β is the thermodynamic forces associated the adhesion intensity β. The states laws give the reversible parts of the RCCM model.

- Irreversible part of the behavior: choice of the dissipation potential Φ. This potential agrees with the Clausius Duhem inequality.

$$\Phi\left(\dot{u}_t, \dot{\beta}; \chi_n\right) = \mu \left| R_n - C_n\, u_n \beta^2 \right| \|\dot{u}_t\| + \frac{b}{p+1} \left|\dot{\beta}\right|^{p+1} + I_{C^-}(\dot{\beta}) \qquad (16)$$

with $C^- = \{\gamma \in W/\gamma \leq 0\}$ and $p \leq 1$. The complementary laws are then written :

$$
\begin{aligned}
R_n^{ir} &= 0 && (17)\\
R_t^{ir} &\in \partial_{\dot{u}_t}\Phi(\dot{u}_t, \dot{\beta}; \chi_n) && (18)\\
G_\beta &\in \partial_{\dot{\beta}}\Phi(\dot{u}_t, \dot{\beta}; \chi_n) && (19)
\end{aligned}
$$

And we obtain the non reversible parts of the RCCM model (controling friction and adhesion).

Therefore we obtain all the relationships characterizing the interface model given at the beginning of this section.

The model has been constructed in two steps:

- step 1: appropriate choices of the free energy and of the potential of dissipation compatible with the Clausius Duhem inequality.
- step 2: the behavior laws are obtained by application of the state laws to the free energy and by application of the complementary laws to the potential of dissipation

- Application of the RCCM model to the fiber/matrix interface to reduce crack propagation in a SiC/SiC composite. The industrial problem was to improve the resistance of a SiC/SiC composite to crack propagation by controlling the properties of the fiber/matrix interface. The objective was to build a model of the fiber/matrix interface behavior in order to check the influence of its characteristics on the crack propagation and to be able to give recommendations for the elaboration of the composite (enzyme, temperature, etc.) in order to improve its resistance. A model coupling unilateral contact, adhesion, viscosity and friction was elaborated. It was presented in the previous section.

Now in this section we will stress three steps of the modeling process: identification of the constitutive parameters, validation of the model and prediction. More details will be given in Sections 4.2, 4.3 and 4.5.

Identifying the constitutive parameters. Identification of the four constitutive parameters was conducted in a fiber micro-indentation experiment carried out at ONERA (see Fig. 32). On Fig. 33 the good agreement between the experimental measures and the model results for these identified values can be noted.

Validating. The model was validated by conducting other micro-indentation experiments operating on larger fibers, with other fiber volume fractions (single fiber surrounded only with matrix) and with different thermal residual stresses. A good agreement was obtained.

Predicting. The model was then used as a tool to optimize the interface characteristics in order to increase the resistance of the composite to crack

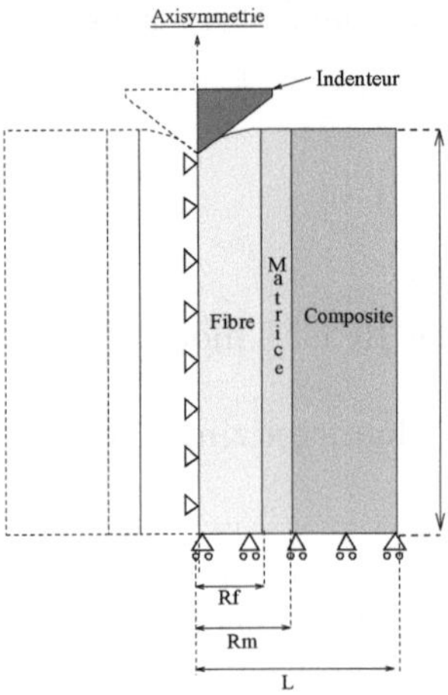

Figure 32. Model of the experiment.

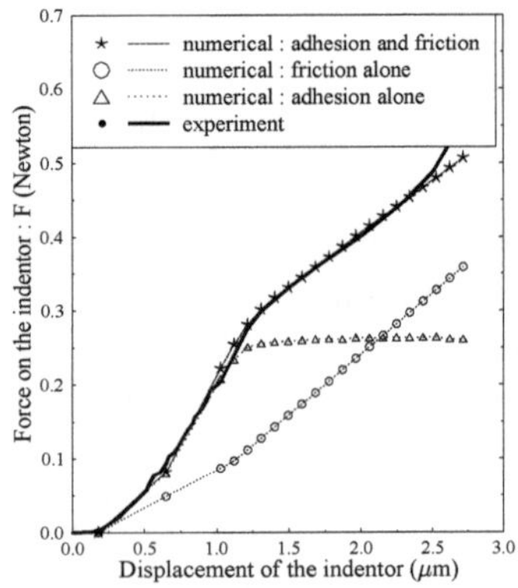

Figure 33. Identification of the constitutive parameters

propagation. The objective was to make recommendations on the suitable interface properties to improve the crack resistance of the material.

When a crack propagates in a composite in a plane orthogonal to the fiber direction, a competition occurs between matrix crack bridging, matrix crack trapping by a row of fibers and fiber breaking. A 3D discretization of a single fiber surrounded with matrix was done. A prescribed displacement was applied on the two ends of the sample. A ductile model was used to describe the propagation of the crack. The global behavior was then evaluated with various values of the model parameters. Very different behaviors were observed and are presented in Fig. 34 and Fig. 35. For symmetry reasons, only half of the sample is represented, the crack propagates in the horizontal plane, the dark zones correspond to the total adhesion breakdown. In Fig. 34, the parameters put in the model correspond to a strong interface. We can observe that crack propagates both in the matrix and in the fiber and there is nearly no rupture along the fiber/matrix interface. Total rupture is obtained when the prescribed displacement reaches $0.12mm$. In Fig. 35, the parameters correspond to a soft interface. In this case, the crack propagates in the matrix, goes around the fiber but does not break the fiber and the interface between the fiber and the matrix progressively collapses (energy dissipation). This time the total rupture occurs when the prescribed displacement reaches $0.4mm$. Thus the composite is three times more resistant. This shows that a soft fiber/matrix interface will improve the composite resistance to crack propagation. Precise values can be given. Details can be found in Raous-Monerie (99).

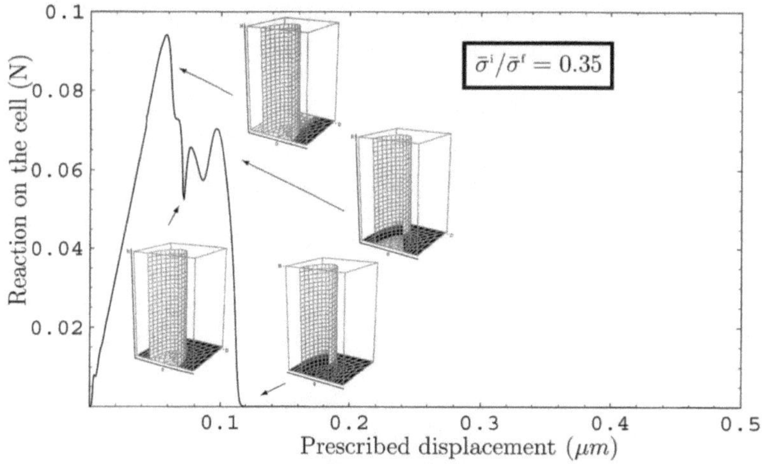

Figure 34. Crack progression (in black) and interface debonding in the case of strong interface

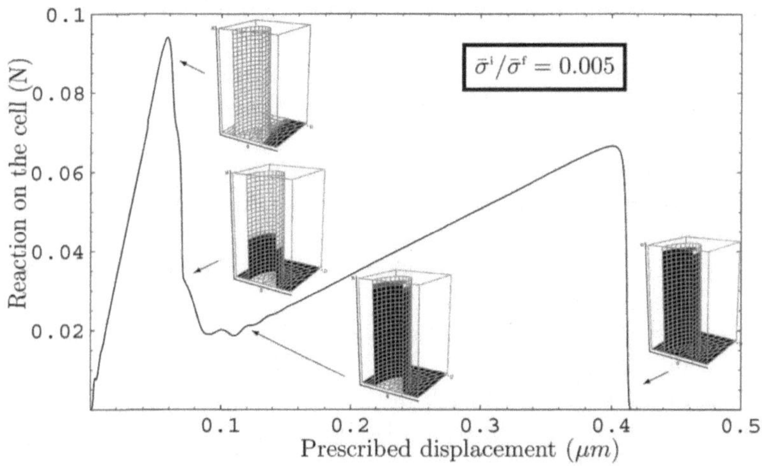

Figure 35. Crack progression (in black) and interface debonding in the case of soft interface

2.5 A unified model for adhesion: inductive-deductive modeling process.

This work was published in Del Piero-Raous (35) (103) . Only the main features are given here. In that work, we showed how a set of various models (Cohesive Zone Model - CZM) can be gathered in a general unified model capturing the main features of the phenomena and based on fundamental concepts (thermomechanics, state space, choice for the energies and the dissipations).

To refer to the general comments given in Section 1, this work presents an up and down process:

- first an inductive process (giving sense to a corpus of data) gives the unified model
- then various models can be obtained as particular cases of the unified model , which is a deductive process (from general to particular).

The purpose of this work is to model a complicated interface response (unilateral contact, friction, adhesion, viscosity, etc.) with the smallest number of variables. This is conducted:

- by considering general laws, typically those of conservation of energy and dissipation of energy, that is, the mechanical version of the first two laws of thermodynamics,
- and choosing
  - a set of state variables, that is, an array of independent variables which fully determine the response to all possible deformation processes,
  - a set of an elastic potential and dissipation potentials, which are functions of state in terms of which the general laws take specific forms,
  - a set of constitutive assumptions.

The unified adhesion model is built as follows. Be given :

- a set of constitutive assumptions
  - a given loading curve (which could be either an experimental one or a prescribed behavior)
  - elastic behavior (with damage)
  - unilateral conditions
- a set of an elastic potential and dissipation potentials related to the effects that one likes to put in the model:
  - damage dissipation
  - viscous dissipation
  - friction dissipation
  - etc

And by using the power equation and its derivatives when necessary, we determined the evolution of the damage variable.

Then, the usual adhesion models (and new ones) can be built from this general formulation by choosing conveniently the loading curve and the elastic and dissipation potentials. This was done in Del Piero-Raous (35) for the RCCM model.

3 Formulations, mathematical aspects, solvers

In Section 1, we discussed the choices of formulations and of the associated mathematical analyses as being important features presented in Section 1. In this section, we want to stress the difficulties encountered in contact mechanics due to the non-smooth character of the laws. There are different kinds of difficulties:

- contact laws are multivalued (non-differentiable) mappings and not functions,
- the friction law is non-associated (no normality rule),
- the variables are defined on the boundary (trace space),
- shocks may occur, time discontinuities may arise and the solutions will be not differentiable.

Thus, formulations are not mathematically elementary. We need to use variational inequalities, differential measures, differential inclusions, etc. Also, solvers have to deal with this non-smooth character of the contact laws and more difficult problems arise.

However - and this may be the most important point to be emphasized in this section - this non-smooth character is essential in the model to deal with the complex phenomena which can be observed with friction. Modeling some phenomena observed in contact mechanics requires that these theoretical specificities associated with this non-smooth character were considered and dealt with correctly. This means that for an accurate description of the contact behavior, regularization should be avoided as often as possible. This is illustrated by the example of the squeal of waist seal sliding on a window glass presented Section 3.11. However, there are of course cases where a simplified model of contact behavior may be sufficient (for example modeling crash tests). We present first the static problem which is a displacement formulation. It has no physical meaning, as friction should be expressed in terms of velocity, but it will be an intermediate problem, very useful for solving the problem expressed on the velocities and moreover the main difficulties can be identified on this formulation.

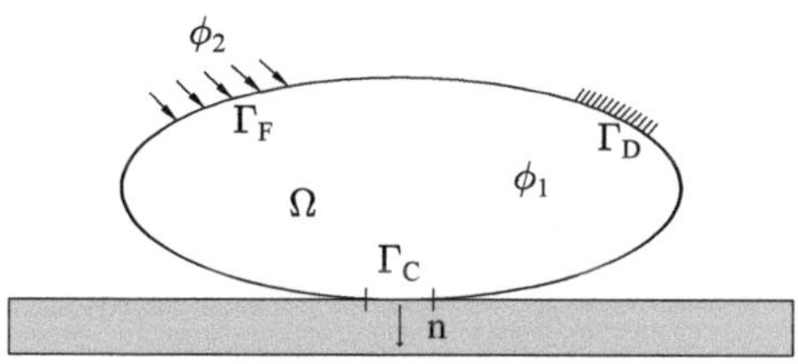

Figure 36. The solid

3.1 Formulations of the "static" problem

As presented in Fig. 36, the solid occupies the domain Ω and is submitted to boundary conditions on Γ_D and to volume and surface loadings in Ω and on Γ_F. On the part Γ_C of the boundary Γ we have unilateral and friction conditions. The solid is here supposed to be elastic.

Problem P_{stat}: Let Φ_1, Φ_2 be the given forces, find the displacement field u, the stresses σ, the strains ϵ and the contact force R such that :

$$\left\{\begin{array}{l}
\text{Elastic behavior and the equilibrium} \\
\left.\begin{array}{lcl} \varepsilon & = & \mathrm{grad}_s\, u \\ \sigma & = & K\varepsilon \\ div\sigma & = & -\phi_1 \end{array}\right\} \text{ on } \Omega \\
\text{Boundary conditions} \\
\quad u = 0 \quad \text{on } \Gamma_D \\
\quad \sigma.n = \phi_2 \quad \text{on } \Gamma_F \\
\text{Unilateral contact with friction} \\
\left.\begin{array}{l} \sigma.n = R \\ u_n \leq 0 \\ R_n \leq 0 \\ u_n R_n = 0 \\ \|R_t\| \leq \mu \mid R_n \mid \text{ with} \\ \text{if } \|R_t\| < \mu \mid R_n \mid \text{ then } u_t = 0 \\ \text{if } \|R_t\| = \mu \mid R_n \mid \text{ then } \exists \lambda > 0 \text{ such that } u_t = -\lambda R_t \end{array}\right\} \text{ on } \Gamma_C
\end{array}\right.$$

3.2 Variational formulation

For a classical elasticity problem without contact conditions, the variational formulation is written under the form of a variational equation. For frictional contact problems, we get an implicit variational inequality (see Duvaut-Lions (40), Raous (95)) or a quasi-variational inequality when the dual formulation is considered (see Panagiotopoulos (83)).

Let K be the convex of the admissible displacements (2D formulation) $K = \{v \in U/\ v_n \leq 0 \text{ on } \Gamma_C\}$ with $U = \left\{v \in \left[H^1(\Omega)\right]^2 / \ v = 0 \text{ on } \Gamma_D\right\}$.

Problem P_{var}: let ϕ_1, ϕ_2 be given as previously defined in *Problem P_{stat}*, find u $\in$ K such that :

$$a(u, v - u) + J_1(u, v) - J_1(u, u) \geq L(v - u) \qquad \forall v \in K \tag{20}$$

with :

$$a(u,v) = \int_\Omega \sigma(u)\varepsilon(u)\, dx = \int_\Omega E_{ijkl}\varepsilon_{ij}(u)\varepsilon_{kl}(v)\, dx \qquad \forall u,v \in U \tag{21}$$

$$L(v) = \int_\Omega \phi_1 v\, dx + \int_{\Gamma_F} \phi_2 v\, ds \qquad \forall v \in U \tag{22}$$

$$J_1(v,w) = \int_{\Gamma_C} \mu|F_n(v)|\|w_t\|\, ds \tag{23}$$

where $a(.,.)$ is the classical bilinear form associated with the elasticity operator E and $L(.)$ is the linear form associated with the loadings (it represents the work of the loads in the virtual displacement v). The form $J_1(.,.)$ is associated with the friction (it represents the work of the tangential contact force; note that the normal force does not work because of the complementarity condition).

3.3 A minimization problem (fixed point method).

It is not possible to associate a minimization problem (minimum of the potential energy) with the variational problem *Problem* P_{var} as can be done in classical elasticity. This is due to the non-associated character of the Coulomb's law, since the sliding velocity does not satisfy the normality rule (see section 2.1). For both mathematical and numerical reasons, it will be helpful to set an equivalent form of the previous problem by using a fixed point on the sliding limit, associated with a Tresca problem. For the Tresca friction, the sliding limit does not depend on the normal force and the sliding velocity satisfies the normality rule, because the Coulomb cone has been replaced by the Tresca cylinder (see Fig. 16). Thus we obtain the following problem.

Problem $\boldsymbol{P_{fp}}$: Find the fixed point of the application S :

$$S(g) = -\mu F_n(u_g) \tag{24}$$

with u_g solution of the following problem $P_{varTresca}$:

Problem $\boldsymbol{P_{varTresca}}$: For a given g, find $u_g \in$ K such that :

$$a(u_g, v - u_g) + j(v) - j(u_g) \geq L(v - u_g) \qquad \forall v \in K \tag{25}$$

with: $j(v) = \int_{\Gamma_C} g\|v_t\|\, ds$

The previous Tresca problem is equivalent to the following minimization problem :

Problem P_{mini}: For a given g, find $u_g \in K$ such that

$$J(u_g) \leq J(v) \qquad \forall v \in K \tag{26}$$

with $J(v) = \frac{1}{2}a(v,v) + j(v) - L(v)$

The problem is now set as a minimization problem under constraints of a non-differentiable functional which needs be solved for each value of the sliding threshold obtained at every step of the fixed point application.

3.4 Other formulations

Many other formulations can be given for the initial *Problem P_{stat}* and each will need different kinds of solvers. It is now easy to understand why the choice of the formulation is one of the important choices in the art of modeling, as was said in Section 1.3. Without going into details, we can cite the following formulations.

Complementarity problem Another approach consists in writing the problem in the form of a linear complementarity problem (see Cottle et al (31)), introducing two new friction variables by separating the tangential displacement into left and right sliding parts (see Raous (95)). In the 2-dimensional case, it is then written, after FEM discretization and condensation in reducing the problem to the contact variables (partial inversion of the linear parts), as *Problem P_{compl}*.

Problem P_{compl}: Find $F \in \Re^p$, $u \in \Re^p$ such that

$$\left.\begin{array}{l} Mu = F^* + R \\ R_i \leq 0, \; u_i \leq 0 \qquad i = 1 \ldots p \\ R_i u_i = 0 \qquad i = 1 \ldots p \end{array}\right\} \tag{27}$$

where M and F^* are respectively a non-symmetric matrix and a loading vector deduced from the FEM problem by condensation (and taking into account the change of variables associated to the choice of new variables), R and u are the contact forces and the contact displacements, and p is the number of contact degrees of freedom. This 2D formulation has been extended to 3D problems using a polygonalization of the Coulomb cone by Klarbring (61).

- Penalization formulation for the frictionless contact. The penalization principle consists in adding an extra force G_ϵ defined on the contact boundary to enforce the conditions $u_n \leq 0$; a penalization parameter ϵ is introduced and a nonlinear variational equation is obtained.

Problem P_{penal}: Find $u \in U$ such that $\forall\, v \in U$

$$a(u, v) = L(v) + G(v) \tag{28}$$

with $G(v) = \int_{\Gamma_C} G_\epsilon \; v \; ds$

Normal penalization can be considered as a numerical form of the compliance law previously introduced in Section 2.1.

- **Lagrange multipliers.** In that case, the contact force is kept as a variable (Lagrange multiplier). It is a mixed formulation. We get a saddle point formulation which is a min/max problem.

- **Augmented Lagrangian.** It is a combination of the penalty and the Lagrange multiplier formulations.

Details of all these formulations can be found in the books by Laursen and by Wriggers (63) (128) (129).

3.5 Formulations of the quasi-static problem

Coulomb friction law has to be expressed on velocities. It was shown in Cocou-Pratt-Raous (22) (23) (24) (27) and Shillor et al (110) that the problem is then written as the coupling of two variational inequalities (one of which is implicit). The problem can be written as follows.

Problem P_{qs}: For t belonging to $[0, T]$ and with presribed initial conditions, find $u(t) \in K$ such that :

$$\begin{aligned} &a(u(t), v - \dot{u}(t)) + J_1(u(t), v) - J_1(u(t), \dot{u}(t)) \geq L(v - \dot{u}(t)) \\ &\qquad\qquad + < R_n(u(t)), v_n - \dot{u}_n(t)) > \qquad \forall v \in V \\ &< R_n(u(t)), z_n - u_n(t)) > \; \geq 0 \qquad \forall z \in K \end{aligned} \tag{29}$$

3.6 Formulation of a dynamics problem.

The main contributions to this topic are from Jean-Jacques Moreau (75) (76) for finite dimensional problems (granular medium). In *Problem* P_{stat}, the equilibrium equation should be replaced by the equation of motion:

$$\rho\, \ddot{u}(x, t) = \mathrm{div}_x \sigma(u(x, t)) + \phi_1(x, t) \tag{30}$$

As was noted before, in the case of a contact problem, the occurrence of impacts and shocks has to be considered and thus the velocities are not continuous (and not differentiable). The acceleration cannot be defined in the usual sense, the notion of differentiable measures has to be used. For the sake of simplicity, we give directly the discrete formulation of the problem.

Problem P_{dyn}: Find U such that
$\forall t \in [0,T]$ $U(t) \in V_h$, $U(0) = U_0$, $\dot{U}(0) = V_0$ and :

$$M.d\dot{U} + K.U + C.\dot{U} = F + Rd\nu \tag{31}$$

and should be satisfied the Signorini and Coulomb conditions for the contact nodes (Section 2.1).
$d\dot{U}$ is a differential mesure representing the discretized acceleration and $d\nu$ is a nonnegative real mesure relative to which $d\dot{U}$ happens to possess a density function. The differential measure is a generalization of the notion of derivative which takes into account the jumps. The derivative $\dot{u} = du/dt$ is replaced by a differential measure du (Stieltjes measure). In the smooth case (u is continuous), we have $du = \dot{u}dt$ where dt is the Lebesgue measure, which is in fact the differential measure of the (real) function t. In the general case, for any compact sub-interval $[a,b]$ we have :

$$\int_{]a,b]} d\dot{U} = \dot{U}^+(b) - \dot{U}^+(a) \tag{32}$$

with right continuity : $\dot{U} = \dot{U}^+$
This formulation in terms of differential measure is the convenient formulation to be used for dealing with the jumps and the shocks which may occur in contact dynamics. Convenient numerical methods to solve this problem set under this sophisticated formulation will be given in Section 3.8.

3.7 Mathematical analysis

In this section, we want to stress that some problems could be very stiff, considering the mathematical properties of the operators and of the solutions.

The art of modeling is then to cope with some choices:

- either transforming the initial problem (regularization, etc.) in order to get a simpler model, easier to solve, but , as was underlined before, the regularized problem is a problem different from the initial one; we should be aware of that a feature because regularization is currently used in most computer codes,
- or addressing the real difficulties of the initial problem, which is not easy to do, but is the right way to get the correct solutions, even if this way is complex.

The mathematical analysis of the non-smooth problems set when modeling contact phenomena is very important. It is essential for a good understanding of the solutions that can be obtained and for overcoming the difficulties which arise both in the formulations and in the solvers. As was said in chapter 1, a large scientific culture is necessary (either individually or collectively as a team). It is impossible to be specialist of everything but it is important to have some knowledge of these various topics, including the mathematical topics presented in this section.

- Main difficulties and some alternatives.
The situation can be summarized as follows.

- The behavior laws are represented by multivalued mapping
  Consequence: variational inequalities
  Alternative: regularization, but we get a very different problem
- The Coulomb law is not associated
  Consequence: no minimum principle
  Alternative: Tresca problem (as intermediate problem) + fixed point on the sliding threshold
- The contact force is a distribution belonging to $H^{-\frac{1}{2}}$ (this point will be addressed below)
  Consequence: compactness issues arising in the mathematical analysis
  Alternative: new definition of the contact forces defined by a convolution with a smooth function with compact support (non-local friction)
- For the dynamic problem : impacts result in velocity discontinuities
  Consequence: differential measure formulation
  Alternative: no alternative in the general case; for a rigid body dynamics restitution coefficients can be used.

- Mathematical framework and difficulties. The following scheme (in 3D) gives a synthetic representation of the mathematical framework for contact problems.This is very helpful to understand the reasons for the difficulties encountered in the mathematical analysis of contact problems.

		Duality pairings		
Contact	$U_c = H^{\frac{1}{2}}(\Gamma_C)^3$	(R, w)	$\Phi_c = H^{-\frac{1}{2}}(\Gamma_C)^3$	
Trace operator	↑		↓	
Displacement	$U = H^1(\Omega)^3$	$< \Phi, v >$	$\Phi_F = L^2(\Omega)^3 \mathrm{x} L^2(\Gamma_F)^3$	
Deformation operator $\epsilon = \mathrm{Grad}_s u$	↑		↓	Equilibrium equations $\mathrm{div}\sigma = -\phi_1$ in Ω, $\sigma.n = \phi_2$ on Γ_F
Strains	$E = L^2(\Omega)^9$	$<< \sigma, \varepsilon >>$	$S = L^2(\Omega)^9$	Stress

$$<< \sigma, \varepsilon >>= \int_\Omega \sigma(u)\varepsilon(v)\, dx \qquad \forall v \in U \tag{33}$$

$$< \Phi, v >= \int_\Omega \phi_1 v\, dx + \int_{\Gamma_F} \phi_2 v\, ds \qquad \forall v \in U \tag{34}$$

$$(R, w) = \int_{\Gamma_C} \mu |R_n(u)| \|w_t\|\, ds \qquad \forall w \in U_c \tag{35}$$

This scheme has 3 levels. Levels 2 and 3 (at the bottom) present the classical scheme for an elasticity problem.

- At level 3, the stress space S and the strain space E are $L^2(\Omega)^9$; they are in duality according to the duality product $<< \sigma, e >>$ which represents the work of the stress σ in the strain ϵ,
- At level 2, the displacement space U is $H^1(\Omega)^3$ and the loading space ϕ_F should be the dual of $H^1(\Omega)^3$ for the duality product $< \phi, v >$ (which represents the work of the loadings (ϕ_1, ϕ_2) in the displacement u); however, as the loadings are given, we are free to choose them in a more regular space and we choose $L^2(\Omega)^3 \mathrm{x} L^2(\Gamma_F)^3$. This is a key point for the study of the existence and the uniqueness of the solution to the mathematical problem set for a small deformation elasticity problem.
- At level 1 - the level of the contact variables - the situation is similar but, as the contact forces are unknown, it is not possible to choose a more regular space for them. The contact displacement space U is $H^{\frac{1}{2}}(\Gamma_C)^3$ (trace of u on the boundary Γ_C) and the contact force is in $H^{-\frac{1}{2}}(\Gamma_C)^3$, which is the dual of U for the duality product (R_c, u_c). $H^{-\frac{1}{2}}(\Gamma_C)^3$ is a distribution space, which will make the mathematical analysis more complicated .

To complete this schema, we have indicated the mappings acting from one space to another. Let us note again that each duality product represents a work. This gives a good picture of the mathematical situation.

- Introduction of non local friction. In classical elasticity, existence and uniqueness theorems are based on coerciveness and continuity of the operators, and the proof is based on the Lax-Milgram theorem and Cauchy-Schwarz inequality (equivalence of norms).

In contact mechanics, the first difficulty which arises is that the contact force is a distribution belonging to $H^{-\frac{1}{2}}(\Gamma_C)$, as explained in the previous section. This implies some compactness difficulties, and a regularization R^{reg} using a convolution product is used to overcome this difficulty.

$$R^{reg} = R * \psi \tag{36}$$

where ψ is a very smooth function with compact support. Thus, the space for the contact force is much smoother and this will be very helpful for the mathematical analysis:

- R is in $H^{-\frac{1}{2}}(\Gamma_C)$, it is a distribution,
- R^{reg} is in $L^2(\Gamma_C)$, it is a function.

However, from a mechanical point of view, it should be noted that the contact forces are now defined by using a notion of non-local forces.

- Overview of the main results about existence and uniqueness of solutions. With this overview of the main mathematical results on contact problems, we want to call attention to the consequences of the non-smooth character of the law. The mathematical difficulties which are pointed out have direct consequences on the art of modeling, both on the kind of phenomena that can be observed and on the computational difficulties that may be encountered.

- Static problem (no mechanical meaning but interesting intermediate problem). The implicit variational inequality needs to be solved - inequalitiy (20) in *Problem* P_{var}.
  - Signorini problem (no friction) Signorini (112): existence and uniqueness of the solution - Fichera (44);
  - Signorini + Coulomb friction: existence if μ is small and no uniqueness - Necas-Jarusek-Haslinger (79), Jarusek (50), Eck-Jarusek (43);
  - Signorini + Coulomb (non-local friction): existence and uniqueness if μ is small - Cocou (21), Duvaut (40), Demkowicz-Oden (37);
  - Normal compliance + Coulomb friction: existence and uniqueness if μ is small - Klarbring-Mikelic-Shillor (59);
- Quasi-static problem. Two coupled variational inequalities (one of them is implicit) need be solved - see Problem P_{qs}.
  - Signorini + Coulomb: existence if μ is small (both in L^∞ and as a multiplier in $H^{-\frac{1}{2}}$) and no uniqueness of the solutions - Andersson (5), Cocou-Rocca (26) (105) (106);
  - Signorini + Coulomb (non-local friction): existence if μ is small (condition only in L^∞) and no uniqueness - Cocou-Pratt-Raous (22) (23);
  - Compliance + Coulomb: existence if μ is small and no uniqueness (only a few works) - Andersson (4), Klarbring-Mikelic-Shillor (59);
  - Examples of non uniqueness for a discrete problem - Ballard (10).

- Dynamic problem. The *Problem* P_{dyn} is formulated in terms of differential measures. There are very few mathematical results in elasticity, more results in viscoelasticity.
  - Continuous problem
    - frictionless in elasticity:
      normal compliance: existence - Martins-Oden (70) (71) Signorini: a few results on specific geometries (axial symmetry) - Munoz-Rivera-Racke (78);
    - normal and tangential compliance in viscoelasticity:
      existence and uniqueness - Martins-Oden (70) (71), Kuttler (62);
    - Signorini problem + non local friction in viscoelasticity:
      existence - Cocou (28), Cocou-Scarella (29);
    - Signorini + Tresca friction in viscoelasticity:
      existence - Jarusek (51);
  - Discrete problem
    - existence and uniqueness for analytical loading in 1D
      - frictionless - Ballard (11);
      - with friction - Ballard-Basseville (12).

- Concluding remarks on mathematical formulations. Contact and friction relate to non-smooth mechanics: the interface laws are multivalued mappings, the formulations are written in terms of implicit variational inequalities or differential measures for dynamic problems, in most cases there is no-uniqueness - and sometimes nonexistence - of solutions, and some problems remain unsolved (no results).

At this point, it is essential to note that the occurrence of multiple solutions is not a fantasy of mathematicians: it can really occur. Klarbring (60) and other authors (Janovsky, Alart-Curnier, Mitsopoulos-Doudoumis) constructed very simple examples with a few masses and springs showing the existence of two solutions, the occurrence of which depends on the values of μ. Unfortunately, the mathematical conditions set on μ depend on mathematical constants such as the coerciveness or the continuity constants that cannot be easily evaluated in terms accessible to the designer.

What should be remembered is that multiple solutions may occur and that the bad conditioning of the problem increases when μ increases. This is also observed with the computational resolution. All this is directly connected with the mathematical results given above.

From a mechanical point of view, this bad conditioning for large values of μ will be observed during the studies on stability. In Section 3.11, this will be illustrated with the presentation of a study on the stability analysis of a mechanical system and the search for occurrence of unstable solutions to modeling squeal phenomena.

In closing this section, we want to underline the strong relationship between mechanics and mathematics, in both directions, that is, from mechanics to mathematics and vice versa. In mechanics, mathematical tools are often used while important developments in mathematics have been motivated by the mathematical study of mechanical problems and especially of contact mechanics problems; for example, variational inequalities, convex analysis, Γ-convergence, etc.. These advances are due to specialists in theoretical mechanics and mathematicians (mostly a French-Italian school and some famous representatives from the USA and Greece). Without trying to be exhaustive, let us cite Jacques-Louis Lions, Georges Duvaut, Jean-Jacques Moreau, Panagiotis Panagiotopoulos, Tyrell Rockafellar, Enrico Magenes, Guido Stampacchia, Umberto Mosco, Bernard Nayroles, Pierre Suquet, Marius Cocou, Patrick Ballard, Yves Renard, Patrick Hild, Michelle Schatzman, Laetitia Paoli, etc.). Let us mention the reference book by R. Dautray and J.L. Lions (33).

Highly theoretical work is still in progress in contact dynamics, along with sophisticated mathematical developments.

3.8 A short overview of the solvers

The topic is broad and work addressing this topic is still in constant development. The idea is obviously for this course not to provide an exhaustive presentation of the solvers used in contact mechanics but to provide some indications and to stress the assumptions that they implicitly make. In fact, we want primarily to draw the attention of the engineer on the care that must be taken in choosing a solver or in choosing a computer code to solve a contact problem. It is essential to check what kind of methods are used, what numerical parameters are involved and to make some control tests on the numerical solution.

Details on numerical methods for contact problems can be found for example in the books by Wriggers (128), Laursen (63), Wriggers-Panagiotopoulos (Eds) (127), Wriggers-Laursen (129), Kikuchi-Oden (58). Let us first have a look at the methods that are currently used to solve quasi-static problems. Some of these methods are commonly used in commercial computer codes.

- **Penalty formulation.** As presented in Section 3.4, penalization is a regularization of the strict contact conditions. An extra force G_ϵ, defined on the contact boundary, is added to enforce the unilateral condition $u_n \leq 0$ and a condition $f_s \leq 0$ controling the friction. Two penalization parameters ϵ_n and ϵ_t (shortly noted by ϵ) are introduced. Normal penalization can be considered as a numerical form of the compliance law previously introduced. The functions G_ϵ are mostly nonlinear functions with a stiff dependence on the normal penetration into the obstacle (or sometimes on the vicinity of the obstacle in case of external penalization) or on the tangential sliding displacement. A nonlinear variational equation (28) is then obtained. After discretization, a nonlinear problem needs to be solved and Newton Raphson methods are usually used (see Alart-Curnier (1)).

$$Au = F + G_\epsilon(u) \tag{37}$$

Two remarks have to be made. First, to conveniently ensure the contact conditions the computing parameters ϵ_n and ϵ_t have to be chosen in such a way that the penalization functions are often stiff and thus the numerical problem remains hard to solve (bad behaviors of the algorithms). When penalization is soft, computations are easier but contact conditions are then often roughly fulfilled. Secondly, as noted in Section

In commercial computer codes, regularization parameter values are often proposed as default values, but specific choices of the function $G_\epsilon(u)$ (both the function shape and the parameters ϵ_c) are generally offered and it is recommended to use them. It will be observed that the choices of the penalization parameters improving the quality of the solutions often lead to large computational times. In any cases great care is recommended when regularization is used.

- **Lagrange multiplier formulation** In that case, the contact force is kept as a variable (Lagrange multiplier). It is a mixed formulation. This method permits an accurate determination of the contact forces. We get a saddle point formulation which is solved by using the Uzawa algorithm (min/max optimization). It should be noted that additional variables (the contact forces) have to be introduced in that case (larger number of DOF).

- **Augmented Lagrangian formulation (widely used).** (see Simo-Laursen (114)) It is a combination of the penalty and the Lagrange multiplier formulations. An iterative process is used to reach the correct value for the multiplier (number of augmentations). The augmentation consists in adding a penalization term. The case with only one augmentation corresponds to the usual penalization. We get a variational equation coupled with the conditions of contact and friction (Kuhn Tucker conditions). A Newton Raphson method, which depends on the number of augmentations, is combined with a radial return process for the contact. For friction, stick prediction is done and correction is conducted for sliding when it is needed.

This method is very powerful and very often used in the computer codes developed in academic research laboratories.

We will present now numerical methods solving the strict conditions for unilateral contact and friction. Details can be found in Raous (95), Raous et al (92), Chabrand et al (16), Lebon-Raous (64), Klarbring-Björkman (61).

- Lemke method (complementarity problem). The complementarity formulation of the contact problem complying with the strict contact conditions was given in section 3.4 (see Klarbring-Björkman (61), Raous (95) (90), Cottle et al. (31)).

Methods derived from mathematical optimization, such as the Lemke method or the interior point method (see (20) (61)) can be used. The Lemke algorithm is a mathematical programming method. It is a direct method based on pivoting techniques, similar to the Simplex method.

In elasticity, a condensation of the problem can be conducted in order to write the problem only on the contact variables. This is done by a partial inversion of the total system that can be performed not by inverting the global matrix but by solving (only once at the beginning) a set of linear problems. Details are given in Raous (95), an extension to treat dynamic problems is given in Vola et al (122). Then the Lemke method is applied on the reduced system whose rank depends only on the number of contact nodes. It is however a full sized matrix which is non-symmetric (because of the friction). Using a direct method is very comfortable; the number of pivotings is less than the rank of the matrix (and in practice much more less). However, because of the full sized matrix, this process cannot be used for huge numbers of DOFs. In that case, when subdomain or multigrid methods are used, the Lemkes method turns out to be still very efficient for coarse grids (see Lebon et al. (65)).

- Fixed point on the sliding limit and minimization problem. It has been shown that the variational inequality problem (strict contact conditions) can be set as a sequence of minimization problems combined with a fixed point method on the sliding limit. At each step, we solve a Tresca problem, i.e., a frictional problem where the sliding limit is given. The Trescas law being an associated law, a minimum principle can be associated. The iteration on the sliding limit converges quite fast (less than 10 iterates and often less than 5) depending on the size of the system and on the prescribed accuracy. Details can be found in Raous (95).

We have to solve *Problem* P_{mini} which is a minimization problem under constraints (u belongs to the convex K) of a non-differentiable functional. Various minimization methods with projection can be used:

- Successive Over Relaxation and Projection (SORP) An optimal parameter of relaxation has to be determined using a trial procedure. The method is very robust but can be costly when extension to non-linear problems is considered.

- Gauss Seidel with Aitken acceleration. No numerical parameter is needed.
- Pre-conditioned conjugate gradient with projection (see Raous-Barbarin (93)). This is a very powerful method but a regularization of the friction term has to be done in order to evaluate the gradient of the functional to determine the descent directions.

- **Multigrid methods.** see Lebon-Raous-Rosu (65). As noted before, we have developed multigrid methods for solving contact problems without regularization of the contact, i.e., using the strict Signorini's conditions and the strict Coulomb law.

Multigrid methods operate at several levels of meshes (usually 2 to 5 levels) which are coarser than the initial mesh where the solution will be calculated. This is an iterative process using complete resolutions on the coarsest grid and a few smoothings conducted on the default of equilibrium on the intermediate grids. At each iteration, the Lemkes method can be used on the coarsest grid to solve the small sized problem, and the projected Gauss Seidel method will be a very efficient smoother on the other finer grids.

In all the examples presented in this course, for both quasi-static and dynamic problems, no regularization of the contact conditions was used and the algorithms fitted the non-smooth character of the laws. This is fundamental for modeling specific phenomena such as squeal, which is due to the occurrence of unstable solutions for the theoretical problem.

- **Dynamics problem.** As presented before, because of the non-derivability of the solutions (shocks), the problem has to be formulated in terms of differentiable measures: equation (31).

A classical Newmark method which is based on a limited development of the solution needs their derivability and cannot be used. However, some improvements have sometimes been developed to adapt the Newmark method to the problem (usually some numerical damping). But, we should be aware that the use of numerical damping will kill not only numerical oscillates but also some real solution oscillates (flutter for example).

A specific method called the Non-Smooth Contact Dynamics (NSCD) method was developed in Montpellier by Moreau-Jean-Dubois (52) (53) for discrete problems (granular materials). A version adapted to continuous formulations and FEM was conducted at the LMA in Marseille - Jean et al (54). The system on differential measure (31) can be written in the following equivalent form: $\forall t \in [0, T]$

$$M(\dot{U}(t) - \dot{U}(0)) = \int_0^t (F - K.U - C.\dot{U})ds + \int_{[0,t]} Rd\nu \qquad (38)$$

$$U(t) = U(0) + \int_0^t \dot{U} ds \tag{39}$$

where ds represents the Lebesgue measure. Be given the time discretization: $i = 0...N, ti = i.h$ (h is the time step), (38) is written:

$$M(\dot{U}(t_{i+1}) - \dot{U}(t_i)) = \int_{t_i}^{t_{i+1}} (F - K.U - C.\dot{U})ds + \int_{[t_i,t_{i+1}]} Rd\nu$$

and

$$\bar{R}^{i+1} = \frac{1}{h}\int_{[t_i,t_{i+1}]} Rd\nu.$$

To complete the time discretization the two following Lebesgue integrals must be approximated :

$$\int_{t_i}^{t_{i+1}} (F - K.U - C.\dot{U})ds \text{ and } \int_{t_i}^{t_{i+1}} \dot{U} ds.$$

The choice of the integration methods must be influenced by the fact that the velocity is discontinuous. We have used the following three methods:

- θ-Method : both integrals are approximated by the classical θ-method i.e. where:
  $$\int_{t_i}^{t_i+1} f ds \approx h(\theta f(t_{i+1}) + (1-\theta) f(t_i)) \ ,$$
- θ-Euler-Method: the first integral is approximated by the θ-method and the second one by the Euler implicit method,
- modified θ-Method: both integrals are approximated by the θ-method but in the contact relations the displacement $u(t_{i+1})$ is replaced by $\hat{u}(t_{i+1}) = u(t_{i+1}) + h(1-\theta)\dot{u}(t_{i+1})$.

In contact and multibody dynamics, we have to mention the reference books by Pfeiffer-Schindler (88), Pfeiffer (87) and Pfeiffer-Glocker (86).

3.9 Numerical analysis

In the art of modeling, it is important to keep in mind the three steps between the initial mathematical problem and the final solution given by the computer code.

- The initial continuous problem is set in functional spaces (Hilbert spaces in our case; see Section 3.1);
- This problem is then approximated by an approached problem set in finite dimensional spaces; when FEMs are used these finite dimensional spaces are generated by a basis of functions constructed on the finite element mesh; this problem is still written in functional spaces (L^2);

- The discrete problem is deduced from the previous one, which is written in $\Re^N$.

Numerical analysis is applied to the study of the properties of the solutions to these various problems and the relationship between these solutions. It is a mathematical task. An important feature is the convergence (and the order of convergence) of the approximated problem towards the initial continuous problem when the size of the refinement (element size) goes to zero.

Numerical analysis is applied to the study of the convergence of the algorithms and the evaluation of the computational errors.

Numerical analysis is of great importance in the art of modeling and constitutes a huge part of applied mathematics. It will help the engineer to choose and control the numerical methods, the algorithms and the numerical parameters.

Once again, it is not possible to be a specialist in all these topics but bearing the main numerical analysis results in mind will be very helpful to conduct a coherent modeling process.

In contact mechanics, among a number of studies, we can cite those by Glowinski- Lions-Tremolires (46) and Barbara Wohlmut (125).

3.10 Conclusions

The interaction between mechanics and mathematics is of great importance in the art of modeling mechanical systems. It concerns not only the way to set and to write the problems (choice of the formulation) but also the way to solve it. Mathematical results which could sometimes be very tough to analyze are in fact of great interest for understanding some subtleness of the models and of the solutions, and also very helpful for choosing and controlling the numerical methods.

In the light of the mathematical analysis briefly presented in this section, we can stress a few points in contact mechanics modeling:

- Contact problems are relevant to non-smooth mechanics. The non-smooth character of the basic laws (Signorini, Coulomb), beyond the simplicity of these laws, contains the ingredients for modeling many mechanical phenomena observed with frictional contact. This non-smooth character should be preserved and studied properly; regularization should be avoided as much as possible because using regularization may result in some fundamental mechanical properties being lost.
- When the friction coefficient μ is large, the problem becomes ill-posed and things take a bad turn, both for theory and for computations. Multiple solutions may occur and it is even possible to construct examples with a few degrees of freedom showing the existence of several

solutions when μ is large. The convergence of some of the algorithms is distorted when μ is large.

- Therefore, when it comes to friction modeling it is very important to keep in mind that: - when the friction coefficient μ tends to zero, the problem will tend towards a tangential free boundary condition, as intuition suggests, - but when friction μ tends to infinity, the problem does not tend towards a tangential clamped boundary condition (as intuition could suggest) but towards a very bad conditioned problem where strange phenomena may occur (multiplicity of solutions, flutter, divergence, etc.). For example, we will show in the next section that a "large μ" makes easier the occurrence of instabilities

3.11 Friction instabilities: a model to reduce squeal of a rubber-like waist seal sliding on a car window

With this example we would like to emphasize various points earlier addressed about the art of modeling:

- how much theoretical and mathematical analysis is helpful to solve industrial problems,
- making it as simple as possible when choosing the model (a constant friction coefficient is used),
- avoiding regularization in order to preserve the non-smooth character of the laws which is fundamental here (strict contact conditions are prescribed),
- using convenient tools for solving the Non-Smooth Contact Dynamic problem (the NSCD method).

- The industrial problem. The problem is to optimize the design of a waist seal in order to reduce or to avoid the squeal phenomena observed during the sliding of this waist seal on a lateral car window. Mainly, we have three constitutive characteristics to optimize: shape of the waist seal, rubber, varnish (that is, the friction coefficient). Because of the strong non-linearities of the problem, optimization through an experimental trial/error process like the one mentioned in Section 1.1 cannot be used (too many experiments to perform).

Modeling has two goals: understanding the reasons for the occurrence of the nuisance and constructing a simulation tool to help the engineer in the design of a new seal.

- Model and assumption. Details on this study conducted within a collaboration with J.A.C. Martins at the IST of Lisbon can be found in Raous et al (98), Martins et al (72), Vola et al (123). See also the work conducted by N Guyen Quoc Son (80) and Franck Moirot and Xavier Lorang on stability analysis with application to brake squeal.

Our work was done in several stages.

- First a mathematical analysis of friction stability was conducted in (72). Theorems giving the sufficient or necessary conditions for instability to occur (under a set of assumptions) in linear elasticity were established.
- Secondly, an extension of the analysis of the stability in finite deformation and nonlinear elasticity (Mooney-Rivlin) was established in (123). This was again a theoretical work. Computational tools for solving this nonlinear and non-smooth problem were developed on the basis of the NSCD method (see Section 3.8).
- The theoretical conditions for the instability to occur was interpreted on the discrete problem as generalized eigenvalue analysis on certain matrices which depend on the contact condition and have to be updated all along the solution evolution.
- Finally, the model was applied to the industrial problem.

- Mathematical analysis of instability for frictional contact. The analysis of instability for frictional problems is a difficult and still very open problem. In the present contribution, we set a basic assumption which is that there is no change in the sliding condition when instability occurs. This means that a point sliding in one direction can oscillate but without reverse sliding, only with changes in the velocity amplitude. This assumption permits a local linearization for conducting the stability analysis. This assumption seems to be very restrictive but it will be observed finally that when the dynamic solution is computed, instability occurrence satisfies the condition given by the following theorems. This suggests that these conditions could be considered quite optimal despite this initial very restrictive assumption.

The regular solutions are computed. In (72), we are interested in the initial condition problem, i.e., a Cauchy problem. In (123), we are interested in the steady sliding solution for modeling the sliding of a waist seal on a window glass. In order to study the stability of these solutions we introduce a perturbation and we evaluate the perturbed solution. The stability analysis depends on the contact status (no-contact nodes, sliding nodes, stuck nodes) either of the steady sliding solution or at every time steps of the regular solution. In every case, the possible growth of the perturbed solution is computed by considering the admissible directions depending on the contact conditions. Generalized eigenvalue problems are obtained. Details of the stability analysis can be found is the papers previously referenced. A series of theorems giving either necessary or sufficient conditions for flutter or divergence instability are then established.

- Numerical analysis of the stability. These theorems are interpreted on the discrete problem (FEM) as the analysis of generalized eigenvalue problems set on matrices which depend on the contact status of the solution (non-contact nodes, sliding nodes and stuck nodes).

The discussion is conducted both on the rate growth of the perturbed solution (real positive part of the complex eigenvalues) and on the frequency of the flutter (imaginary part of the eigenvalues).

Let us summarize the modeling process:

- let us first emphasize that we use contact and friction models without regularization and with a constant friction coefficient
- we compute the regular solutions, either Cauchy solution or steady sliding solution (small or finite deformations, linear or nonlinear elasticity, etc.)
- stability analysis is conducted on the generalized eigenvalue problems set on matrices depending on the contact condition.
- in order to evaluate the optimality of the conditions, computation of the dynamic solution using the Non Smooth Contact Dynamics method is conducted in the situation where instability is predicted to occur by the theory.

- Application to the waist seal Two specimens of waist seals are considered (see Fig. 37). For each, the possible occurrence of squeal is studied relative to the variation of the main constitutive parameters (especially the friction coefficient). This consists in analyzing the generalized eigenvalue spectra. Validation of the model is conducted by comparing the model results and the results of experiments conducted by Renault. We will compare the two specimens presented in Fig. 37 and evaluate for each of them the possibility of squeal to occur by analyzing the eigenvalue spectrum.

First the steady sliding solution is computed for a glass window moving up and moving down for various values of the friction coefficient (finite deformations, rubber nonlinear elasticity, unilateral contact and friction) given in Fig.37. For each we need to build some matrices depending on the contact condition (see (96), (72), (123) for details). Generalized eigenvalue problems are solved. To analyze the flutter instability occurrence, the imaginary part is interpreted in terms of frequencies (see Fig. 38 and 39), the real part in terms of growth rates. The eigenmodes are also computed. A large number of results are given in (123). We only present on Fig. 38 and 39 the spectra of the flutter instability which may occur for each of the two geometries in function of the values of the friction coefficient. It can be noted that in the acoustic range (100Hz 14 000Hz), geometry 1 generates a flutter vibration (2000Hz and 14 000 Hz) for $\mu > 0.3$, although in the same range $0.3 < \mu < 0.45$ geometry 2 is stable, flutter does not occur.

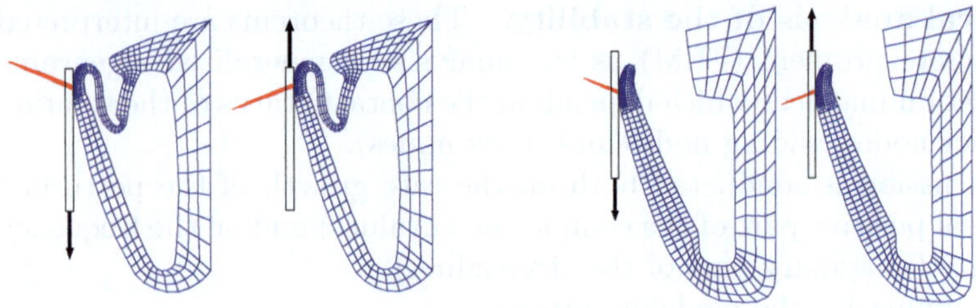

Figure 37. The two specimens: steady sliding solutions (deformation and forces) for a glass window moving up and moving down

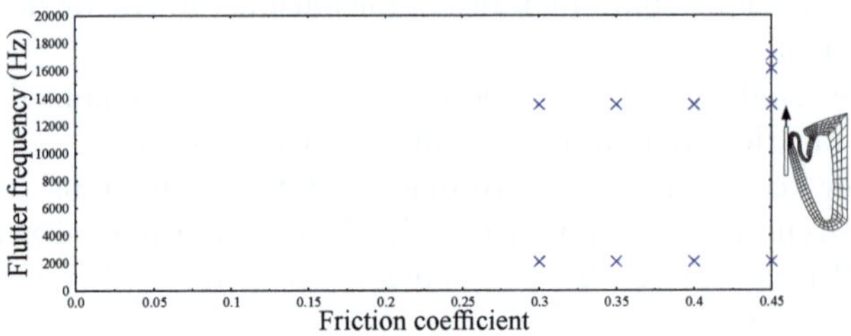

Figure 38. Flutter frequencies (geometry 1, glass moving up, $\mu = [0, 0.45]$).

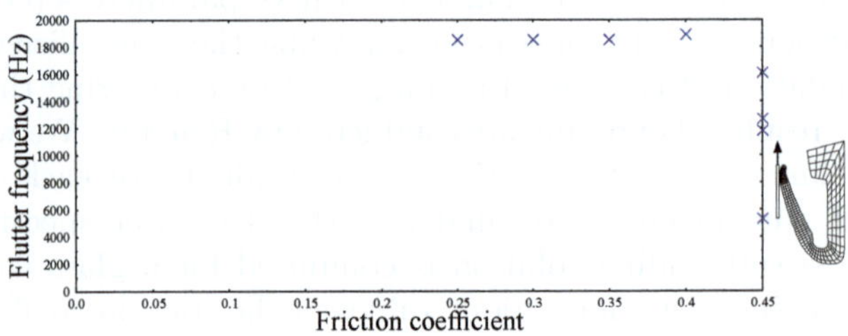

Figure 39. Flutter frequencies (geometry 2, glass moving up, $\mu = [0, 0.45]$).

Then, when the dynamics evolution of the perturbed solution is computed using the NSCD method for $\mu = 0.4$ for the two geometries, a growing oscillating solution (flutter) for geometry 1 and a stable solution for geometry 2 can be observed (see Fig. 40 and Fig. 41. This confirms that despite the very strong assumption made at the beginning of the stability analysis, the stability conditions seem to be quasi optimal. This is validated by the experiments which show that geometry 2 is more stable (no squeal) than geometry 1 and when flutter occurs with geometry 1, the frequency of the squeal noise is about 2000Hz.

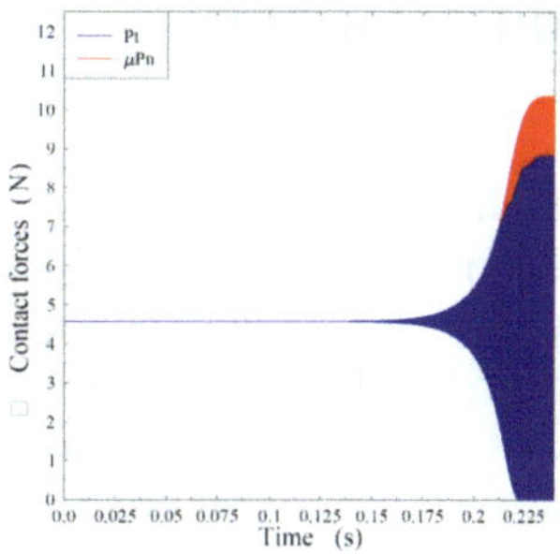

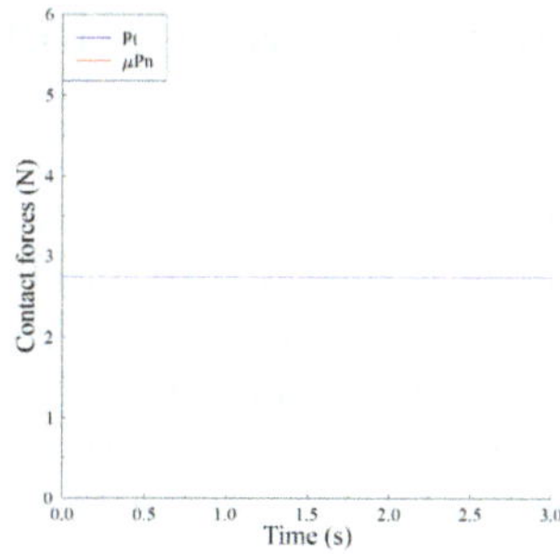

Figure 40. Evolution of the contact forces (geometry 1, glass moving down, $\mu = 0.4$, $\theta = 0.55$).

Figure 41. Evolution of the contact forces (geometry 2, glass moving down, $\mu = 0.4$, $\theta = 0.55$).

Conclusion With this example, we show that a mathematical analysis of stability with a model with constant friction coefficient, associated with a convenient formulation (differentiable measures for the dynamics) and with a convenient numerical method (NSCD) makes it possible to characterize friction instabilities. We show that considering a variable friction coefficient is not a necessary condition for modeling stick-slip or instability, as it has very often been asserted. We showed that the ingredients contained in the Coulomb law are sufficient to characterize this phenomenon, provided that the formulation and the numerical resolution would comply with the non-smooth character of the law.

Using this model with constant friction coefficient refers to the idea of making it as simple as possible when building a model. Introducing a variable friction coefficient would include extra parameters that are poorly known and would have disturbed the analysis.

Another important point to note is that regularizations are risky. In particular stability analysis would have been impossible if either a penalization for the frictional contact or a time regularization for the shocks had been used. One reason is that regularizations kill the oscillations and another is that the stability analysis would be completely dependent on the values of the regularization parameters!

It was necessary to preserve the non-smooth character of the contact law and to treat it correctly - which refers to the previous assertion: using the rights tools appropriate to the solution properties.

4 Identification, validation and validity domain - examples

Model validation constitutes a fundamental step in the model-building process before using the model for predicting new behaviors. Parameter identification is an intermediate step which only shows that there exists at least one set of parameters that provides a good description of a reference experiment, no more. The validation must show that the model using the previously determined constitutive parameters can give a good simulation of other experiments than the one used for the identification. Then, prediction consists in using the model in situations where no experiments have been carried out. The examples considered (geometries, load amplitude, load velocities, etc.) have to be compatible with the founding model assumptions (validity domain). As presented below, contact mechanics often requires specific procedures. This will be illustrated by several examples.

4.1 Validation of the numerical methods

The first issue is the validation of the numerical method used to solve the problem. It consists in verifying that the numerical tools have been correctly chosen and implemented, and evaluating the errors they introduce.

In contact mechanics, we need to check first if the solution fulfills the contact conditions: no penetration (or small penetration if penalization is used), only traction on the contact boundary, condition $\|R_t\|/|R_n| = \mu$ when sliding occurs, etc. Besides that, as usual, a more complete analysis has to be conducted to check the ability of the numerical method to correctly solve the problem.

The validation of a numerical method could be conducted using:

- an analytical solution when it is possible (it is the best way but is not always possible with a complex model),
- a Benchmark which is a reference example chosen by a user community which permits comparisons of the results obtained using various computational codes as well as discussions between developers. This means that a consensus has emerged within the scientific community to consider a solution as a reference solution. The characteristics of a Benchmark should be the same as those of the problem under consideration: kinematics, material behavior, load type, etc.

Analytical solutions for contact problems are rare and limited to basic cases: small deformations, elastic behaviors, simple geometries (sphere, cylinder, contact with a half-plane), simple loadings (static contact, indentation, etc.) and often frictionless contact. In contact mechanics, very often we will have to use benchmarks.

This validation of the numerical method is essential. It is absolutely required when developing a new method, and it is also recommended and very useful when using an existing method (in a commercial code for example). Furthermore it also constitutes a way to evaluate the performances of the algorithms:

- to test the convergence,
- to evaluate the errors (and to compare them with the theoretical estimates),
- to adjust the computational parameters and test the sensitivity of the solutions to their variation,
- to compare the efficiency of various methods.

Example: validation of the computational methods for solving the contact problem of a rubberlike structure with unilateral contact and friction. An extension of the NSCD method presented in Section 3.8 was developed to compute the dynamic solution of the problem presented in Section 3.6. This new method was built to solve problems of unilateral contact with Coulomb friction (without regularization) for a rubberlike material with large deformations and non-linear elasticity (with incompressibility condition). The problem is too complex to provide a reference analytical solution which can be used to validate the method. Two Benchmarks were considered, one for quasi-static loading and one for dynamic loading.

- Computation of the quasi-static compression of a cylinder. This Benchmark was proposed and used by Simo-Taylor (113), Sussman-Bathe (117) and Liu-Hofsetter-Mang (68). Cases without friction and with friction ($\mu = 0.2$) were considered (see Fig. 42, 43 and 44)

There is very good agreement between our results and those obtained by Sussman-Bathe (117) (displacement/hydrostatic pressure formulation) and Simo-Taylor (113).

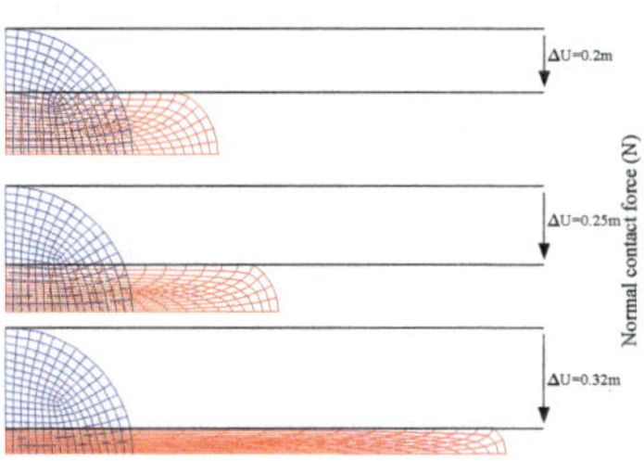

Figure 42. Deformations for the frictionless case

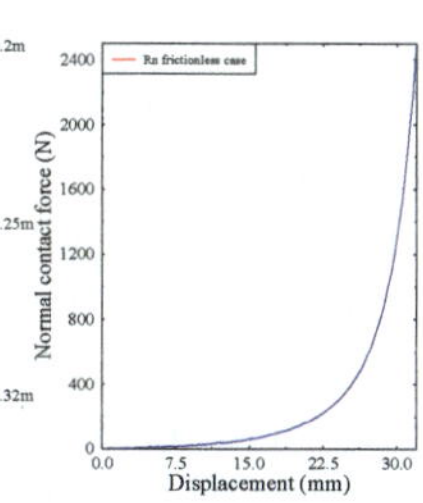

Figure 43. Resultant contact forces for μ=0

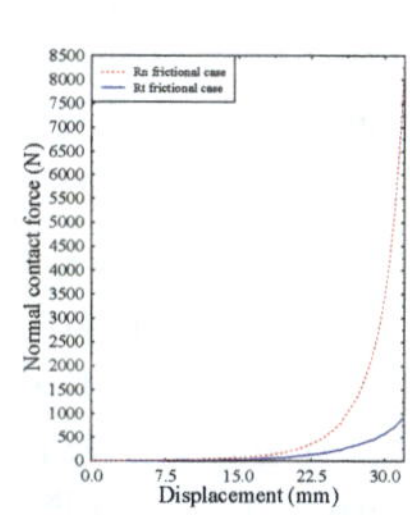

Figure 44. Resultant contact forces for μ=2

- Dynamic impact of a cylinder inside a cone. This Benchmark was proposed by Wriggers et al (126). It is the impact of a rubberlike cylinder inside a rigid cone. With these examples, we stressed the importance of validating the numerical methods associated with a model and the difficulties finding a reference test in the case of complex problems.

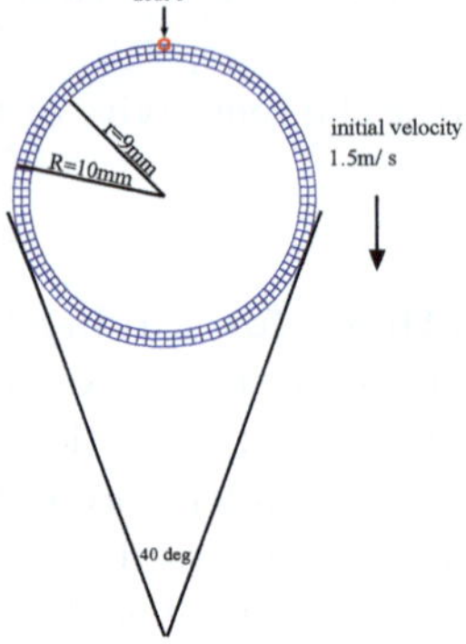

Figure 45. impact of a cylinder into an angle

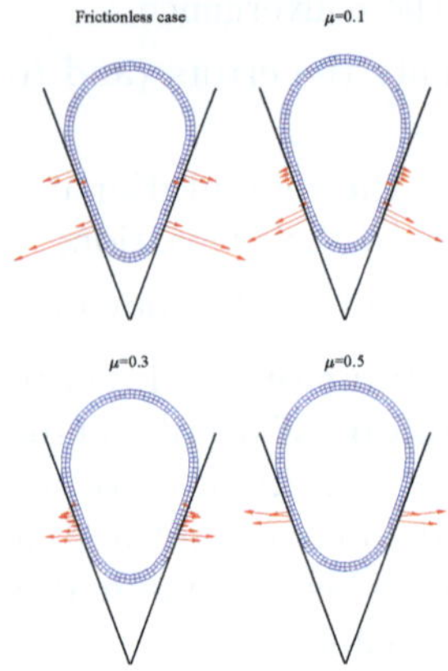

Figure 46. Deformation of the cylinder and contact forces for various friction coefficients.

4.2 Identification of the constitutive parameters.

As earlier noted, mechanics of materials and structures, it is usually possible to determine the constitutive parameters with preliminary and standard experiments conducted on specimens. However, for contact problems, the identification of the constitutive parameters (and especially the friction coefficients) is very difficult because of the strong dependence on the environmental conditions (temperature, surface condition, residual lubricant, etc.). Special methodologies are needed; for example conducting the identification on the mechanical problem itself (the complete structure) taking a given loading situation as reference. Once this is done, the validation of the model will be conducted on other cases of loadings by using the parameters determined during the identification.

It should be kept in mind that identification of the constitutive parameters is an important and difficult task. It is a key point in the art of modeling. It is a problem of minimization of the error between the experiment and model results. As said earlier, the choice of error norm is very important (L^2 or L^∞ for example). This minimization problem does not have a unique solution (a priori no convexity property of the functional to minimize). We have to deal with local minima and so the search for the correct parameters has to be enriched by estimations of ranges of values based on mechanical considerations. We will illustrate this with a few examples in Sections 4.4, 4.5 and 4.6.

4.3 Validation of the model.

Now that the numerical methods are validated and the constitutive parameters evaluated, model validation will consist in verifying whether the model using the previously determined parameters is able to describe other experiments. The experimental conditions have to satisfy the constraints created by the validity domain conditions of the model. The norm used to evaluate the error between simulation and experimental results should be the same as the one used for the identification. This will be commented and illustrated with the following examples.

4.4 Example in Mechanics of Materials: cyclic behavior of polymeric foam.

The construction of a model for the cyclic behavior of polymeric foam was presented in Section 1.9. Here, we give details about the identification step and highlight some specific ways to proceed (see Pampolini-Raous (82)). The model uses a number of parameters (to characterize viscosity, nonlinear elasticity and damage) and we will stress two points:

- the number of parameters should be minimized (as few as possible and as many as necessary): to characterize the viscosity, we will minimize the number of relaxation times to be taken into account,
- mechanical considerations should be included in selecting the parameters of the nonlinear elastic model, instead of conducting a blinded minimization process.

- Identification of the viscosity parameters. The relaxation times characterizing the viscosity were identified using extra relaxation experiments conducted on foam specimens. The relaxation curves give the evolution of the stress when an initial deformation is prescribed and kept constant. The problem is to determine the number of Zener/Maxwell moduli and the values of the associated constitutive parameters, which will make it possible to simulate these relaxation experiments properly. It can be shown that the problem reduces to approximate the experimental curves by a series of exponentials. The corresponding minimization problem is solved by combining an interpolation method (Prony method) and a minimization method (Hooke and Jeeves). For a model to be as simple as possible, we start with one modulus and make the identification, i.e., we determine the best values of the parameters to minimize the gap between theory and experiment. This gap remains large when only one modulus is used and so we progressively increase the number of moduli in order to take the smallest

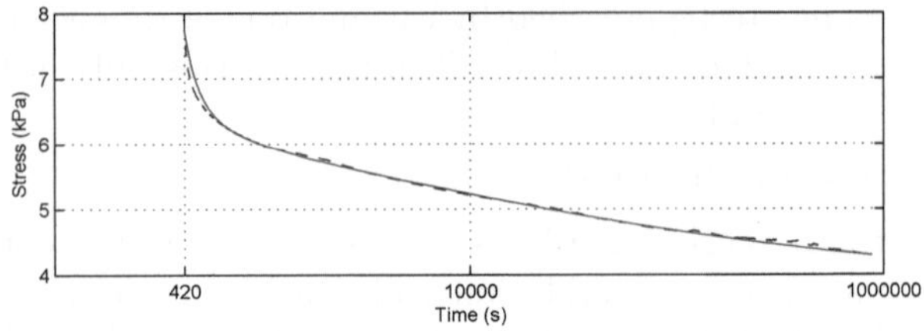

Figure 47. Relaxation experiment lasting 10 days: comparison of a model with 4 exponentials (dotted line) and 5 exponentials (full line) with the experimental curve which is indistinguishable from the full line one.

number of moduli needed to get a good approximation of the relaxation curve. This is presented in Fig. 47. Long period effects (recovery effect after a resting period) and short ones (during a cycle) coexist. Thus, very different relaxation times were identified. It turned out that five Zener/Maxwell moduli (i.e. ten parameters) were sufficient (and necessary!) to get a good description of the viscoelastic effect.

- Identification of the parameters of the nonlinear elasticity. Once the viscosity parameters were identified using the relaxation experiments, those of the nonlinear elastic springs were identified for the first loading cycle. Six parameters characterize the shape of the strain energy given in Fig. 6 : c, μ, m, β, k, a. Instead of conducting a blind optimization, we analyzed first the role of each constant in the response curve (σ_e, ϵ) of the nonlinear spring (Fig. 6 in Section 1.9). It can be seen from Fig. 48 that:

- c determines the initial slope and the value of the local maximum - Fig. 48(a);
- μ determines the second ascending branch - Fig. 48 (b);
- m determines the descending branch, the position and the value of the local maximum - Fig. 48 (c);
- when β increases, σ_e decreases when $\epsilon \simeq a$ - Fig. 48 (d);
- k determines the slope of the curve when $\epsilon \simeq a$ - Fig. 48 (e);
- a determines the position of the local minimum - Fig. 48 (f).

After this preliminary analysis, the identification procedure was performed on numerical simulations of a complete loading/unloading cycle (Fig. 49). The following procedure was used for this purpose:

1. select c to obtain the initial slope of the experimental curve;
2. select m (with c being fixed) to obtain the appropriate value of the force at the beginning of the plateau regime;
3. select μ (with c and m being fixed) to obtain the appropriate value of the force at the end of the loading process ($\epsilon = 0.7$);
4. perform an optimization routine with the Hooke and Jeeves method to determine the values of a, β and k.

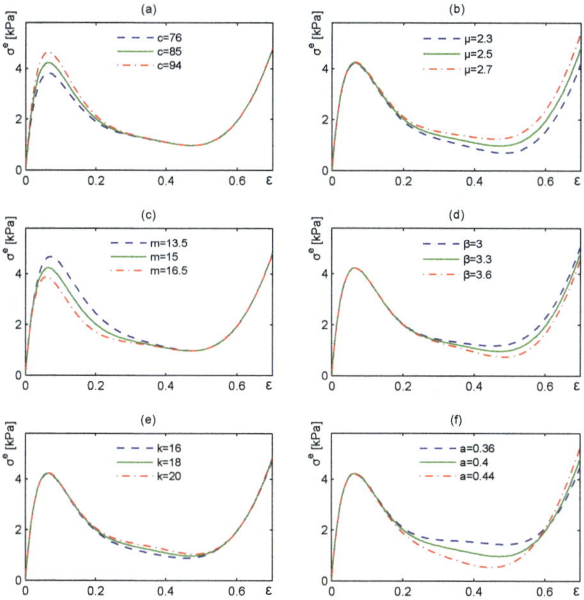

Figure 48. Nonlinear elasticity model: influence of the constitutive parameter variation on the stress/strain curve.

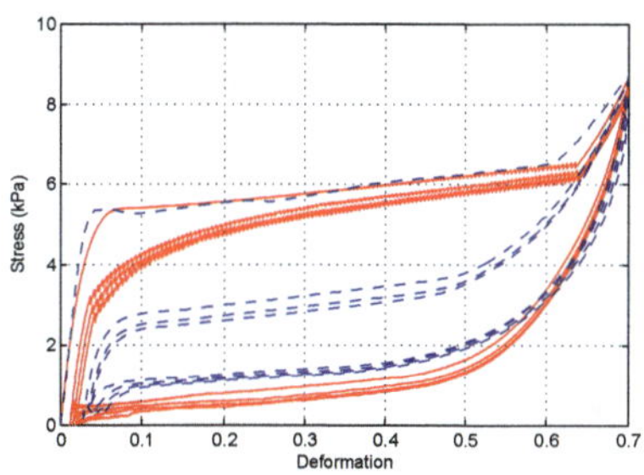

Figure 49. Theory/experiment comparison when using the nonlinear elasticity model with viscosity

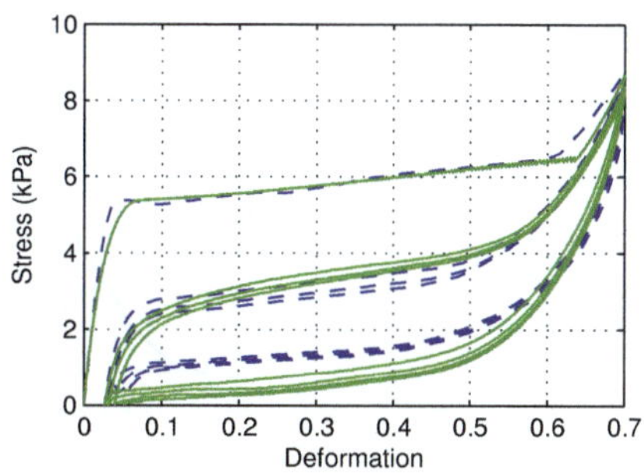

Figure 50. Theory/experiment comparison when using the model with nonlinear elasticity, viscosity and damage.

The parameter characterizing the damage evolution is then identified from multi-cycle experiments. The final results are given in Fig. 49 and Fig. 50 (model results in full line; experimental ones in dotted line).

- Validation of the model In this example, model validation was achieved through simulation of complex loadings using the same values of the constitutive parameters (those resulting from the identification): cycles with various amplitudes, cycles with intermediate unloadings, etc. Results were presented in Section 1.9.

4.5 Example in Mechanics of Materials : fiber/matrix interface in a composite material

With this example of micro-indentation of a single fiber in a SiC/SiC composite, which was presented in Section 2.4, we stress the importance of the mechanical analysis during the identification process.

This experiment was conducted in order to identify the constitutive parameters of the RCCM model which was developed to describe the behavior of a fiber/matrix interface. As presented in Section 2.4, the RCCM model has four parameters: the friction coefficient μ, the initial stiffness of the interface C ($C_n = C_t$), the adhesion energy ω (the Dupré energy) and the interface viscosity b.

Again, the identification procedure was not performed as a blind optimization. Preliminary studies were conducted to get a range of admissible values for each parameter. Extra experiments and energetic analysis were conducted: creep analysis (when the prescribed displacement keeps a constant value at the end of the indentation) and cyclic loading/unloading displacement of the indenter. These preliminary studies based on mechanical

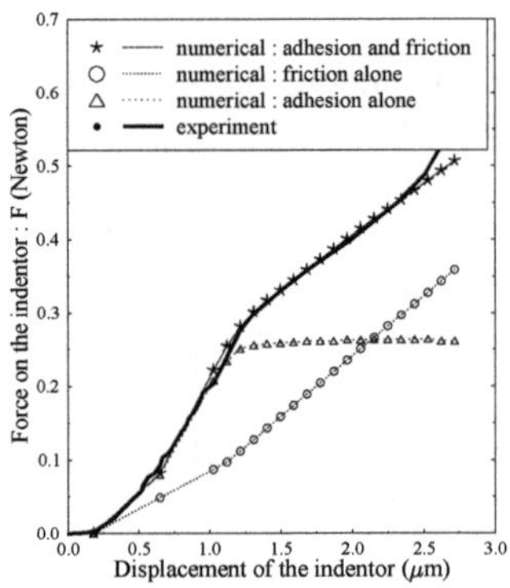

Figure 51. Identification of the constitutive parameters on an indentation experiment - Experiment (full line) - RCCM model (star line)

considerations and some elementary computations provide:

- an estimate of the initial stiffness of the interface based on the thickness and the properties of the third body present in the interface (the fiber envelop which is a pyrolytic carbon);
- an estimate of the friction coefficient based on the analysis of dissipation during a cycle;

- a range of possible values for the adhesion energy ω based on the values of Dupré energy for these materials and energetic analysis;
- an estimate of the order of magnitude of b through experiments with different values of the indentation velocity coupled with the creep analysis.

On the basis of these preliminary estimates, the precise values of the parameters were identified using the experimental curve given in Fig. 51. Details can be found Raous-Cangémi-Cocou (95) (96).

- Validation. Validation of the model was conducted through experiments with fibers of various diameters and different loadings.

4.6 Example in Civil Engineering: steel/concrete interface in reinforced concrete

With this example, we illustrate the ultimate step of the construction of a model (see Section 1.7): when the best simulation that can be obtained after identification of the parameters is not satisfactory, the model needs to be improved, i.e., new ingredients have to be introduced, and extra phenomena taken into account.

This is a civil engineering problem: modeling the behavior of the steel-concrete interface in reinforced concrete. The RCCM model was used to simulate the interface and an experiment consisting in pulling out a steel bar embedded in a concrete specimen was conducted (see Raous-Karray (101) (57)). The mesh is presented in Fig. 52.

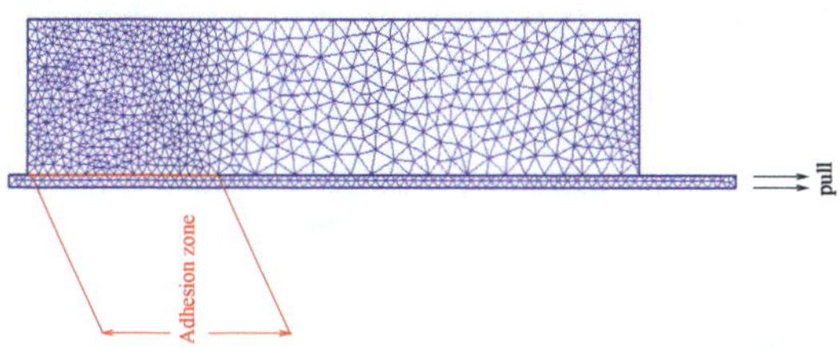

Figure 52. Pull-out of a steel bar from a concrete specimen experiment: the mesh

- Identification. During the identification process, using the standard RCCM model it was impossible to determine the values of the four parameters (μ, C, ω, b) which could give a correct simulation of the experiment: we got either a good description of the peak or a good description of the asymptote (see Fig.54). The friction coefficient μ was the key parameter. Therefore, we introduced a friction coefficient depending on the sliding displacement (see Fig. 53). From a mechanical point of view, this corresponds to taking into account a wear phenomenon corresponding to the grinding of the interface when sliding occurs, due to the powdery nature of concrete. The debris generated at the interface act as lubricant and the friction coefficient decreases when sliding occurs.

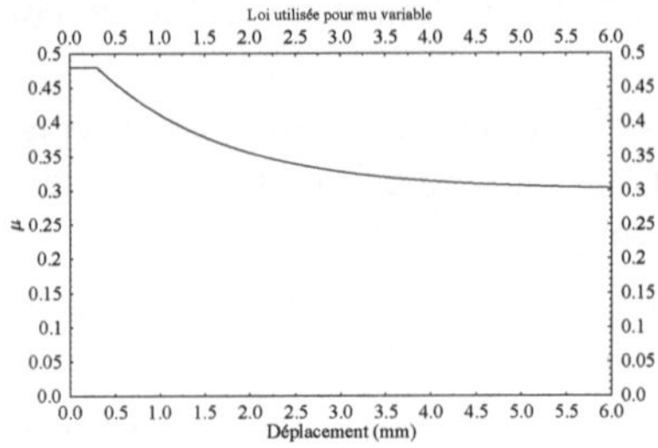

Figure 53. Variation of the friction coefficient related to the sliding displacement

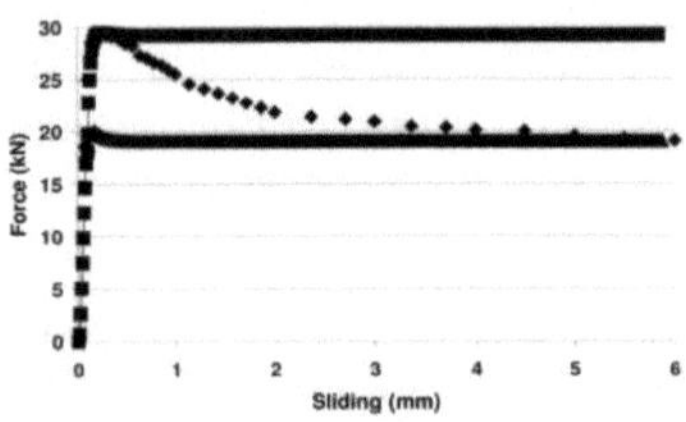

Figure 54. Approximation of the experimental curve (cross line) by the regular RCCM model with $\mu = 0.28$ and $\mu = 0.46$.

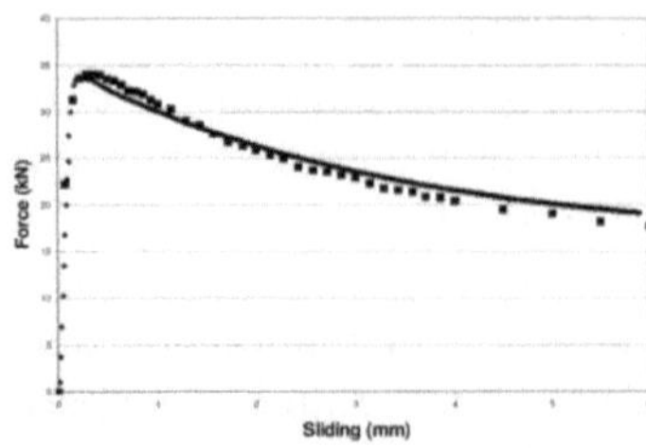

Figure 55. Approximation of the experimental curve (square line) by the model with variable friction coefficient

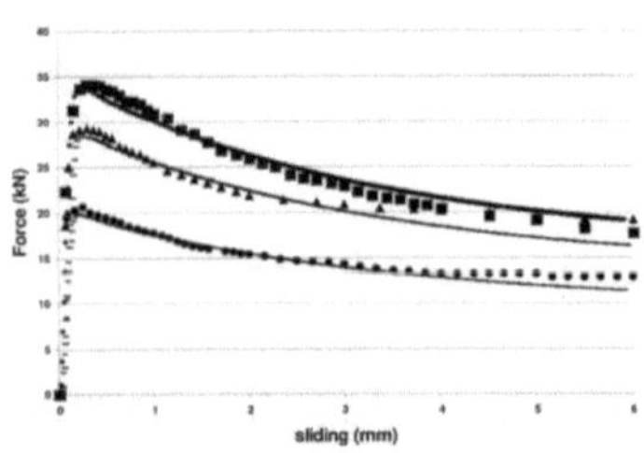

Figure 56. Simulations (full line) and experimental results for 3 rods of diameter 14mm (squares), 12mm (triangles) and 10mm (circles).

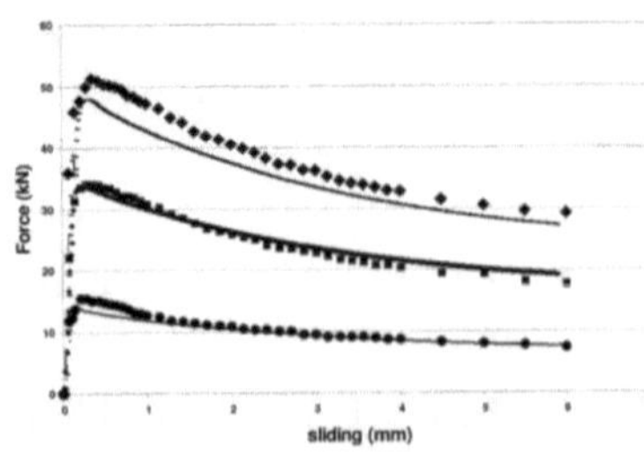

Figure 57. Simulations (full line) and experimental results for 3 lengths of the adhesive zone: 15cm (diamonds), 10cm (squares) and 5cm (circles).

Then, it was possible to find values of the constitutive parameters C, ω, b and those defining the variation of μ given in Fig. 53 to obtain a very good approximation of the experimental result (see Fig. 55).

- **Validation.** Validation of the model was performed by conducting experiments on rods of different diameters and on specimens with contact interfaces of different lengths . The results presented in Fig. 56 and 57 show the good agreement between the model simulation and the experimental results. This validates the model and in particular the choice of the function $\mu(u_t)$ given in Fig. 53.

4.7 Conclusion

Using these examples, we stressed that identification of the constitutive parameters of a model could be a difficult task (especially in contact mechanics). It is not a simple optimization problem. It is very important to take into account the mechanical meaning of the parameters as well as the range of their possible values. Complementary experiments and specific protocols may be needed (in Section 4.4, both relaxation and cyclic loadings were used and an ordered sequence was defined for the parameter determination).

In contact mechanics, identification and validation are often conducted with the same kind of experiments and special care should be brought to the processes which are conducted and to the experiments which are chosen in the two cases.

As said at the very beginning, identification of the constitutive parameters and validation of the model are two key steps to get an efficient model for simulating and predicting the behavior of a mechanical system. It is sometimes considered that the only noble part in the art and craft of modeling is the construction of the model.

What we have tried to show in this last chapter is that not only are constitutive parameter identification and model validation two key points in the art of modeling, they also are real scientific approaches.

Bibliography

[1] P. Alart, A. Curnier, *A mixed formulation for frictional contact problems prone to Newton like solution methods*, CMAME, 92(3),353-375, 1991.

[2] G. Anciaux, J.-F. Molinari, *Contact mechanics at the nanoscale, a 3D multiscale approach*, IJNME, 79(9), 1041-1067, 2009.

[3] G. Anciaux, S. B. Ramisetti, J.-F. Molinari, *A finite temperature bridging domain method for MD-FE coupling and application to a contact problem*, CMAME, 205-208(0), 204-212, 2012.

[4] L.-E. Andersson, *A quasistatic frictional problem with normal compliance*, Nonlinear Anal. Theory Methods Appl., 16, 347-369, 1991.

[5] L.E. Andersson, *Existence results for quasistatic contact problems with Coulomb friction*, Appl. Math. Opt., 42(2), 169-202, 2000.
[6] J.F. Archard, *Contact and rubbing of flat surface*, J. Appl. Phis., 24(8), 981-988, 1953.
[7] J. F. Archard, W. Hirst, *The wear of metals under unlubricated conditions.* Proc. Royal Society, A-236, 397-410, 1956.
[8] J. F. Archard, *Elastic deformation and the laws of friction*, Proc. Royal Society of London, A-243, 190-205, 1957
[9] J. F. Archard, *Surface topography and tribology*, Tribology International, 7, 213-220, 1974.
[10] P. Ballard, *A counter-example to uniqueness in quasi-static elastic contact problems with friction*, Int. J. Eng. Sci., 37, 163-178, 1999.
[11] P. Ballard, *The dynamics of discrete mechanical systems with perfect unilateral constraints*, Arch. Rational Mech. Anal., 154, 199-274, 2000.
[12] P. Ballard, St. Basseville, *Existence and uniqueness for dynamical unilateral contact with Coulomb friction: a model problem*, Math. Model. Num. Anal., 39(1), 59-77, 2005.
[13] Y. Berthier, *Third body reality - Consequences and use of the third body concept to solve a friction and wear problems*, in *Wear, Materials, Mechanisms and Practice*, Wiley, 2005.
[14] A. Bizzarri, M. Cocco, *Slip-weakening behavior during the propagation of dynamic ruptures obeying rate and state dependent friction laws*, J. Geophys. Res., 108(B8), 2373, 2003.
[15] G. Bouchitte, A. Lidouh, J.-C. Michel, P. Suquet, *Might boundary homogenization help to understand friction*, in Proceedings of Contact Mechanics International Symposium, Curnier (Ed.), Presses Polytechniques, Lausanne, 1992.
[16] P. Chabrand, F. Dubois, M. Raous, *Comparison of various numerical methods for solving unilateral contact problems with friction*, Mathematical and Computer Modelling, 28(4-8), 97-108, 1998.
[17] M. Campillo, I. R. Ionescu, *Initiation of antiplane shear instability under slip dependent friction*, J. Geophys. Res., 102 (B9), 20363-20371, 1997.
[18] G. P. Cherepanov, A.S. Balakin, V.S. Ivanova, *Fractal fracture mechanics A review*, Engineering Fracture Mechanics, 51(6), 997-1033, 1995.
[19] J. Cho, T. Junge, J.-F. Molinari, G. Anciaux, *Toward a 3D coupled atomistic and discrete dislocation dynamics simulation: dislocation core structures and Peierls stresses with several character angles in FCC aluminum*, Adv. Mod. Simul. Engng, 2(12), 2015.
[20] P. W. Christensen, A. Klarbring, J. S. Pang, N. Stromberg, *Formulation and comparison of algorithms for frictional contact problems*, Int. J. Numer. Meth. Eng., 42, 145-173, 1998.
[21] M. Cocou, *Existence of solutions of Signorini problems with friction*, Int. J. Engng. Sci., 22(5), 567-575, 1984.
[22] M. Cocou, E. Pratt, M. Raous, *Existence d'une solution du problème quasi statique de contact unilateral avec frottement non local*, CRAS Paris, 320 Serie I, 1413-1417, 1995.

[23] M. Cocou, E. Pratt, M. Raous, *Formulation and approximation of quasistatic frictional contact*, Int. J. Engng. Sci., 34(7), 783-798, 1996.

[24] M. Cocou, E. Pratt, M. Raous, *Constructive aspects of functional analysis for the treatment of frictional contact*, Math. Comp. Mod., 28(4-8), 109-120, 1998.

[25] M. Cocou, L. Cangémi, M. Raous, *Approximation results for a class of quasistatic contact problems including adhesion and friction*, in *Proc. IUTAM Symposium on Variations de domaines et frontières libres en mécanique des solides, Argoul-Frémond-Nguyen (Eds)*, Kluwer, 211-218, 1999.

[26] M. Cocou, R. Rocca, *Existence results for unilateral quasistatic contact problems with friction and adhesion*, Math. Mod. Num. Anal., 34, 981-1001, 2000.

[27] M. Cocou, M. Raous, *Implicit variational inequalities arising in frictional contact mechanics : analysis and numerical solutions for quasistatic problems*, in *From convexity to non-convexity dedicated to memory of Prof. G. Fichera, Gilbert-Panagiotopoulos-Pardalos(Eds)*, Kluwer, Dordrecht, 255-267, 2001.

[28] M. Cocou, *Existence of solutions of a dynamic Signorinis problem with nonlocal friction in viscoelasticity*, ZAMP, 53, 1099-1109, 2002.

[29] M. Cocou, G. Scarella, *Analysis of a dynamic unilateral contact problem for a cracked viscoelastic body*, ZAMP, 57, 523-546, 2006.

[30] M. Cocou, M. Schryve, M. Raous, *A dynamics unilateral contact problem with adhesion and friction in viscoelasticity*, ZAMP, 61, 721-743, 2010.

[31] R. W. Cottle, F. Giannessi, P.-L. Lions (Eds), *Variational Inequalities and Complementary Problems in Mathematical Physics and Economics*, John Wiley, 1979.

[32] C. A. Coulomb, *Théorie de machines simples*, Mémoire de Mathémathiques et de Physique de l'Académie Royale, 10, 161-342, 1785.

[33] R. Dautray, J.-L. Lions, *Analyse mathématique et calcul numérique pour les sciences de l'ingénieur*, Masson, Paris, 1302 pages, 1987.

[34] L. De Laurenzis, P. Wriggers, *Computational homogenization of rubber friction on rough rigid surfaces*, Comp. Mat. Sci., 77, 264-284, 2013.

[35] G. Del Piero, M. Raous, *A unified model for adhesive interfaces with damage, viscosity and friction*, Europ. J. Mech. - A/Solids, 29(4), 496-507, 2010.

[36] G. Del Piero, G. Pampolini, *The influence of viscosity on the response of open-cell polymeric foams in uniaxial compression: experiments and theoretical model*, Continuum Mech. Thermodyn., 24, 181-199, 2012.

[37] L. Demkowicz, J.T. Oden, *On some existence and uniqueness results in contcat problems with non local friction*, Nonlinear Analysis - Theory Methods Applications, 6(10), 1075-1093, 1982.

[38] R. Descartes, *Discours de la méthode pour bien conduire sa raison et chercher la vérité dans les Sciences*, Imprimerie Ian Maire, La Haye, 8 juin 1637

[39] S. Dumont, F. Lebon, M.L. Raffa, R. Rizzoni, H. Welemane, *Multiscale Modeling of Imperfect Interfaces and Applications*, In *Computational Methods for Solids and Fluids, Ibrahimbegovic(Ed)*, CMAME, 41, 81-122, 2016.

[40] G. Duvaut, J.-L. Lions, *Les inéquations en mécanique et en physique*, Dunod, Paris, 1972.

[41] G. Duvaut, *Equilibre d'un solide élastique avec contact unilatéral et frottememt de Coulomb*, CRAS, Paris, 290A, 263-265, 1980.

[42] G.A. Drosopouilos, P. Wriggers, G. Stavroulakis, *A multi-scale computational method including contact for the analysis of damage in composite materials*, Computational Materials Science, 12(95), 522-535, 2014.

[43] C. Eck, J. Jaruseck, *Existence results for the static contact problem with Coulomb friction*, Math. Mod. Meth. Appl. Sci., 8, 445-468, 1998.

[44] G. Fichera, *Problemi elastostatici con vincoli unilaterali: il problema di Signorini con ambigue condizioni al contorno*, Memorie della Accad. Naz.dei Lincei, Classe di Scienze Fisiche, Matematiche e Naturali, 8(7) n2, 91-140, 1964.

[45] M. Fremond, *Adherence des solides*, J. Mec. Theor. Appl., 6(3), 383-407, 1987.

[46] R. Glowinski, J.-L. Lions, R. Trémolieres, *Analyse numérique des inéquations variationnelles*, Dunod, Paris, 1976.

[47] J.A. Greenwood, J.B. Williamson, *Contact of nominally flat surfaces*, Proc. Royal Soc. of London, A 255, 300-319, 1966.

[48] S. Hyun, L. Pei, J.-F. Molinari, *Finite-element analysis of contact between elastic self-affine surfaces*, Physical Review E, 70(2), 2004.

[49] Y. Ida. *Cohesive force across the tip of a longitudinal shear crack and Griffiths specific surface energy*, J. Geophys. Res., 77, 3796-3805, 1972.

[50] J. Jarusek, *Contact problems with bounded friction coercive case*, Czechoslovak Math. J., 33(108), 237-261, 1983.

[51] J. Jarusek, *Dynamic contact problems with given friction for viscoelstic bodies*, Czechoslovak Mathematical Journal, 46 (121), 475-487, 1996.

[52] M. Jean, J.-J. Moreau, *Dynamics in the presence of unilateral contact and dry friction: a numerical approach*, in *Unilateral problems in structural analysis, Del Piero-Maceri (eds)*, CISM Course, 304, Springer-Verlag, Wien, 1987.

[53] M. Jean, *The Non Smooth contact dynamics method*, Comp. Meth. Appl. Mech. Engng, 177, 235-257, 1999.

[54] M. Jean, V. Acary, Y. Monerie, *Non-smooth contact dynamics approach of cohesive materials*, Phil. Trans. Royal Soc. London (A), 359, 2497-2518, 2001.

[55] K. L. Johnson, *Contact mechanics*, Cambridge University Press, 1985.

[56] T. Junge, J.-F. Molinari, *Plastic activity in nanoscratch molecular dynamics simulations* of pure aluminium. Int. J. Plast., 53, 90-106, 2014.

[57] M. A. Karray, S. Barbarin, M. Raous, *Traitement de la liaison béton-acier par un modèle d'interface couplant adhésion et frottement*, Annales Maghrébincs de l'Ingénieur, 18(2), 2004.

[58] N. Kikuchi, J.T. Oden, *Contact problems in Elasticity : a study of variational inequalities and finite element methods*, SIAM, Philadelphia, 1988,.

[59] A. Klarbring, A. Mikelic, M. Shillor, *On friction problems with normal compliance*, Nonlinear Anal. Th. Meth. Appl., 13, 935-955, 1989.

[60] A. Klarbring, *Examples of non uniqueness and non existence of solutions to quasistatic contact problem with friction*, Ingenieur-Archiv, 60, 529-541, 1990.

[61] A. Klarbring, G. Björkman, *A mathematical programming approach to contact problems with friction and varying surfaces*, Comput. Struct., 30(5), 1185-1198, 1988.

[62] K.L. Kuttler, *Dynamic friction contact problems for general normal and friction laws*, Nonlinear Anal. Theory Methods Applications, 28(3), 559-575, 1997.

[63] T.A. Laursen, *Computational contact and impact mechanics*, Springer, 2003.

[64] F. Lebon, M. Raous, *Multibody contact problem including friction in structural assembly*, Computer and Structures, 43(5), 925-934. 1992.

[65] F. Lebon, M. Raous, I. Rosu, *Multigrid methods for unilateral contact problems with friction*, in *IUTAM-Symposium on Computational Methods in Contact Mechanics, Wriggers-Nackenhorst (Eds)*, Springer, 1-16, 2007.

[66] Ch. Licht, E. Pratt, M. Raous, *Remarks on a numerical method for unilateral contact including friction*, Int. Series Num. Math., 101, Birkhuser Verlag Basel, 129-144, 1991.

[67] Ch. Licht, G. Michaille, *A modelling of elastic adhesive bonded joints*, Adv. Math. Sci. Appl., 7, 711-40, 1997.

[68] C.H.Liu, G. Hofstetter et H.A. Mang, *3D finite element analysis of rubber-like materials at finite strains.*, Eng. Comp., 11, 1994.

[69] B. Lorenz, B. N. J. Persson, *Interfacial separation between solids with randomly surfaces: comparison of experiment with theory*, J. Phys. Condens. Matter, 21, 1-6, 2009.

[70] J.A.C. Martins, J.T. Oden, *Existence and uniqueness results for dynamic contact problems with nonlinear normal and tangential interface laws*, Nonlinear Anal. Theory Methods Applications, 11(3), 407-428, 1987.

[71] J.A.C. Martins, J.T. Oden, *Corrigendum of (70)*, Nonlinear Anal. Theory Methods Applications, 12(7), 747, 1988.

[72] J. A. C. Martins, S. Barbarin, M. Raous, A. Pinto da Costa, *Dynamic stability of finite dimensional linearly elastic systems with unilateral contact and Coulomb friction*, Comp. Meth. Appl. Mech. Engng., 177(3-4), 289-328, 1999.

[73] N. Metropolis, S. Ulam, *The Monte Carlo Method*, J. American Stat. Assoc., 44(247), 335-341, 1949.

[74] Y. Monerie, M. Raous, *A model coupling adhesion to friction for the interaction between a crack and a fiber/matrix interface*, ZAMM, 80, 205-209, 2000.

[75] J.-J. Moreau, *Standard inelastic shocks and the dynamics of unilateral constraints*, in: *Unilateral Problems in Structural Analysis, Del Piero-Maceri (Eds.)*, CISM Courses, 288, Springer-Verlag, Wien New York, 173-221, 1985.

[76] J.-J. Moreau, *Unilateral contact and dry friction in finite freedom dynamics*, in [Moreau-Panagiotopoulos (Eds) (77)], 1-82, 1988.

[77] J.-J. Moreau, P. D. Panagiotopoulos (Eds), *Non Smooth Mechanics and Applications*, CISM Courses and Lectures, 302, Springer-Verlag, Wien, 1988.

[78] J. Munoz-Rivera, R. Racke, *Multidimensional contact problems in thermoelasticity*, SIAM J. Appl. Math., 58(4), 1307-1337, 1998.

[79] J. Necas, J. Jarusek, J. Haslinger, *On the solution of the variational inequality to the Signorini problem with small friction*, Boll. U.M.I., 5 (17-B), 796-811, 1980.

[80] Q.S. NGuyen, *Bifurcation and stability in dissipative media (plasticity, friction, fracture)*, Appl. Mech. Rev., 47, 1994.

[81] G. Pampolini, G. Del Piero, *Strain localization in open-cell polyurethane foams: experiments and theoretical model*, J. Mech. Mater. Struct., 3, 969-981, 2008.
[82] G. Pampolini, M. Raous, *Nonlinear elasticity, viscosity and damage in open-cell polymeric foams*, Arch. Appl. Mech, 84, 1861-1881, 2014.
[83] P.D. Panagiotopoulos, *Inequality problems in Mechanics, convex and non convex energy functions and Hemivariational inequalities*, Birkhuser Verlag, Boston Basel, 1985.
[84] L. Pei, S. Hyun, J.-F. Molinari, *Finite element modeling of elasto-plastic contact between rough surfaces*, J. Mech. Phys. Sol., 53 (11), 2385-2409, 2005.
[85] B. N. J. Persson, *Relation between Interfacial Separation and Load: A General Theory of Contact Mechanics*, Physical Review Letters, 99(12), 2007.
[86] F. Pfeiffer, Ch. Glocker, *Multibody Dynamics with Unilateral Contacts*, John Wiley and Sons, New York, 1996.
[87] F. Pfeiffer, *Mechanical System Dynamics*, Springer, Berlin, Heidelberg, 2009.
[88] F. Pfeiffer, Th. Schindler, *Introduction to Dynamics*, Springer, Berlin Heidelberg, 2015.
[89] S. B. Ramisetti, G. Anciaux, J.-F. Molinari, *A concurrent atomistic and continuum coupling method with applications to thermo-mechanical problems*, Int. J. Num. Meth. Engn, 97, 707-738, 2014.
[90] M. Raous, *On two variational inequalities arising from a periodic viscoelastic unilateral problem*, in [*Cottle-Gianessi-Lions (Eds), 1979*], 285-302, 1979.
[91] M. Raous (Ed), *Numerical methods in mechanics of contact involving friction*, Special Issue J. Méca. Th. Appl., 7(suppl. n1), 1988.
[92] M. Raous, P. Chabrand, F. Lebon, *Numerical methods for frictional contact problems and applications*, in [*Raous (Ed.)(91)*], 111-128, 1988.
[93] M. Raous, S. Barbarin, *Preconditioned conjugate gradient method for a unilateral problem with friction*, in *Contact Mechanics, Curnier (Ed)*, Press. Polytech. Univ. Romandes, 423-432, 1992.
[94] M. Raous, M. Sage, *Numerical simulation of the behavior of surface asperities for metal forming*, in *Numerical Methods in Industrial Forming Processes, Chenot-Wood- Zienkiewicz (eds)*, Balkema, 75-180, 1992.
[95] M. Raous, *Quasistatic Signorini problem with Coulomb friction and coupling to adhesion*, in *New developments in contact problems, Wriggers-Panagiotopoulos (Eds)*, CISM Courses and Lectures, 384, Springer Verlag, Wien-New York, 101-178, 1999.
[96] M. Raous, L. Cangemi, M. Cocou, *A consistent model coupling adhesion, friction and unilateral contact*, Comp. Meth. Appl. Mech. Engng., 177(3-4), 1999.
[97] M. Raous, *Constitutive models and numerical methods for frictional contact*, in *Handbook of Materials BehaviorNon linear Models and Properties, Lemaitre (Ed.)*, Academic Press, 777-786, 2001.
[98] M. Raous, S. Barbarin, D.Vola, *Numerical characterization and computation of dynamic instabilities for frictional contact problems*, in *Friction and instabilities, Martins-Raous (Eds)*, CISM Courses and Lectures, 457, Springer Verlag, Wien - New York, 233-292, 2002.

[99] M. Raous, Y. Monerie, *Unilateral contact, friction and adhesion in composite materials: 3D cracks in composite material*, in *Contact Mechanics, Martins-Monteiro Marques (Eds)*, Coll. Solid Mech. Appl., Kluwer, 333-346, 2002.

[100] Raous, M., Schryve, M., Cocou, M., *Restorable adhesion and friction*, in *Nonsmooth/Nonconvex Mechanics with Applications in Engineering, Baniotopoulos (Ed.)*, Ziti Publisher, Thessaloniki, 165-172, 2006.

[101] M. Raous, M.A. Karray, *Model coupling friction and adhesion for steel-concrete interfaces*, Int. J. Comp. Appl. Tech., 34 (1), 42-51, 2009.

[102] M. Raous, G. Festa, J.-P. Vilotte, C. Henninger, *Adhesion and friction for fault interfaces in geophysics*, Keynote lecture in *Mini-Symp. Comput. Contact Mech., IV Europ. Conf. Comput. Mech. ECCOMAS (Sol. Struct. Coupled Pbs Engng.)*, Paris, May 17-21, 2010.

[103] M. Raous, *Interface models coupling adhesion and friction*, Them. Issue: Surface mechanics: facts and numerical models, CRAS Paris, 339, 491-501, 2011.

[104] R. Rizzoni, S. Dumont, F. Lebon, E. Sacco, *Higher order model for soft and hard elastic interfaces*, Int. J. Solids Struct., 51, 4137-4148, 2014.

[105] R. Rocca, M. Cocou, *Existence and approximation of a solution to quasistatic problem with local friction*, Int. J. Engng. Sci., 39(11), 1233-1255, 2001.

[106] R. Rocca, M. Cocou, *Numerical Analysis of quasi-static unilateral contact problems with local friction*, SIAM J. Numer. Anal., 39(4), 1324-1342, 2001.

[107] A. L. Ruina, *Slip instability and state variable friction laws*, J. Geophys. Res., 88 (10), 359-370, 1983.

[108] A. Rutherford, *Mathematical modelling techniques*, Dover Publications, inc, New York, 269 pages, 1978.

[109] R. A. Sauer, *A Survey of Computational Models for Adhesion*, J. Adhesion, 92(2), 81-120, 2016.

[110] M. Shillor, M. Sofonea, J.J. Telega, *Models and Analysis of Quasistatic Contact*, Lect. Notes Phys., 655, 2004.

[111] M. Serpilli, *Mathematical Modeling of weak and strong piezoelectric interfaces*, J Elast, 121, 235-254, 2015.

[112] A. Signorini, *Questioni di elasticita non linearizzata e semi-linearizzata*, Rend. di Matem. delle sue appl., 18, 1959.

[113] J.C. Simo et R.L. Taylor, *Quasi-incompressible finite elasticity in principal stretches. Continuum basis and numerical algorithms.* Comp. Meth. Appl. Mech. Eng., 85, 1991.

[114] J.C. Simo, T.A. Laursen, *An augmented Lagrangian treatment of contact problems involving friction*, Computers Structures, 42 (1), 97-116, 1992.

[115] P. Spijker, G. Anciaux, J.-F. Molinari, *The effect of loading on surface roughness at the atomistic level*, Comp. Mech., 50, 273-283, 2012.

[116] P. Spijker, G. Anciaux, J.-F. Molinari, *Relations between roughness, temperature and dry sliding friction at the atomic scale*, Trib. Int., 59, 222-229, 2013.

[117] T. Sussman et K.J. Bathe, *A finite element formulation for nonlinear incompressible elastic and inelastic analysis.*, Comp. Struc., 26(1-2), 1987.

[118] D. Tabor, *Friction-The present state of our understanding*, J. of Lubri. Techn., Trans. ASME, 103, 169-179, 1981.

[119] I. Temizer, P. Wriggers, *A multiscale contact homogenization technique for the modeling of third bodies in the contact interface*, Comp. Meth. Appl. Mech. Engng., 198, 377-396, 2008.

[120] K. Uenishi and J. R. Rice. *Universal nucleation length for slip-weakening rupture instability under nonuniform fault loading*, J. Geophys. Res., 108 (B1), 2003.

[121] J.-P. Vilotte, G. Festa, M. Raous, C. Henninger, *Earthquake rupture with scale-dependant friction and damage interface law*, in *American Geophysical Union Fall Meeting*, USA, Dec 2009.

[122] D. Vola, E. Pratt, M. Jean, M. Raous, *Consistent time discretization for a dynamical frictional contact problem and complementarity techniques*, Rev. Europ. Elêments Finis, 7(1-3), 149-162, 1998.

[123] D. Vola, M. Raous, J.A.C. Martins, *Friction and instability of steady sliding: squeal of a rubber/glass contact*, Int. J. Num. Meth. Engng., 46, 1699-1720, 1999.

[124] P. Wagner, P. Wriggers, C. Klapproth, C. Prange, *Multiscale FEM approach for hysteresis friction of rubber on rough surfaces*, Comp. Meth. Appl. Mech. Engng., 296, 150-168, 2015.

[125] B. Wohlmuth, *Variationally consistent discretization schemes and numerical algorithms for contact problems*, Acta Numerica, 569-734, 2011.

[126] P. Wriggers, T. Vu Van et E. Stein, *Finite element formulation of large deformation impact-contact problems with friction.*, Comp. Struct., 37(3), 1990.

[127] P. Wriggers, P. D. Panagiotopoulos (Eds), *New developments in contact problems*, CISM Courses and Lectures, 384, Springer Verlag, Wien-New York, 1999

[128] P. Wriggers, *Computational contact mechanics*, John Willey sons, 2002

[129] P. Wriggers, T.A. Laursen (Eds), *Computational contact mechanics*, CISM Courses and Lectures, 498, Springer Verlag, Wien-New York, 2007.

[130] P. Wriggers, J. Reinelt, *Multi-scale approach for frictional contact of elastomers on rough rigid surfaces*, Comp. Meth. Appl. Mech. Engng., 198(21-26), 1996-2008, 2009.

[131] V. A. Yastrebov, G. Anciaux, J.-F. Molinari, *From infinitesimal to full contact between rough surfaces: Evolution of the contact area*, Int. J. Solids Struct., 52, 83-102, 2015.

Modeling for Non-linear Vibrational Response of Mechanical Systems

Steven W. Shaw, Oriel Shoshani, and Pavel M. Polunin

Department of Mechanical Engineering
Michigan State University
East Lansing, MI, 48824, USA

Abstract This chapter describes the modeling of mechanical systems in order to describe their non-linear vibratory response, focusing on free vibration and frequency response near resonance. Some general thoughts about modeling are offered first, along with some archetypical models for non-linear vibration, motivated by physical examples. We focus on considerations related to the inclusion of non-linearities in system models and how non-linearity affects the ability of a model to describe system response. Methods for analyzing non-linear vibration systems are not emphasized, and only the minimum tools required to demonstrate the results of interest are introduced. Models with one and two modes with polynomial non-linearities are considered, since these offer descriptions of generic behaviors that are quite common and qualitatively distinct from linear system response. Sample physical systems that are used to demonstrate the main ideas include the simple pendulum, transverse vibrations of beams, capacitively driven micro-systems, the spherical pendulum, and a spring-pendulum system. The chapter closes with some general thoughts about non-linear system modeling and its use in design.

1 Introduction

The models used by engineers to describe the dynamics of mechanical systems have the form of mathematical equations that involve idealizations of the physical system of interest. These models are usually in the form of differential or differential/algebraic equations that arise from first principles such as Newton's laws of motion, Lagrange's equations of motion, or a variational principle. In this chapter we consider topics related to the modeling of oscillatory motions of mechanical systems, focusing on situations in which the model must account for non-linear behavior. Before

F. Pfeiffer and H. Bremer (eds.), *The Art of Modeling Mechanical Systems*,
CISM International Centre for Mechanical Sciences 570,
DOI 10.1007/978-3-319-40256-7_5

turning to that specific topic, we first offer some general comments about the mathematical modeling of physical systems.

First and foremost, there is no such thing as a "correct model" in an absolute sense. A better designation is that of a "good" model, or a "successful" model, as being the "simplest" model that captures the behavior of interest. The concept of "simplest" in terms of a mathematical model can be measured in a number of ways, such as a minimal number of degrees of freedom (DOF), and/or a minimal number of independent parameters, and/or its amenability to mathematical or computational analyses. Two famous quotes that are relevant to this concept are the following: "with four parameters I can fit an elephant, and with five I can make him wiggle his trunk," which was attributed to John von Neumann by Enrico Fermi (Dyson, 2004); and "everything should be made as simple as possible, but not simpler," which is widely (but perhaps inaccurately) attributed to Albert Einstein (Website). Both these quotes remind us that complicated models are not necessarily better models. The development of a successful model requires physical insight, at least partial knowledge about the expected response, experience with the class of problems at hand, and typically several iterations. The ultimate test for any model is, of course, to compare its output with measurements from controlled experiments, although in some cases it is meaningful to test a reduced-order model against computer simulations of a more complex, typically higher order, model.

Another important clarification of terms relevant to the topic at hand is that no physical system is, in itself, "linear" or "non-linear." These terms apply only to mathematical models, since they are mathematical definitions. In fact, a separate classification for the study of non-linear dynamics should really be unnecessary, as noted by mathematician Stanislav Ulam, who remarked: "[it] would be like defining the bulk of zoology by calling it the study of non-elephant animals" (Campbell, 1987). So, in this respect it only makes sense to have a special designation for linear models. Another relevant comparison comes from mechanics, which considers models for both rigid and non-rigid, i.e., flexible, bodies. Of course, all bodies are more accurately described by non-rigid models, since no object is absolutely rigid, but it is often useful and convenient to consider the idealization of a rigid body, which provides the required information about a given system. The situation with non-linear and linear models is analogous.

These observations point to the recognition that linear models are very special. Engineers and scientists know that one must generally start with a non-linear model in order to obtain a valid linear model, and that the limitations of a linear model can be understood only in light of the underlying non-linear model. However, linear systems are widely used since they are very useful and attractive for analysis and prediction, based largely on the principle of superposition (as described below). However, the lack of superposition for a dynamical system does not mean one should abandon all hope of understanding its behavior, and, in fact, non-linear dynamics has, over the past fifty or so years, become a standard field of study across many disciplines, including engineering, physics, chemistry, and biology; see, for example Strogatz (2014) for an introductory treatment and wide-ranging examples.

For the class of problems of interest here, namely, the non-linear vibrations of mechanical systems, a typical model involves inertial terms that can be described using kinetic energy, conservative forces that can be described using a potential, and non-conservative forces that include effects such as dissipation and time-varying inputs from external sources. These non-conservative terms account for the influence from the surrounding environment on the system of interest, and typically add or subtract energy from the system. When modeling vibrations about a stable equilibrium, the equations of motion can be expanded in a Taylor series which provides convenient forms that are often amenable to analysis and provide information about linear system behavior, as well as some quite generic non-linear behaviors. This is the focus of the present chapter.

The first level of expansion about equilibrium provides a linear vibration model which possesses well-known properties and abundant methods for analysis. The fundamental property that defines a linear system is *superposition*, which states that if the system response to input $u_i(t)$ is $x_i(t)$, for $i = 1, 2$, then the response to $c_1u_1(t)+c_2u_2(t)$ is $c_1x_1(t)+c_2x_2(t)$, where the c_i are arbitrary constants. This property includes scaling as a special case, that is, when the system input is scaled by a factor c, the response also scales with c. Superposition imposes severe limitations on the types of responses a model can exhibit, due to the form of its solutions. One consequence is that a linear system model with constant coefficients and a time-periodic external input at a frequency ω has a unique steady-state response at that same frequency, with no additional harmonics, and if the equilibrium is asymptotically stable, the periodic response is similarly stable. Fortunately, the behavior of many systems is adequately described by such linear models. This, along with the fact that the mathematical tools available for studying linear systems are readily accessible to undergraduate students, is why they are widely studied in engineering. However, too often students are not exposed to the limitations of linear models, or to the availability and behavior of non-linear models. This shortcoming is less widespread now, since the personal computer allows students to easily explore the behavior of non-linear system models.

A quite general model for the dynamics of a mechanical system with n DOF can be expressed in terms of a Lagrangian formulation using generalized coordinates $q_i(t)$ and the Lagrangian $L = T - V$ defined in terms the system kinetic energy $T(q_i, \dot{q}_i, t)$ and potential energy $V(q_i, t)$. The resulting model is a set of n differential equations expressed as

$$\frac{d}{dt}\frac{\partial L}{\partial \dot{q}_i} - \frac{\partial L}{\partial q_i} = Q_i \qquad i = 1, 2, \cdots, n \tag{1}$$

where the $Q_i(q_i, \dot{q}_i, t)$ are generalized forces not derivable from the potential, which typically include damping, external excitation, etc. These differential equations are linear only when the kinetic and potential energies are quadratic in the $\dot{q}_i$'s and q_i's, respectively, and the Q_i's are linear in the $\dot{q}_i$'s and q_i's. Such models can be obtained by the series expansion approach described above. However, in general the equations are non-linear due to effects from T (inertial non-linearities), V (conservative, restoring force non-linearities), and/or Q_i (non-linear damping, friction, etc.). For vibratory motions about a stable equilibrium, one can expand both T and V about the equilibrium of interest, as described above, or one can generate the equations of motion and then expand. Such expanded models are considered in this chapter, where our interest is in the modeling of vibratory mechanical systems for which non-linear behavior is important. We focus on modeling aspects rather than on techniques for solving the model equations, for example, by perturbation or numerical methods, or otherwise gaining an understanding of the system dynamics, for example, by topological methods. The reader should look to other sources for such information, of which there are hundreds of books; for mechanical systems these include (Nayfeh and Mook, 2008; Nayfeh, 2008; Guckenheimer and Holmes, 1983; Thompson and Stewart, 2002; Hagedorn, 1981; Bogolyubov and Mitropolskii, 1955; Schmidt and Tondl, 1986; Stoker, 1950; Fidlin, 2005; Moon, 2008; Virgin, 2000).

The chapter is organized as follows. In the next section, we offer some simple examples of systems with a single DOF in order to introduce some basic elements of non-linear vibrations that apply for situations when the system response can be captured using a single vibrational mode. The discussion starts with properties of free vibration and then considers system response to harmonic excitation near the main resonance, and we consider the conditions for which non-linearity needs to be accounted for in a vibration model. We also consider the case of parametric resonance, which leads to a sub-harmonic response that arises from an instability of the equilibrium of a system whose stiffness is fluctuating in a periodic manner. In Section 3 we expand the discussion to consider cases when two modes are needed to describe the system response, focusing on situations when non-linearity comes into play in such systems, namely, the case of internal resonances. Again, we focus on the essential aspects of modeling such systems, and introduce tools only as needed to that end. While only two modes are considered here, the results presented will make it clear that multiple modes can interact in a similar manner under the right conditions.

Our discussion focuses on models that describe non-linear behavior of free vibration and forced vibration near resonance conditions. We employ the method of averaging to describe the behavior of these models, but, of course, many methods exist, including the popular method of multiple scales(Nayfeh and Mook, 2008). The immense subject of chaotic dynamics is not considered here, but information about that topic is abundant, for example, in Strogatz (2014); Guckenheimer and Holmes (1983); Thompson and Stewart (2002); Moon (2008); Virgin (2000).

2 Single Mode Vibrations

We start by considering free undamped vibrations of four systems modeled by a single DOF that lead to similar equations of motion, from which we proceed to distill some important information about modeling for non-linear free vibration. We then turn to a brief consideration of modeling for the effects of non-linear damping. Harmonically forced vibration for this class of models is then considered, focusing on features of the system frequency response near resonance, and we determine conditions on the system and input parameters for which a non-linear model is required to correctly capture the features of the frequency response. We then consider the case of sub-harmonic response from parametric resonance, which arises when the system stiffness is periodically modulated in time.

It should be noted that we consider in detail only systems with symmetric non-linearities, modeled by odd functions of the state variables, since these are most easily amenable to analysis. Non-symmetric models are equally important, perhaps even more so, but are omitted for the sake of brevity; the reader can consult the non-linear vibration references cited above to learn about non-symmetric system models.

2.1 Free Vibrations.

The Simple Pendulum. The simple pendulum is idealized by a point mass m fixed to one end of a light rigid bar of length ℓ whose other end is constrained by a revolute joint so that the mass is restricted to move in a circle centered on its pivot in a fixed plane, as shown in Figure 1.

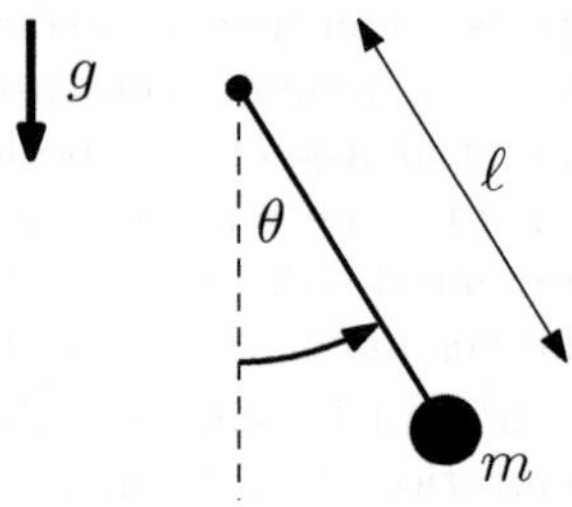

Figure 1. Simple planar pendulum.

This model assumes that the pendulum bob is small but massive such that $m\ell^2$ is much larger than the moment of inertia of the support bar about the pivot and of m about its center of mass. The well-known equation of motion for the pendulum angle θ, as measured from the downward vertical position, is found from the kinetic energy $T = \frac{1}{2}m\ell^2\dot{\theta}^2$ and potential energy $V = -mg\ell\cos\theta$, where g is gravitational acceleration. Dividing the result through by $m\ell^2$ yields

$$\ddot{\theta} + \frac{g}{\ell}\sin\theta = 0. \tag{2}$$

This equation represents the fully non-linear model and has an explicit solution (in the language of dynamical systems, it is integrable), expressed in terms of Jacobi elliptic functions (Baker and Blackburn, 2005). For small angles $|\theta| \ll 1$ the response can be approximated by considering a model obtained from a series expansion of the non-linear term $\sin\theta$, given by

$$\ddot{\theta} + \frac{g}{\ell}\left(\theta - \frac{1}{6}\theta^3 + \cdots\right) = 0. \tag{3}$$

The linearized model ignores the cubic and higher-order terms, yielding a natural frequency $\omega_n = \sqrt{g/\ell}$, and it is of interest to consider the circumstances under which one can use the linear model, or when one should include the cubic term, which results in a so-called *weakly non-linear* model, defined more specifically below. For free vibration both models, linear and weakly non-linear, exhibit periodic motion, and the main issue is the level of acceptable approximation for the period of oscillation and the magnitude of higher harmonics in the response. Note that in this case, since the non-linear coefficient is negative, that the effective stiffness decreases as a function of amplitude, resulting in a decrease of frequency as the amplitude is increased. We consider the frequency of oscillation and its amplitude dependence in a quantitative manner after introducing more examples. For a discussion of higher harmonics and their effects on the response, we refer to Nayfeh and Mook (2008).

Free Vibration of a Fixed-Fixed Beam. We next consider the transverse vibrations of a beam with both ends clamped, accounting for the stretching that occurs along the axis due to transverse deformations. As we show here, the mid-line stretching has the effect of increasing the tension of the beam through a term that arises from non-linear kinematics, leading to frequency shifts that depend on the shape and amplitude of beam vibration. Our formulation is similar to that found in Lifshitz and Cross (2008).

The transverse vibrations for a slender uniform beam of length ℓ are described by the partial differential equation (Meirovitch, 2010)

$$\rho \frac{\partial^2 u}{\partial t^2} - \tau \frac{\partial^2 u}{\partial x^2} + EI \frac{\partial^4 u}{\partial x^4} = 0 \tag{4}$$

where $u(x,t)$ is the transverse displacement at a point x along the beam, ρ is the mass per unit length, τ is the axial tension, E is the Young's modulus of the beam material, and I is the second moment of cross-sectional area relative to the neutral axis. For the clamped-clamped case, boundary conditions $u(0,t) = u(\ell,t) = 0$, $\frac{\partial u}{\partial x}(0,t) = \frac{\partial u}{\partial x}(\ell,t) = 0$ apply. We consider an infinitesimal beam element of undeflected and non-elongated length dx, as shown in Figure 2, which experiences an incremental transverse deflection du. It is seen that the element is stretched by an amount

$$\sqrt{dx^2 + du^2} - dx = \left(\sqrt{1 + \left(\frac{du}{dx}\right)^2} - 1 \right) dx = \frac{1}{2} \left(\frac{du}{dx}\right)^2 dx + \ldots,$$

which induces an increase in the local tension proportional to the axial strain and expressed to leading order by $\Delta\tau = \frac{EA}{2\ell}\left(\frac{du}{dx}\right)^2 dx$, where A is the beam cross sectional area. This tension is assumed to be instantaneously achieved, in which case every beam element sees a tension obtained by integrating this effect over the length of the beam, resulting in a total instantaneous uniform tension of

$$\tau = \tau_o + \frac{EA}{2\ell} \int_0^\ell \left(\frac{du}{dx}\right)^2 dx.$$

Substituting this into Eq. (4) yields a non-linear model that accounts for this stretching, which, as expected, can only increase the tension, and thus also the vibrational frequencies, of the beam.

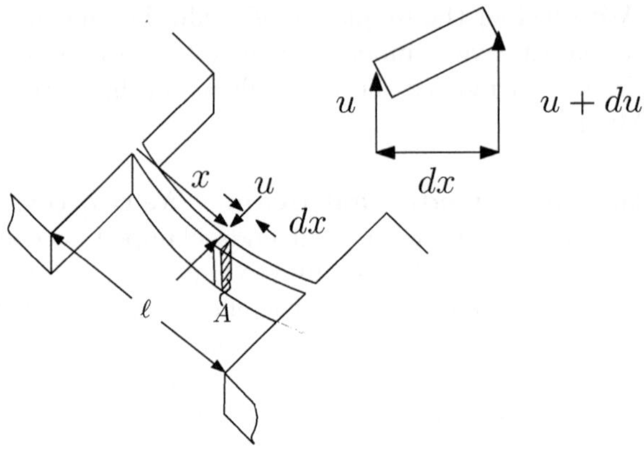

Figure 2. Schematic representation of the flexural vibration of a clamped-clamped beam, shown in a deformed configuration. The element depicted in the beam, and shown top-view in the inset, is used to derive the non-linear effects of mid-line stretching.

Eq. (4) describes general transverse motion of the beam. To consider the vibrations of an individual eigen-mode, under the assumption that it does not interact with other modes, we assume a solution of the form $u(x,t) = \phi(x)q(t)$ where $\phi(x)$ is a mode shape normalized so that its maximum is unity along the beam. This is substituted into the beam equation and projected onto that mode by multiplying the resulting equation by $\phi(x)$ and integrating over the length of the beam, leading to an ordinary differential equation governing $q(t)$, given by

$$\ddot{q} + \omega_o^2(1+\alpha)q + \gamma q^3 = 0 \tag{5}$$

where $\omega_o = \sqrt{\frac{EI\lambda_2}{\rho\ell^4\lambda_0}}$ is the natural frequency of the mode with zero tension ($\tau_o = 0$), $\alpha = -\tau_o \frac{\ell^2\lambda_1}{EI\lambda_2}$ describes the change in natural frequency induced by tension τ_o, $\gamma = \frac{EA\lambda_3}{\rho\ell\lambda_0}$ is the non-linear coefficient from mid-line stretching, and the (non-dimensional) modal coefficients are given by $\lambda_0 = \frac{1}{\ell}\int_0^\ell \phi^2\, dx$, $\lambda_1 = \ell\int_0^\ell \phi\, \phi_{xx}\, dx$, $\lambda_2 = \ell^3\int_0^\ell \phi\, \phi_{xxxx}\, dx$, $\lambda_3 = \ell\int_0^\ell \phi_x^2\, dx$, where a subscript x denotes a partial derivative with respect to x (note that some of these coefficients can be simplified using integration by parts, but that is not important for present purposes). Note that $\lambda_1 < 0$ and $\lambda_{0,2,3} > 0$ for nondegenerate modes, that is, those with positive eigen-frequency, so that the stiffness non-linear coefficient γ is positive, so that the frequency increases with increasing amplitude, similar to the effects of added initial tension, since $\alpha > 0$. Again, for free vibrations the issue of whether or not to include the non-linear term is related to the accuracy of the frequency of vibration. Before turning to that question we consider a model consisting of this beam in an electrostatic field, in which case higher order non-linearities can come into play.

Free Vibration of an Electrostatically Actuated Vibratory System. Here we consider the dynamics of a micro-scale resonator in the presence of electrostatic fields, typical of micro-electro-mechanical systems (MEMS) that are widely used for actuation and sensing in micro-scale devices, and which readily exhibit non-linear behavior (Rhoads et al., 2010; Lifshitz and Cross, 2008). In order to account for the interaction of the mechanical system with its electromagnetic environment, one has to construct an associated electrostatic potential $V_{el}(q_i, t)$ and include it in the Lagrangian L. Or, equivalently, one can use the attendant force in a Newtonian formulation. An accurate form of the electrostatic potential depends on the geometry of the resonator and the attendant electrodes, as well as on the character of the relative motion of the resonator with respect to the driving/sensing electrodes, which can take the form of comb-fingers or parallel surfaces (Lu and Fedder, 2004; Lemkin, 2001). In Figure 3 we illustrate a simple model for such an electrostatic actuation/sensing scheme, by considering again the transverse vibrations of a beam with both ends fixed, and surrounded on both sides by parallel capacitive drive and sense electrodes. In this case, with symmetrically placed electrodes, we can write the electrostatic force (per unit length) acting on the beam as

$$F_{el}(x,t) = \frac{\epsilon_0 w}{2}\left[\frac{V_{b1}^2}{(d-u)^2} - \frac{V_{b2}^2}{(d+u)^2}\right], \tag{6}$$

where $u = u(x,t)$ is the physical displacement of the resonator body (the beam), w is the width of the beam, and V_{bi} is the bias voltage applied between the resonator and i^{th} electrode. While generally one has $V_{b1} \neq V_{b2}$, symmetrically biased vibratory systems can be of particular interest. Such a device has been proposed as a tunable resonator element, wherein equal DC bias voltages applied to both electrodes allow one to vary the resonant frequencies of symmetric modes (Kaiser et al., 1993), and these models have been widely studied (Rhoads et al., 2006b,a; Batra et al., 2007; Nayfeh and Mook, 2008).

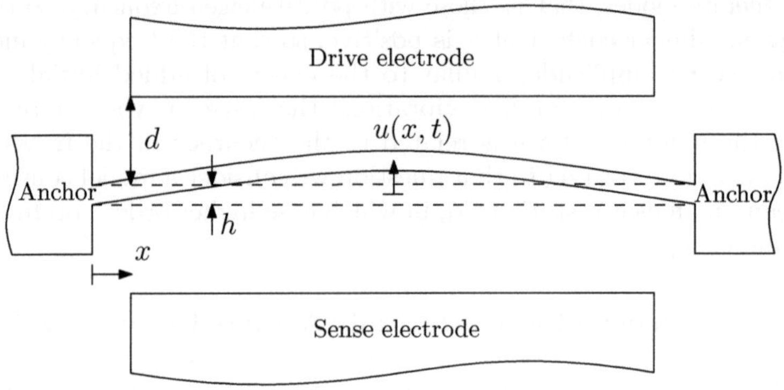

Figure 3. Schematic representation of a clamped-clamped beam resonator with parallel surface electrodes used for capacitive actuation and sensing. The relative scale of the gap d is exaggerated for clarity.

In order to prevent sticking and electrical shorting (which can result from the so-called pull-in instability) of the beam to one of the electrodes, the beam vibration amplitude must satisfy $u < \delta d$ where δ is a factor, usually 1/3 or less (Nayfeh et al., 2007; Alsaleem et al., 2010). This constraint allows one to expand $F_{el}(x,t)$ in Eq. (6) in a Taylor series about $u = u_0$, where u_0 is the beam displacement profile in the absence of periodic forcing. In the symmetric case, $V_{b1} = V_{b2} = V_b$, it is clear that $u_0 = 0$, and as a result, the Taylor expansion of the electrostatic force becomes

$$F_{el}(x,t) \approx \frac{2\epsilon_0 w V_b^2}{d^2} \sum_{n=1}^{N} n \left(\frac{u}{d}\right)^{2n-1}, \tag{7}$$

where N determines the number of terms that are retained in $F_{el}(x,t)$. From Eq. (7) it follows that N is mainly determined by the resonator vibration amplitude as compared with d, as well as by the desired level of accuracy of the analytical model. It follows further that while the beam thickness h dictates the appearance of mechanical non-linearities due to the beam stretching, the electrode gap size d is an additional length scale that regulates the presence of electrostatic non-linear stiffness effects. The competition between these effects can lead to situations where one must include up through quintic terms from electrostatic effects.

In order to include the effect of electrostatic force on the dynamics of a particular vibration mode $\phi(x)$, one has to project $F_{el}(x,t)$ on $\phi(x)$ using, for example, a Galerkin approach (Meirovitch, 1997). The calculation of precise mode shapes is generally a challenging task that requires to account for non-ideal "clamped-clamped" boundary conditions and electrostatic stiffness effects. In this case it is convenient, and quite accurate, to project $F_{el}(x,t)$ onto the primary flexural mode $\phi_1(x)$ in the form shown in Eq. (6) using the method of assumed modes with a function $\phi(x)$ that satisfies the boundary conditions (Meirovitch, 1997). The main idea of this method is to approximate an exact mode shape for the primary flexural mode by a relatively simple expression, e.g., $\phi_1(x) \approx \sin^2(\pi x/\ell)$, which closely resembles the deformation profile of a clamped-clamped beam. Since the resulting expression for the modal electrostatic force in this case is cumbersome(Polunin et al., 2016a), we follow here an alternative path: we first expand $F_{el}(x,t)$ in the Taylor series shown in Eq. (7) and then project this approximated expression onto $\phi_1(x)$. In this case, the governing modal equation of motion becomes

$$\ddot{q} + \omega_{0m}^2(1+\alpha)q + \gamma_m q^3 = \kappa \sum_{n=1}^{N} c_n \left(\frac{q}{d}\right)^{2n-1}, \tag{8}$$

or

$$\ddot{q} + \omega_{0m}^2(1+\alpha)q + \gamma_m q^3 = \kappa \left[c_1 \frac{q}{d} + c_2 \left(\frac{q}{d}\right)^3 + c_3 \left(\frac{q}{d}\right)^5 \right] \tag{9}$$

for the quintic truncated model, where a subscript m indicates a coefficient from purely mechanical (elastic) forces. Where the left-hand side is identical to the one shown in Eq. (5), $\kappa = \frac{2\epsilon_0 w V_b^2}{d^2}$ is the strength of the electrostatic potential, and the projection coefficients are given by $c_n = \frac{n}{l}\int_0^l \phi_1^{2n} dx$. Eq. (8) shows that electrostatic force acting on the vibratory system renormalizes the resonator mechanical parameters, derived in the previous section, such as the natural vibration frequency and the Duffing non-linearity. Furthermore, the c_n's are positive definite from their definition, thus, electrostatic effects alway reduce, that is, *softens*, the corresponding mechanical parameters. The natural frequency and Duffing (cubic) non-linear coefficients are modified by electrostatics to become

$$\omega_0^2 = \omega_{0m}^2(1+\alpha) - c_1\frac{\kappa}{d}, \quad \gamma = \gamma_m - c_2\frac{\kappa}{d^3}. \tag{10}$$

Note that by varying the bias voltage one can modify the the system natural frequency, via parameter κ, and that there exists a voltage, corresponding to $\omega_0 = 0$, that will result in electrostatic buckling of the beam.

It is clear that mechanical and electrostatic effects have different physical origins and that their relative strengths are determined by two physically different length scales, the beam thickness h and the electrode gap size d, respectively. Furthermore, since electrostatic forces always soften the corresponding hardening mechanical coefficients of the resonator, mechanical and electrostatic restoring forces represent competing sources of non-linearity in this type of vibratory system. Generally, h and d can differ in their orders of magnitude, and the overall dynamic behavior of the resonator is determined by an interplay of mechanical and electrostatic effects. For example, high-frequency MEMS resonators used in sensing applications are frequently designed in a way that $d \ll h$. Note that small d is necessary to create sufficiently strong electrostatic field between the resonator body and attendant electrodes, while the beam thickness h strongly influences the resonant frequency of the system. In this case, when the resonator amplitude is small, the dynamics of the vibration mode is essentially determined by the mechanical stiffness, $\omega_{0m}^2 \gg c_1\kappa/d$ and $\omega_0^2 > 0$. As the resonator amplitude increases, the cubic term in the resonator stiffness starts to affect the modal dynamics and, depending on the relative strengths of γ_m and γ_e $(= c_2\frac{\kappa}{d^3})$, the Duffing parameter γ can be either positive (hardening frequency response) or negative (softening frequency response), as depicted by the dashed ($\gamma > 0$) and the dot-dashed ($\gamma < 0$) lines, respectively, in Figure 4. As the resonator vibration amplitude increases further, one may have to include the effects of the quintic term in the resonator stiffness. It can be shown that when $d \ll h$, the electrostatic contribution to the resonator quintic stiffness term is dominant over the mechanical quintic effects, and if $\gamma > 0$ the resonator frequency response will exhibit a hardening-to-softening transition, as shown as the solid curve in Figure 4. Note that near the turning point the frequency is locally independent of the amplitude, akin to a linear resonator, and such conditions have been proposed as operating points in self-sustained oscillating MEMS devices used for timing applications, due to favorable noise properties at such points (Yurke et al., 1995; Kenig et al., 2012).

For these models one does not need to consider terms of order higher than quintic in the electrostatic force, since all higher-order terms will be very small if $|q| \ll \omega_{0m}^{1/2} d^{5/4}(\kappa c_3)^{-1/4}$, just as one need not consider terms of order higher than cubic in the mechanics, since $|q| \ll \omega_{0m}|\gamma|^{-1/2}$ will render them small. In fact, the only sensible reason to keep multiple orders of non-linear terms is if there are such competing effects. Also note that at some amplitude the response will become unstable, due to the electrostatic attraction to the electrodes, and pull-in will occur, at which point the present model ceases to be valid.

As was the case for the previous examples, since the non-linear forces acting on these systems are conservative, the response of this single DOF model is a periodic oscillation in time with an amplitude-dependent

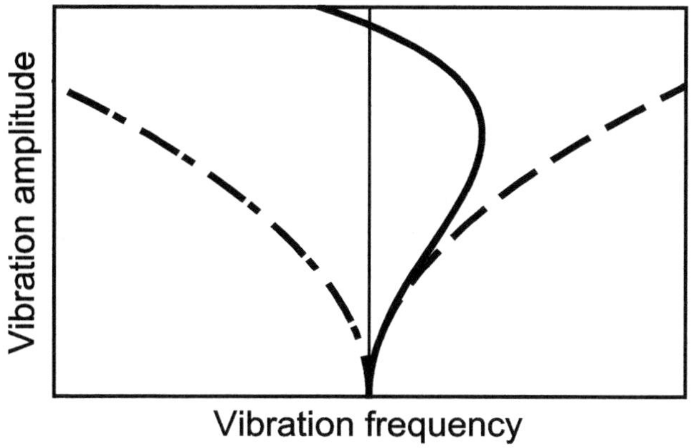

Figure 4. Amplitude versus frequency plots for three cases of vibratory systems undergoing free (unforced and undamped) oscillations. The thin vertical line represents the linear resonator model for which the frequency is independent of the vibration amplitude. The dot-dashed line represents a softening frequency response, such as the pendulum, the dashed line represents a hardening frequency response, such as the clamped-clamped beam, and the thick solid line represents a hardening-to-softening frequency response with competing sources of non-linearity, as in the clamped-clamped beam in the presence of DC electrostatic bias.

frequency; this holds even if one retains the full mechanical and electrostatic potentials, without expanding them.

A Simple System with Inertial Non-linearity. As a final example we very briefly consider the single DOF system shown in Figure 5, which exhibits an inertial non-linearity. This system consists of a solid uniform cylinder of mass m and radius R which has embedded in it a point mass M at a distance a from its center point C. The cylinder is assumed to roll without slip, an idealization that should be examined in terms of contact mechanics, as considered in other chapters of this volume. The configuration of the cylinder is described by the coordinate θ, taken to be zero when M is directly below C. The kinetic and potential energies are given by $T = \frac{1}{2}\left(J_C + M(R^2+a^2) - 2MRa\cos\theta\right)\dot{\theta}^2$ and $V = -Mga\cos\theta$, resulting in an equation of motion

$$\left(J_C + M(R^2+a^2) - 2MRa\cos\theta\right)\ddot{\theta} + MRa\dot{\theta}^2\sin\theta + Mga\sin\theta = 0 \quad (11)$$

where $J_C = \frac{1}{2}mR^2$. For small oscillations, keeping leading order non-linear effects, the model is

$$\left(J_C + M(R-a)^2 + MRa\theta^2\right)\ddot{\theta} + MRa\theta\dot{\theta}^2 + Mga\theta\left(1 - \frac{1}{6}\theta^2\right) = 0 \quad (12)$$

which clearly has both stiffness and inertial non-linearities at the same (cubic) order. Whether the inertial non-linearities are hardening or softening is not immediately clear. A systematic treatment of this question can be carried out by rearranging the equation to isolate the acceleration term and expanding again to obtain the form

$$\ddot{\theta} + \omega_n^2\theta(1 + c_1\theta^2 + c_2\dot{\theta}^2) = 0 \quad (13)$$

where, for the present model, $c_1 < 0$ and $c_2 > 0$. The response of this system is also periodic, generally with an amplitude dependent frequency. By applying the method of averaging it can be shown that the system is hardening (softening) if $3c_1 + \omega_n^2 c_2 > 0$ (< 0), and by grinding out the calculations for this example it is found that $3c_1 + c_2\omega_n^2 < 0$, rending the system softening, as expected, since its large amplitude behavior is qualitatively similar to that of a pendulum.

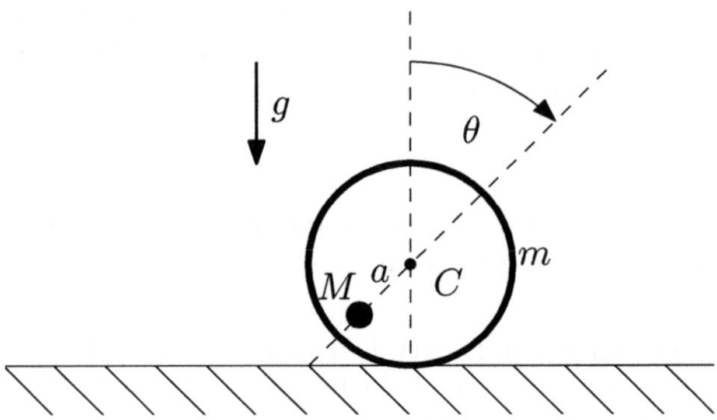

Figure 5. A uniform cylinder with an embedded mass that rolls without slip.

2.2 Validity of Weakly non-linear Duffing Models

At this point we note that the pendulum equation, Eq. (3), the beam mode equation, Eq. (5), and the electrostatic beam equation Eq. (8) (retaining only the cubic term), have the same form, namely

$$\ddot{x} + \omega_n^2 x + \gamma x^3 = 0, \quad (14)$$

which is the classic Duffing equation, the prototypical model for vibrations in which the non-linear effects are symmetric in x and weak relative to the linear terms. Specifically, the model describes vibrations for which $|\gamma x^3| \ll |\omega_n^2 x|$, which implies $|x| \ll \omega_n|\gamma|^{-1/2}$. For the pendulum this restricts the angle to be $|\theta| \ll \sqrt{6}$, which is certainly satisfied by angles up to 30 degrees, and may be valid quite a bit further in amplitude, depending on the required precision in frequency, as quantified below. For a mechanical beam with zero pretension ($\tau_o = 0$ or, equivalently, $\alpha = 0$) and rectangular cross section of transverse thickness h, this restricts the maximum deflection to be $|q| \ll h\left|\frac{\lambda_2}{6\lambda_1\lambda_3}\right|^{1/2}$, which for the fundamental mode of the beam reduces approximately to $|q| \ll 0.25h$, which implies that the beam vibration amplitude must be much less than the beam thickness (in the absence of electrostatic effects).

The shift in the frequency of oscillation for the weakly non-linear case can be approximated by the method of averaging, which is described below. For a vibration amplitude a, such that $x(t) \approx a\cos(\omega_a t)$, the amplitude dependent frequency is given by

$$\omega_a(a) = \omega_n\left(1 + \frac{3\gamma}{8\omega_n^2}a^2 + \cdots\right). \tag{15}$$

where it is seen that the correction $\frac{3\gamma}{8\omega_n^2}a^2$ is small when $a \ll \omega_n\sqrt{\frac{8}{3\gamma}}$, consistent with the scaling $a = |x| \ll \omega_n|\gamma|^{-1/2}$, derived above for weak non-linear effects.

While these conditions imply relatively *weak* non-linearity, we shall see that one must retain these non-linear terms in some cases in order to capture important *qualitative* aspects of the response when the system is driven into certain types of resonance conditions. For example, for harmonically forced vibration with damping, the behavior predicted by the linear and weakly non-linear models can be dramatically different, as we show subsequently.

2.3 Non-linear Damping

To this point we have considered only conservative non-linearities, focusing on their effects on the frequency of oscillation. Of course, damping causes oscillations to decay, and for a weakly non-linear model this will result in frequency changes as the motion dies out. When this decay is slow, that is, when it takes many oscillations to appreciably reduce the amplitude, one can track the frequency in a quasi-static manner using Eq. (15).

Here we consider models for vibration decay that result from linear and non-linear damping, given by

$$\ddot{x} + 2\Gamma(1 + \beta x^2)\dot{x} + \omega_n^2 x = 0, \tag{16}$$

where Γ is a linear damping coefficient, specifically expressed as the inverse of a time scale that governs decay of free vibration at small amplitudes (as shown below), and β is a non-linear damping coefficient. Note that $\dot{x}x^2$ form of non-linear damping is not unique. The form of damping $\dot{x}^3$, for example, arises from the same microscopic origins in micro/nano-scale devices and leads to similar results (Zaitsev et al., 2012).

To determine an approximation of the response the method of averaging is employed, which is more fully developed below. The method assumes a response of the form $x(t) = a(t)\cos(\omega_n t + \phi(t))$, where the amplitude a and phase ϕ are, for small damping, slowly varying functions of time. In this case the phase remains constant (since there is no stiffness non-linearity to cause a frequency shift) and the result of averaging is an equation that governs the slowly varying amplitude (details about the coordinate transformation are provided in the next subsection),

$$\dot{a}(t) = -\Gamma\left(1 + \frac{\beta}{4}a(t)^2\right)a(t). \tag{17}$$

If the damping model is linear ($\beta = 0$), this is easily solved to obtain $a(t) = a_0 \exp(-\Gamma t)$, the classic exponential decay for an initial amplitude a_0. The non-linear case also has a closed form solution, but it is rather cumbersome (Polunin et al., 2016b), so suffice it to note that one can think of an amplitude-dependent effective value of Γ, crudely given by $\Gamma_{eff}(a) = \Gamma(1 + \frac{\beta}{4}a^2)$, which shows that for $\beta > 0$ the decay rate is faster at larger amplitudes, and it approaches Γ as the amplitude goes to zero, that is, in the range of validity of the linear model. Figure 6 shows the amplitude decay versus time for different values of β, demonstrating the effects of non-linear damping.

One can estimate the range over which the linearized damping model is valid by the condition $|\frac{\beta}{4}a^2| \ll 1$, which implies that the amplitude must satisfy $a \ll 2|\beta|^{-1/2}$. Thus, again the valid amplitude range scales with the inverse of the square root of the non-linear coefficient. There are many forms of non-linear damping, all of which similarly lead to amplitude-dependent rates of decay. We next turn to behavior of non-linear systems driven near resonance.

2.4 Response to Harmonic Excitation - Primary Resonance

As noted in the introduction, a key feature of linear system models is that when they are driven with a frequency ω, their steady-state response

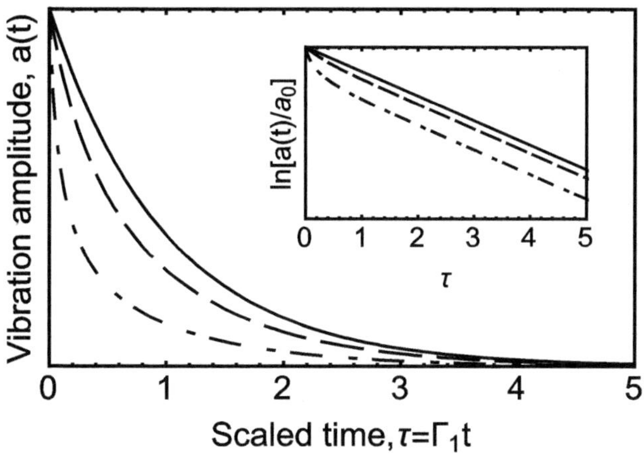

Figure 6. Effect of non-linear dissipation on the decay of unforced vibrations. The solid, dashed, and dot-dashed curves correspond to $\beta = 0, 1, 10$ respectively. Inset: decay of unforced vibrations on a logarithmic scale, where Γ_{eff} is represented by the slope of the decay, which approaches Γ for $a \ll 2\beta^{-1/2}$.

is unique and is also at frequency ω. It is clear that this is not the case with even simple non-linear models, like the Duffing equation, since the cubic term will necessarily result in over-tones at 3ω, etc. However, for weakly non-linear models these overtones are generally small and one can approximate the response using a single harmonic at the main frequency. A more interesting feature of the steady-state response of non-linear models is that if one scales the input amplitude by some factor c, the steady-state response will not scale similarly, and in fact, it can be drastically different, as we show here.

To demonstrate these features we consider a weakly non-linear system with linear damping and direct harmonic forcing, described by the model

$$\ddot{x} + 2\Gamma\dot{x} + \omega_n^2 x + \gamma x^3 = f\cos(\omega t), \tag{18}$$

where f is the magnitude of an external direct drive normalized by the system mass, ω is the drive frequency, and Γ is the linear damping coefficient considered above. This model describes, for example, the motion of the pendulum subjected to a harmonic torque or when the support moves horizontally in a harmonic manner, or the clamped-clamped beam when it is subjected to a transverse harmonic force or it is subjected to transverse harmonic support motion, or the capacitively driven beam when the voltage on one of the electrodes is harmonically modulated $V(t) = V_{dc} + V_{ac}\cos(\omega t)$ (which leads to driving frequencies ω and 2ω, since the force depends on the square of the applied voltage).

For small forcing amplitudes, the response of this system is well approximated by the linearized model ($\gamma = 0$) except when damping is small and it is driven near resonance, in which case large amplitudes result, bringing the possibility of non-linear effects into play. The most dramatic non-linear effect is the appearance of multiple steady-state responses for a given set of input conditions near resonance. There are several approaches to analyzing this behavior; here we use the method of averaging, introduced above, but more fully considered here, following Guckenheimer and Holmes (1983). The method employs a coordinate transformation to rotating amplitude and phase coordinates that will be slowly varying when the system is lightly forced near resonance with small damping. The transformation is given by $(x(t),\ \dot{x}(t)) = (a(t)\cos(\omega t + \phi(t)),\ -a(t)\omega\sin(\omega t + \phi(t)))$, which requires a constraint equation

$$\dot{a}(t)\cos(\omega t + \phi(t)) - \dot{\phi}(t)a(t)\sin(\omega t + \phi(t)) = 0, \tag{19}$$

similar to that used in the method of variation of parameters in differential equations. Under this coordinate transformation, Eq. (18) yields an equation involving $(a'(t),\ \phi'(t))$ which, together with Eq. (19) can be solved for $(a'(t),\ \phi'(t))$, since the equations are linear in these variables. This amplitude and phase will have fast oscillations on a time scale ω^{-1}, superimposed on a slow drift that occurs on a time scale set by Γ^{-1} and the nearness of the drive frequency ω to the linear natural frequency ω_n. The fast oscillations are eliminated by averaging (by integration) over one period of the excitation, $2\pi/\omega$, which can be viewed as another coordinate transformation that removes fast oscillating terms (Guckenheimer and Holmes, 1983). The result is the following set of equations which govern the slowly varying amplitude and phase, denoted here as $(r(t), \psi(t))$,

$$\dot{r}(t) = -\frac{1}{2\omega}(2\Gamma\omega r + f\sin\psi) \tag{20}$$

$$\dot{\psi}(t) = \frac{1}{8r\omega}(3\gamma r^3 - 4r\sigma - 4f\cos\psi) \tag{21}$$

where we have introduced the frequency detuning $\sigma = (\omega^2 - \omega_n^2)$. Note that for f, Γ, γ, σ, small, these indeed are slow equations, since the terms on the right hand side are small. The key advantage of this approach is that the averaged equations, Eqs. (20) and (21), do not depend explicitly on time, rendering their analysis significantly less complicated than the original equation (18). Steady state responses of $x_{ss}(t)$ near resonance have frequency ω

and can be approximated by $x_{ss}(t) = \bar{r}\cos(\omega t + \bar{\psi})$ where $(\bar{r},\ \bar{\psi})$ are fixed points of the above equations. Furthermore, the stability of a given steady-state response is dictated by stability of its attendant fixed point, and under quite general conditions a bifurcation of a fixed point corresponds to a similar bifurcation of the corresponding steady-state response (Guckenheimer and Holmes, 1983).

Uncoupled equations for the fixed points $(\bar{r},\ \bar{\psi})$ can be found by setting the right hand side of Eqs. (20) and (21) equal to zero and solving for for $(f\cos\bar{\psi}, f\sin\bar{\psi})$ from which an equation for $\tan\bar{\psi}$ can be found, and then eliminating $\bar{\psi}$ using $(f\cos\bar{\psi})^2+(f\sin\bar{\psi})^2 = f^2$ yields an uncoupled equation for $\bar{r}$. These equations are given by

$$\tan\bar{\psi} = \frac{-8\Gamma\omega}{3\gamma\bar{r}^2 - 4\sigma} \tag{22}$$

$$f^2 = \frac{\bar{r}^2}{16}\left(9\bar{r}^4\gamma^2 - 24\bar{r}^2\gamma\sigma + 16\left(4\Gamma^2\omega^2 + \sigma^2\right)\right) \tag{23}$$

Direct solution of these for fixed points as a function of system and input parameters requires solving a cubic equation for $\bar{r}^2$. However, a standard trick for showing amplitude results is to vary $\bar{r}$ and compute the corresponding f^2 (a linear equation) or σ (a quadratic equation) directly from the second equation, the latter of which results in frequency responses such as those shown in Figure. 7 for different levels of excitation f. The most important feature of these frequency responses is that for "small" f (to be quantified below) a usual linear resonance frequency response is recovered, which is (nearly) symmetric about the resonance point given by $\sigma = 0$. As the drive amplitude f is increased, the resonance peak begins to become asymmetric and bend, leftward for $\gamma < 0$ (softening) and rightward for $\gamma > 0$ (hardening). Beyond a critical value of f, denoted by f_{cr} and determined below, the resonance curve bends sufficiently far that the response becomes multi-valued, specifically, it transitions from a unique response over all frequencies (σ values) to one where there are three steady-state response amplitudes over a range of frequencies near resonance. This range of frequencies with multiple steady-state solutions is bounded from both ends by a pair of saddle-node bifurcations in which a stable and an unstable response branch merge and annihilate one another (Guckenheimer and Holmes, 1983). When plotting amplitude $\bar{r}$ verses frequency (or detuning σ), the upper and lower branches are stable and the middle branch is unstable, rendering bistability and hysteresis in the response as the frequency is swept back and forth through resonance.

The conditions on the input parameters (f, σ) for the onset of bi-stability are given by the conditions for which the response becomes multi-valued. A

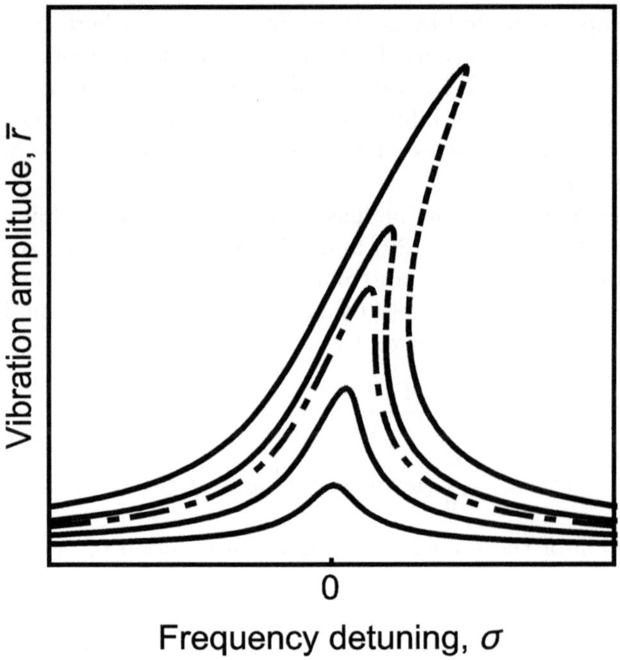

Figure 7. Frequency responses in the form of amplitude versus detuning of a forced Duffing resonator at different levels of excitation f. The dot-dashed curved represents the case of critical forcing amplitude, f_{cr}, beyond which the resonator response becomes multi-valued. Frequency responses below this threshold represent nearly linear resonator response, while those above it lead to bistable steady-state behavior. Solid (dashed) lines represent dynamically stable (unstable) responses.

convenient method for doing this is to consider the dependence of the term on the right hand side of Eq. (23), denoted as $F(\bar{r}^2, \sigma)$, as a function of $\bar{r}^2$, which is a cubic equation. Then solutions of the equation $f^2 = F(\bar{r}^2, \sigma)$ are points where the cubic curve $F(\bar{r}^2)$ cross the horizontal line f^2. For $f^2 = 0$ there is clearly a root at $\bar{r}^2 = 0$ and as f^2 increases this root initially moves positive, approximately along $\bar{r}^2 = f^2 \left(4\Gamma^2\omega^2 + \sigma^2\right)^{-1}$, which is the expression for the linear response. The onset of bi-stability occurs when the cubic curve has an inflection point, since perturbing this case yields either one or three roots. The conditions for the inflection point can be found by simultaneously solving $dF(\bar{r}^2, \sigma)/d(\bar{r}^2) = 0$ and $d^2F(\bar{r}^2, \sigma)/d(\bar{r}^2)^2 = 0$ for the amplitude and frequency at the inflection point. This calcuation yields

$$\bar{r}_{cr}^2 = \frac{16\Gamma\omega_n}{3\sqrt{3}|\gamma|} \qquad \sigma_{cr} = 2\sqrt{3}\Gamma\omega_n \ \mathrm{sgn}(\gamma) \tag{24}$$

where we have substituted ω_n for ω since they are very close to one another near the resonance condition. This substitution allows one to express the threshold conditions in terms of system parameters that describe damping Γ, natural frequency ω_n, and non-linearity γ. It is interesting to compare this critical amplitude with the estimate for weak non-linear behavior in free vibrations, and it is seen that both amplitudes scale with $|\gamma|^{-1/2}$, but that in free vibration without damping the critical amplitude scales with ω_n while in force vibration it scales with $\sqrt{\Gamma\omega_n}$, which accounts for the fact that this is related to a resonance amplitude that depends strongly on the level of damping. The critical value of the excitation amplitude at the onset of bi-stability can then be determined by evaluating $f_{cr}^2 = F(\bar{r}_{cr}^2, \sigma_{cr})$, yielding

$$f_{cr}^2 = \frac{256\ \Gamma^3\omega_n^3}{9\sqrt{3}|\gamma|}. \tag{25}$$

The condition f_{cr} can be evaluated for a given model, such as the pendulum or clamped-clamped beam, providing a threshold level of the excitation amplitude above which the system response is *qualitatively* different from the linear response. Of course, the asymmetry of the resonance curve exists for all amplitudes, but is small until f approaches f_{cr}.

Sub-harmonic and super-harmonic resonances can also occur, which are responses with strong harmonic components that are not equal to, but are rationally related to, ω. These arise when ω/ω_n is near rational values, but only a very few are of practical interest in most cases. These are more fully considered in non-linear vibrations textbooks cited above. We next consider the simplest, and most common, sub-harmonic resonance that occurs in these systems.

2.5 Response to Harmonic Excitation - Parametric Sub-harmonic Resonance

Here we consider a resonance that arises when the drive stems from periodic modulation of the system stiffness at a frequency that is approximately twice the system natural frequency ω_n. In this case the system, whose natural tendency is to respond at or near ω_n, synchronizes with the drive whose frequency is approximately twice that value, resulting in a response at frequency $\omega/2$, that is, it has twice the period of the drive, which is a sub-harmonic of order two. This is the case of primary parametric resonance, and it results when the drive enters the system as a multiplier of the displacement, which can be viewed as a periodically modulation of the system stiffness.

Before turning to details, we consider when this resonance will occur for some of the physical examples described above. The first is the simple pendulum, which experiences parametric excitation when its support point is harmonically oscillated in a vertical manner at a frequency that is nearly twice the natural frequency, or when its length varies harmonically (like a child pumping a swing, although this leads to additional terms in the model); either of these effects leads to an effective harmonic variation in the pendulum frequency, via g or ℓ. These can destabilize the equilibrium of the pendulum and induce an oscillation that looks like its natural swing, whose amplitude is set by non-linear effects. Note that in this oscillation, the excitation executes two full cycles while the pendulum completes one full swing. The second example is the clamped-clamped beam, which experiences parametric resonance when its tension is modulated at a frequency that is twice that of a given mode. For the electro-mechanical oscillator, this resonance occurs when the bias voltages on both sides are harmonically modulated at a frequency nearly twice the natural frequency of the device (which is set by the combination of mechanical and electrostatic restoring forces). In all cases, this form of excitation can destabilize the equilibrium and induce oscillation that is limited by non-linear effects. It is worth noting that this "resonance" is actually an instability, and that the linear model predicts that the instability will lead to unbounded response even in the presence of damping. In fact, the resonant amplitude is set by non-linear effects rather than damping, as we will show below. We now turn to some analysis that allows one to quantify these results for a generic model.

The generic model for these cases is the non-linear Mathieu equation, for which we include both non-linear stiffness (γ) and non-linear damping (β), with parametric excitation at frequency 2ω (expressed in this way for convenience) and amplitude λ,

$$\ddot{x} + 2\Gamma(1 + \beta x^2)\dot{x} + \omega_n^2 \left(1 + \lambda \cos\left(2\omega t\right)\right) x + \gamma x^3 = 0. \tag{26}$$

Note that the trivial response $x = 0$ is a solution of this equation, since this represents a system whose only input is a harmonically varying stiffness. However, the stability of this rest state must be considered. The equation linearized about $x = 0$ has time-periodic coefficients, and analysis requires Floquet theory (Iakubovich and Starzhinskiĭ, 1975; Erugin et al., 1966; Cesari, 2012). Perturbation methods can also be employed to investigate this stability issue, as well as to determine the non-linear response, so that is our preferred approach.

The method of averaging described above must be slightly modified to handle this resonance. The use of driving frequency 2ω in the model allows one to use an assumed form of a response at ω, which is close to ω_n; this is the sub-harmonic of order two in which the system responds with a period twice that of the drive. To capture this response using averaging, one must average over two periods of the forcing, which is $2\pi/\omega$ (the period being π/ω in this case), thereby searching for responses with period double that of the drive. A complication arises when using polar coordinates, however, since the phase ϕ is not defined for the zero amplitude ($a = 0$, that is, trivial) response, so that it is more convenient to employ the transformation using Cartesian coordinates, given by $(x(t),\ \dot{x}(t)) = (u(t)\cos(\omega t) + v(t)\sin(\omega t), -u(t)\omega\sin(\omega t) + v(t)\omega\cos(\omega t))$, which implies the constraint $\dot{u}(t)\cos(\omega t) + \dot{v}(t)\sin(\omega t) = 0$. Substitution of the transformation in Eq. (26) results in an equation involving $\dot{u}$ and $\dot{v}$. That equation and the constraint equation are linear in $\dot{u}$ and $\dot{v}$. These are solved and the result integrated over $2\pi/\omega$, assuming that u and v are constant over the integration period. This process results in the averaged equations that govern the (slow) behavior of (u, v),

$$\dot{u} = -\Gamma u\left(1 + \frac{1}{4}\beta(u^2 + v^2)\right) + \frac{v}{8\omega}\left(-4\sigma + 3\gamma(u^2 + v^2) - 2\lambda\omega_n^2\right) \quad (27)$$

$$\dot{v} = -\Gamma v\left(1 + \frac{1}{4}\beta(u^2 + v^2)\right) + \frac{u}{8\omega}\left(4\sigma - 3\gamma(u^2 + v^2) - 2\lambda\omega_n^2\right) \quad (28)$$

where we have substituted $\sigma = \omega^2 - \omega_n^2$ for the frequency detuning away from the resonance condition. Note that these equations govern the behavior of u and v, which is slow when the system is weakly non-linear, lightly damped, and lightly driven near resonance. The first terms on the right hand sides describe the linear (Γ) and non-linear (β) damping, which affect only the amplitude of oscillation. The two next terms capture frequency shifts (resulting in phase drift) from the the detuning (σ) and Duffing non-linearity (γ), and the final term results from the parametric excitation (λ).

The linearized version ($\beta = \gamma = 0$) of these equations has only the trivial solution. Considering the eigenvalues of the trivial response, it follows that it is stable for small drive amplitudes (except for the very special case $\Gamma = 0$, $\omega = 1$), and becomes unstable for $\lambda > 2\omega_n^{-1}\sqrt{\sigma^2 + 4\Gamma^2(\sigma^2 + \omega_n^2)}$. This threshold is the boundary of an instability region in the system parameter space that is referred to as an Arnold tongue, which provides conditions on the drive amplitude and frequency, for a given level of linear damping, over which the trivial response is unstable. Examples of this tongue for various values of damping are shown in Figure. 8. Note that for zero damping

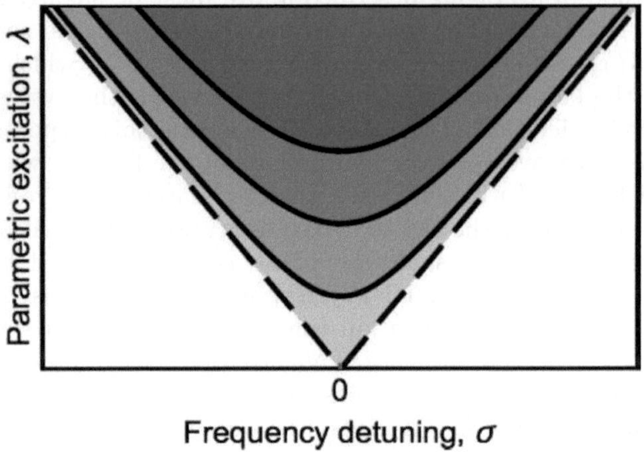

Figure 8. Stability regions for the trivial response of PR for different values of damping Γ. The dashed line represents the undamped case, $\Gamma = 0$, and solid lines (from bottom to top) represent Arnold tongues with $\Gamma = \Gamma_0, 2\Gamma_0, 3\Gamma_0$. Increasing the damping results in a vertical shift of the Arnold tongue, indicating that higher excitation amplitudes λ are required in order to de-stabilize the trivial response.

($\Gamma = 0$) the tongue is a shaped like a "V" that emanates from the point $(\lambda, \omega) = (0, 1)$, and for non-zero damping it lifts off to finite values of drive and the sharp bottom is rounded off.

The system also has non-trivial periodic responses, set by non-linear effects that require solution for the constant values of (u, v) from Eqs. (27) and (28). The stability of these responses is found in the usual manner using the eigenvalues of the system linearized about the steady-state response. This problem is solvable in closed form, since the zero solution factors out and provides a quadratic equation in the amplitude $a = \sqrt{u^2 + v^2}$, but the results are quite cumbersome. For zero damping ($\Gamma = 0$) one can find solutions in a rather simple form. However, in the general case it can be easily programmed for plotting, and analyzed in detail for bifurcation conditions. In all cases, the results show that a pair non-zero responses emerge from each edge of the Arnold tongue in a pitchfork bifurcation(Guckenheimer and Holmes, 1983), one edge being a super-critical and the other sub-critical, with the order depending on the sign of the Duffing non-linearity γ. The responses within a given pair have identical amplitude but a relative phase of π, which makes them physically indistinguishable from one another, since they are simply time-translated versions of each other that have the same relative phase to the drive, which is at twice the frequency of the response (Danzl and Moehlis, 2010). In other words, the responses are physically identical, but have opposite phase. Samples of these responses, shown as frequency response curves in the form of amplitude a versus detuning σ, are provided in Figure 9 for two sets of parameter conditions. The most

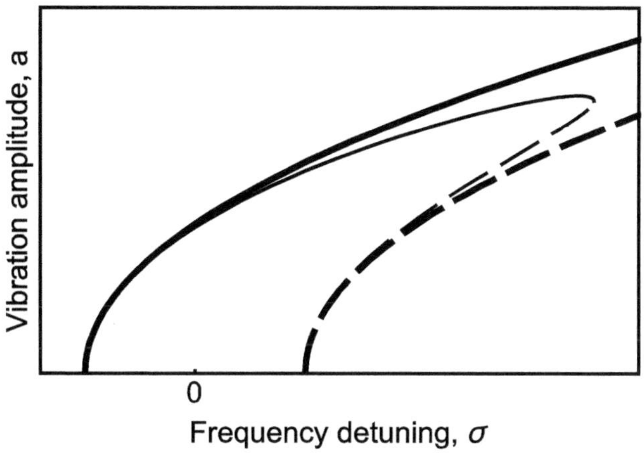

Figure 9. Frequency response in the form amplitude versus detuning of a parametrically-driven vibratory system, noting that zero detuning represents excitation frequency exactly twice the natural frequency. The frequency response drawn in thick lines is obtained when only linear damping is present, $\beta = 0$, while the frequency response for a resonator with dissipative non-linearity $\beta > 0$ is shown in thin lines. Solid and dashed lines represent stable and unstable periodic solutions to Eq. (26), respectively. Note that the zero response always exists, and is unstable between the pitchfork bifurcation points, and stable outside of that region.

interesting features of these frequency responses are: (i) they are hardening (softening) for $\gamma > 0$ (< 0); (ii) one pair (from the super-critical bifurcation) is stable and the other pair (from the sub-critical bifurcation) is unstable; (iii) there exists a range of detuning values where both the trivial and non-trivial responses coexist, leading to a state of bistability in which the trivial response co-exists with both types of non-trivial responses, one pair of which is stable; (iv) the system exhibits hysteresis as the frequency is swept up and down; and (v) the two pairs of non-trivial responses merge and close off at a finite frequency (value of σ) only if one includes non-linear damping in the model, $\beta \neq 0$. More details about this resonance can be found in the standard non-linear vibrations texts cited above.

Before leaving this discussion of parametric resonance, we note some extensions and applications. First, note that some physical models lead to parametric excitation terms of the form $g(x)\cos(2\omega t)$. Traditional scaling approaches suggest that $g(x)$ be expanded with both x and λ small, yielding Eq. (26) as a generic model for this resonance. However, the more general model is required to capture some interesting and subtle phenomena that has been observed in some physical systems, for example, in MEMS resonators (Rhoads et al., 2006b). See Shoshani and Shaw (2016) for analysis and examples of this more general model for parametric resonance.

Another application of parametric excitation is so-called *parametric amplification* in which both direct and parametric resonance excitation are applied to a system. This can be achieved, for example, by oscillating the pendulum support at an angle with an input that combines two frequencies, ω and 2ω, or in the capacitively driven beam by including oscillations in the bias voltages at these two frequencies. The generic model for non-linear parametric amplification is given by

$$\ddot{x} + 2\Gamma\dot{x}(1 + \beta x^2) + \omega_n^2\left(1 + \lambda\cos(2\omega t)\right)x + \gamma x^3 = f\cos(\omega t + \nu) \tag{29}$$

which includes a phase difference between the two harmonics of the input, both of which are near resonant. Typically, one considers the linear model, which shows that the presence of the parametric excitation, so long as its amplitude is below the instability threshold λ_{cr}, alters the system response near resonance and can be used to amplify or attenuate the resonant amplitude, depending on the phase ν. See Rugar and Grütter (1991) for the seminal paper on this subject, and Kumar et al. (2011); Rhoads and Shaw (2010); Rhoads et al. (2008) for applications to mechanical systems and extensions to non-linear parametric amplification.

Finally, it should be noted that additional parametric resonance conditions can also occur, involving ω/ω_n near other rational ratios, yielding additional Arnold tongues of higher order, but these are less commonly encountered since their resonant frequency ranges are inherently narrower and more significantly reduced by damping. An example where up to five tongues have been measured in a MEMS experiment is described in Turner et al. (1998).

3 Multi-Mode Non-linear Vibration

Up to this point we have discussed single mode vibrations, focusing on free vibration and response to harmonic excitation. We now turn to multimode non-linear vibrations, focusing on response to harmonic excitation. To start, it is crucial to note that linear models for multiple DOF have the very special property of *invariance*, which says essentially that modes behave independently of one another. For example, in free vibrations the system initial conditions set the individual modes in motion, and these modes each behave according to a single-DOF model, and never interact with one another. Similarly, in forced vibration, the modes are individually excited and respond accordingly, without interference from other modes. This is clearly the case for systems with distinct natural frequencies, in which case the natural modes are uniquely defined. This property also holds for systems with repeated natural frequencies, although the analysis is more subtle in that case since the modes are not uniquely defined, but once the modes are specified in a convenient manner, the modes behave independently of one another (Meirovitch, 1997). This modal independence follows from a view of the system expressed in terms of modal coordinates, obtained by a similarity transformation of the original system, which leads to a set of N uncoupled single-DOF models describing the dynamics of the N individual modes. This is most easily formulated for systems without damping, or with Caughey damping (Caughey and Okelly, 1965), in which case the modes are standing waves, but is true even more generally, when the modes are traveling waves. This view is very powerful and serves as the basis for the vast majority of practical vibration studies.

The inclusion of non-linearities in a model necessarily destroys this invariance, so it is natural to consider the conditions under which weakly non-linear systems behave in a manner that is *qualitatively* different from the underlying linear model. In other words, under what circumstances might one need to include non-linear effects in multi-mode model, even if the non-linear terms (forces) are small compared to linear terms.[1] In order to intuitively see when these situations occur, the simplest situation is when a system has repeated natural frequencies, which occurs quite naturally in systems with symmetries, for example, circular or square structures. In this case small effects can have large consequences, as a result of high sensitivities near such a degenerate point. In fact, even small linear terms that disturb symmetry can have large effects in such cases. However, as our focus is on non-linear modeling, we maintain focus on the effects of non-linearity in such situations. In fact, we consider such a system for our first example, namely the spherical pendulum, where the symmetry of the system leads to freedom in selection of the vibration modes, and the interaction of the

[1] Of course, in models with strong non-linearities the linearized system modes will always interact, but in such cases the modes are used only for convenience, if at all, since they have no special meaning for the system.

modes is demonstrated by considering the response to harmonic excitation of a single mode. We then consider a spring-pendulum system with frequencies in a two-to-one (2:1) ratio, for which the higher frequency mode drives the lower frequency mode by a type of parametric excitation through non-linear coupling terms.

Other such *internal resonance* conditions exist, and in general, they occur between two modes when their natural frequencies are rationally related ($p : q$) and certain coupling terms exist. However, in systems with damping, only low order $p : q$ resonances of the type considered here are observed, the most common being 1:1, 2:1, and 3:1. Several of these cases are thoroughly explored in the literature, including at least two entire books on the subject (Nayfeh, 2000; Tondl, 2000).

Modeling these resonances requires that one keep non-linear terms that couple the linear vibration modes, and these terms can be captured by the expansion methods outlined previously in this chapter. In order to examine the modal interactions it is convenient to first convert the model to linear modal coordinates so that the linear terms are uncoupled. This is generally a cumbersome process, aided by symbolic computation for specific models, and it leads to a multitude of non-linear terms, even for models with two DOF and quadratic and/or cubic non-linearities. However, not all these terms are important. The important non-linear terms in such models can be found by applying perturbation methods, in which only the essential terms survive. These essential terms are those that cause resonant interactions of the modes. In fact, one can often guess the nature of these terms by considering their harmonic content, since only terms that resonate with a given mode will survive a systematic analysis. The most formal and rigorous approach for this process is to employ the theory of normal forms, in which successive coordinate transformation are used to remove non-resonant terms (Guckenheimer and Holmes, 1983). In most common cases, this approach is equivalent to the more straightforward application of averaging (Sethna, 1995).

Before turning to examples, it should be noted that these non-linear modal interactions can be explored in terms of free vibrations, since internal resonances lead to energy exchange between vibration modes. The nature of this exchange depends on the initial amount of energy provided to the interacting modes, as well as system parameters that include the nature of individual mode (Duffing) non-linearities, the nearness of the modal frequencies to perfect $p : q$ resonance conditions, the level of individual modal damping, and the strength of the non-linear modal coupling terms. While interesting, these free vibrations are not considered here, since the demonstration of non-linear modal interactions is more compelling in the forced response, to which we now turn.

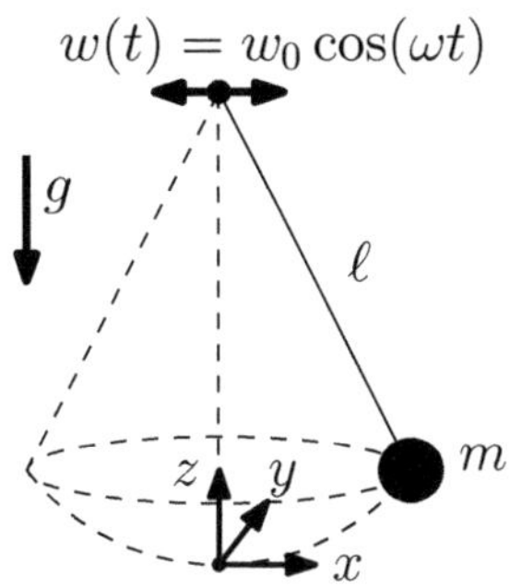

Figure 10. Schematic representation of a spherical pendulum with planar (x direction) forcing from harmonic base motion w.

3.1 A 1:1 Internal Resonance: The Spherical Pendulum

The behavior described here is not particular to the spherical pendulum; it also appears in other systems with repeated frequencies, such as strings, square plates, circular plates, etc. In fact, if one drives such a system near resonance, the non-linear interactions described here are very likely to be observed, and, in fact, for light damping they cannot be avoided except at very small vibration amplitudes.

For our example of 1:1 internal resonance, we consider the spherical pendulum shown in Figure 10, which is one of the simplest examples of a two-DOF symmetric system with repeated natural frequencies. It is simply the non-planar extension of the simple pendulum, with a spherical joint at its root. Our approach closely follows that of Miles (Miles, 1962, 1984a). As shown in Figure 10, the pendulum has length ℓ and mass m, is driven by a base motion $w(t) = w_0 \cos(\omega t)$ in the x direction, and the absolute position of m is given by (x, y, z), with a holonomic constraint given by $(x - w)^2 + y^2 + (z - \ell)^2 = \ell^2$. The system kinetic and potential energies are given by $T = m(\dot{x}^2 + \dot{y}^2 + \dot{z}^2)/2$ and $V = mgz$. We can reduce the number of DOF from three (x, y, z) to two using the constraint, and (x, y) is the convenient set of generalized coordinates for weakly non-linear responses in which the pendulum angle from the vertical is not large. Using Lagrange's equations we obtain the following equations of motion (EOM) for the driven direction x and its perpendicular direction y (Gottlieb and Habib, 2012)

$$\begin{aligned}\ddot{x} + N(x - w) &= 2\Gamma(\dot{x} - \dot{w}) \\ \ddot{y} + Ny &= 2\Gamma\dot{y}\end{aligned}$$

where we have added equal linear damping to both directions, and introduced the following definitions,

$$N = g\sqrt{\ell^2 - R^2} - (x - w)\ddot{w} + (\dot{x} - \dot{w})^2 + \dot{y}^2 + (R\dot{R})^2/(\ell^2 - R^2)$$
$$R^2 = (x - w)^2 + y^2, \qquad R\dot{R} = (x - w)(\dot{x} - \dot{w}) + y\dot{y}$$

Note that due to the symmetry of the system, if we set the drive to zero, $w(t) = 0$, we obtain identical EOM for $\pm x$ and $\pm y$, i.e., the unforced EOM are invariant under the transformation $(x, y) \to -(y, x)$. In addition, the unforced EOM are uncoupled, indicating that x and y are linear modal coordinates (as would any pair of non-parallel direction coordinates of this type). Moreover, even for $w(t) \neq 0$, the out-of-plane DOF y possesses the reflection symmetry $(y \to -y)$. Note that this model has both conservative and inertial non-linearities.

Consideration of the linear model indicates that only the driven mode x will respond to the excitation w, and that y remains zero, as it is unforced. In fact, one sees that such a response satisfies the non-linear model as well, predicting a response with $y = 0$ and x governed by the EOM

$$\ddot{x} + N|_{y\equiv 0}(x - w) = 2\Gamma(\dot{x} - \dot{w}).$$

Clearly this represents the response of a simple planar pendulum to base excitation, so that a softening frequency response is observed when the base motion is harmonic. This is demonstrated below, where we also show that when the suspension magnitude exceeds certain threshold ($w_0 > w_{c_1}$), there is energy transfer from the (driven) in-plane x mode to the (undriven) out-of-plane y mode, resulting in coupled mode responses that are due to the non-linear resonant coupling between the modes.

In order to investigate the system response, we again apply the method of averaging, using the scaled Cartesian coordinates $x = w_0^{1/3}\ell^{2/3}[u_1(\tau)\cos(\omega t) + v_1(\tau)\sin(\omega t)]$, $y = w_0^{1/3}\ell^{2/3}[u_2(\tau)\cos(\omega t) + v_2(\tau)\sin(\omega t)]$, rather than the polar notation which is singular for $y = 0$. We also rescale time by defining the slow time scale $\tau = (w_0/\ell)^{2/3}(\omega t/2)$. Then, by retaining only the leading-order non-linear terms (in this case, cubic non-linearities), we obtain the following non-dimensional evolution equations for the slowly varying modal coordinates (Miles, 1962, 1984a)

$$\frac{du_1}{d\tau} = -\alpha u_1 - (\sigma + E/8)v_1 - 3Mu_2/4 \tag{30}$$

$$\frac{dv_1}{d\tau} = -\alpha v_1 + (\sigma + E/8)u_1 - 3Mv_2/4 + 1 \tag{31}$$

$$\frac{du_2}{d\tau} = -\alpha u_2 - (\sigma + E/8)v_2 + 3Mu_1/4 \tag{32}$$

$$\frac{dv_2}{d\tau} = -\alpha v_1 + (\sigma + E/8)u_2 + 3Mv_1/4 \tag{33}$$

where $\alpha = (2\Gamma/\omega)(w_0/\ell)^{-2/3}$ is the rescaled damping parameter, $\sigma = (w_0/\ell)^{-2/3}(\omega^2 - g/\ell)/\omega$ is the rescaled frequency detuning parameter, and

$$E = u_1^2 + v_1^2 + u_2^2 + v_2^2, \quad M = u_1 v_2 - u_2 v_1$$

are measures of the pendulum energy and angular momentum with respect to the z axis, respectively. Note that the averaged equations (Eqs. (30)-(33)) preserve all the symmetries of the original EOM, i.e., they are invariant under the transformation $(u_2, v_2) \to -(u_2, v_2)$. In the unforced case $w(t) = 0$, which corresponds to the subtraction of the unity term (which is the scaled excitation term) from Eq. (31), they are invariant under the transformation $(u_1, v_1, u_2, v_2) \to -(u_2, v_2, u_1, v_1)$. Also note that these averaged equations can be derived by first averaging the system Lagrangian or Hamiltonian, and then applying Hamilton's principle; see Miles (1984a). Finally, note that the averaged equations are much simpler than the original equations of motion, a result of the averaging process which eliminates all non-resonant terms and yields EOM that do not depend explicitly on time.

In order to find the fixed points representing the steady-state responses, we set the right hand side of Eqs. (30) and (33) to zero, and note that the steady-state response is completely determined by the pendulum energy and angular momentum, that is, $u_i = u_i(E, M)$ and $v_i = v_i(E, M)$. Hence, at steady-state we can reduce the number of algebraic equation from four to two

$$\begin{aligned} E^3 + 16(\alpha^2 - 3\sigma^2)E + 128\sigma(\sigma^2 + \alpha^2) - 24 = 0, \\ M^2 = \tfrac{5}{36}\left[E^2 + \tfrac{32}{5}\sigma E - \tfrac{64}{5}(\sigma^2 + \alpha^2)\right] \end{aligned} \tag{34}$$

These equations simplify considerably for the case of planar motion, $(u_2, v_2) = (0, 0)$, reducing to

$$E_1^3 + 16\sigma E_1^2 + 64(\sigma^2 + \alpha^2)E_1 - 64 = 0, \quad M = 0 \tag{35}$$

where $E_1 = u_1^2 + v_1^2$ is simply the square of the amplitude of vibration in the x coordinate. Note that Eq. (35), which governs the frequency response curve of the planar motion, is identical to Eq. (23) with a softening Duffing non-linearity, as expected. However, the full equations for steady-state contain much more, including non-planar responses.

Figure 11 shows samples of the system frequency response curves, indicating the stability of the fixed points of Eqs. (30)-(33), for three different values of the parameter α, which depends on the excitation amplitude, the damping, and other parameters. These response curves are the real and positive roots for E in the cubic equations given in Eqs. (35),(34), where for the non-planar motion the condition $M^2 > 0$ must also hold for physically meaningful solutions (since the angular momentum is not complex). The transitions in stability of the fixed points are associated with bifurcation points in Eqs. (30)-(33). The conditions for these bifurcation points can be obtained from the quartic characteristic polynomial of this fourth-order system. However, for the sake of brevity we omit details and explain these bifurcations and the attendant transitions in stability in a qualitative way, and we refer the interested reader to Miles (1984a) for more details and quantitative descriptions.

In Figure 11(a) we show the response curve of the pendulum for $\alpha = 0.625$, which yields excitation level of $u_0 = 5.724\ell(\Gamma\sqrt{\ell/g})^{3/2}$. For this relatively small magnitude of excitation the system behaves almost linearly. That is, only the planar solution exists and there are no bi-stability regions. However, the non-linearity of the x mode induces a small level of asymmetry in the response curve. The response for excitation magnitude $u_0 = 7.414\ell(\Gamma\sqrt{\ell/g})^{3/2}$ ($\alpha = 0.526$) is shown in Figure 11(b), where it is seen that in the vicinity of the primary resonance ($\sigma = 0$) the planar solution losses stability due to pitchfork (symmetry-breaking) bifurcations and a pair of stable, non-trivial, non-planar solutions appear. This pair of non-planar solutions is associated with the in-phase ($w_0^{1/3}\ell^{2/3}\sqrt{u_2^2 + v_2^2}\cos(\omega t + \phi)$) and out-of-phase ($w_0^{1/3}\ell^{2/3}\sqrt{u_2^2 + v_2^2}\cos(\omega t + \phi - \pi)$) solutions of the out-of-plane DOF, which correspond to the $y \to -y$ symmetry. The conditions for these two pitchfork bifurcations can be found by setting $M^2 = 0$ in Eq. (34), and solving simultaneously the three equations in Eqs. (35)-(34), yielding an algebraic relation between σ and α which can be shown to have either 0, 1 or 2 real solution(s) for σ, depending on the value of α (see Miles (1984a)). These responses physically represent elliptical rotations of the pendulum mass, which must exist in symmetric pairs with rotation in opposite directions.

As the magnitude of the excitation is further increased to $w_0 = 22.6\ell(\Gamma\sqrt{\ell/g})^{3/2}$ ($\alpha = 0.25$), the response becomes like that shown in Figure 11(c), in which additional saddle-node bifurcations appear in the planar solution which satisfy the condition $d\sigma/dE = 0$ in Eq. (35); these result in bi-stability of the

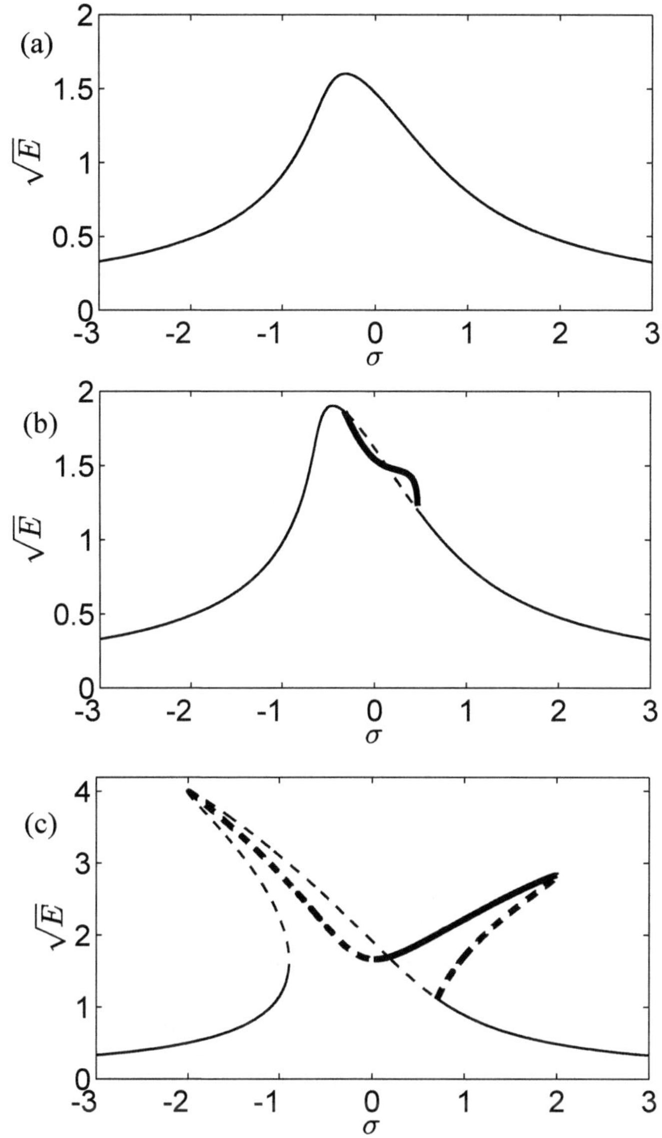

Figure 11. System frequency response curves for: (a) $\alpha = 0.625$, (b) $\alpha = 0.526$, and (c) $\alpha = 0.25$. Planar/non-planar solutions are denoted by thin/thick lines, and stable/unstable solutions are denoted by solid/dashed lines.

single mode response, just as described for the Duffing equation, although here the upper branch can be unstable. In addition, there are also saddle-node and Hopf bifurcations of the non-planar solution, which correspond to $d\sigma/dE = 0$ at the first equation in Eq. (34) and $c_3(c_2c_1 - c_3) - c_4c_1^2 = 0$, respectively, where c_k are the coefficients of the characteristic polynomial $\lambda^4 + c_1\lambda^3 + c_2\lambda^2 + c_3\lambda + c_4 = 0$ of the Jacobian matrix of Eqs. (30)-(33) evaluated at the fixed points. Moreover, the Hopf bifurcation, shown by the change of stability of the non-planar response (the thick line in Figure 11(c)) leads to a limit-cycle in the averaged equations that produces slow periodic oscillations in Eqs. (30)-(33), which results in slow periodic modulation of the amplitudes and phases in the original system variables. The physics of this response corresponds to beating motions in which the response exchanges amplitudes between in-plane and out-of-plane responses, resulting in a beating type of whirling response of the pendulum. The amplitude and the frequency of this limit-cycle can be estimated by applying the method of harmonic balance to Eqs. (30)-(33). Note that for σ values below the Hopf bifurcation threshold shown in Figure 11(c), both the planar and non-planar solutions are unstable, and in this region there are non-stationary responses in which the amplitude of oscillation is slowly modulated. This modulation can be periodic, as described above, but can undergo additional bifurcations resulting in chaotic modulations, which themselves can undergo bifurcations, including global bifurcations that cause sudden and dramatic changes in the system response. Additional details about these dynamics, which are surprisingly complicated for a two DOF system model, yet are completely generic for this internal resonance, and can be found in Johnson and Bajaj (1989); O'Reilly and Holmes (1992).

In closing this section, we note that this simple system model demonstrates that complicated interactions of vibration modes can occur even when non-linear effects are small, if the modal frequencies allow them to resonate with one another. This example has two modes with identical natural frequencies, a feature resulting from the symmetry of the system. In terms of engineering applications, this arises in any circular or square structure. As an example, a common MEMS device with circular symmetry is a disk-ring-gyro (DRG), which also exhibits this internal resonance. It has been observed that non-linear modal interactions in that device result in a type of parametric amplification that is favorable to device sensitivity (Nitzan et al., 2015). Of course, in real systems the symmetry is not perfect and the modes have slightly different natural frequencies. In such cases, the model that accounts for this difference shows modal interactions can still occur, but are eliminated if the inter-modal detuning is too large. Specifically, as noted above, in general the nature and strength of the non-linear

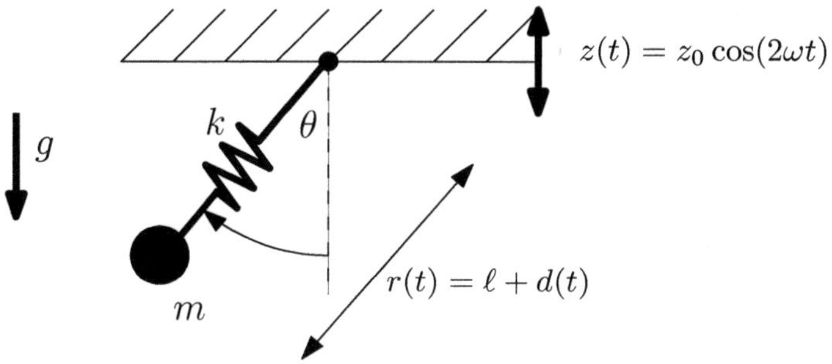

Figure 12. Schematic representation of a spring pendulum with vertical base excitation.

interactions depend on this *internal* detuning, the level of modal damping, the level of excitation (which sets the energy in the modes), the nearness of the drive frequency to resonance, and the nature of the modal coupling terms.

While modal interactions are clear and not surprising for this 1:1 case, they can also occur in more subtle cases, as we show next.

3.2 A 2:1 Internal Resonance: The Spring Pendulum System

The physical system considered in this section is depicted in Figure 12. It consists of an ideal planar pendulum whose mass is suspended via a spring so that the effective pendulum length is given by $r(t) = \ell + d(t)$ where $\ell = \ell_0 + \Delta$ is the static pendulum length, which is dictated by the unstretched spring length ℓ_0 and the static deflection $\Delta = mg/k$. The generalized coordinates used are $d(t)$ and the pendulum angle $\theta(t)$. The system is driven by a vertical base motion $z(t)$, which is assumed to be harmonic, and which directly drives d and parametrically drives θ. The kinetic and potential energies for this system are given by $T = \frac{m}{2}\left[\dot{d}^2 + \dot{z}^2 + (l+d)^2\dot{\theta}^2 + 2\dot{z}\left(\dot{\theta}(l+d)\sin\theta - \dot{d}\cos\theta\right)\right]$ and $V = \frac{k}{2}(\frac{mg}{k} + d)^2 - mg(l+d)\cos\theta + z$. Lagrange's equations are used to derive the equations of motion for θ and d, and with common factors are removed, and with linear damping added to the pendulum pivot and the sliding mass, they are given by

$$\begin{aligned}\ddot{\theta}(\ell+d)+\Gamma_{\theta}\ell\dot{\theta}+2\dot{d}\dot{\theta}+(g+\ddot{z})\sin\theta &= 0\\ \ddot{d}+\Gamma_{d}\dot{d}+\frac{k}{m}d-(\ell+d)\dot{\theta}^{2}+g(1-\cos\theta)-\ddot{z}\cos\theta &= 0.\end{aligned}$$

This model has both conservative and inertial non-linearities. A convenient feature of these equations, shared with the spherical pendulum example, is that the physical coordinates are also modal coordinates, seen by the fact that the linearized equations are uncoupled. However, in this case the modal frequencies, $\omega_\theta = \sqrt{g/\ell}$ and $\omega_x = \sqrt{k/m}$, are not identical. It is simple to show that $\frac{\omega_x}{\omega_\theta} = \sqrt{1+\frac{k\ell_0}{mg}}$, and hence $\omega_\theta < \omega_x$, and the system can take on a full range of frequency ratios, depending on parameter values. Also, note that $\theta = 0$ is a solution of the θ equation, which renders a simple linear equation with base excitation $\ddot{z}$ for the d equation. It is shown below that this response can become unstable under certain resonance conditions, resulting in a nontrivial response with some interesting features. For example, due to the reflection symmetry ($\theta \to -\theta$) any periodic solutions will come in pairs (in-phase and out-of-phase), not surprisingly, just as in the case of parametric resonance.

The non-linear resonance behavior of interest here can be captured by keeping second order non-linear terms in the model. Defining $x = d/\ell$ and $w = z/\ell$, and expanding out to second order in x, θ, and w yields

$$\begin{aligned}\ddot{\theta}(1+x)+\Gamma_{\theta}\dot{\theta}+\omega_{\theta}^{2}\theta+2\dot{x}\dot{\theta} &= -\ddot{w}\theta\\ \ddot{x}+\Gamma_{x}\dot{x}+\omega_{x}^{2}x-\dot{\theta}^{2}+\frac{\omega_{\theta}^{2}}{2}\theta^{2} &= \ddot{w}\end{aligned}$$

We consider harmonic vertical base excitation, modeled by $w(t) = w_0\cos(2\omega t)$ that resonates the mass in its d direction, that is vertically, by taking 2ω close to ω_x. This sets up large motion in x if damping is small, and this results in a harmonic variation of the pendulum length, and hence the effective pendulum frequency. If that variation happens to occur at a frequency that is about twice the frequency of the pendulum swinging motion, it can result in parametric resonance of the θ response. However, the pendulum motion acts back on the sliding motion of the mass, and this sets up the conditions for an internal resonance. Specifically, this resonance condition arises when $2\omega \approx \omega_x$ drives the sliding motion of the mass at a frequency 2ω, and when $\omega_x \approx 2\omega_\theta$, so that the motion of the mass induces swinging of the pendulum at ω, that is, at half the drive frequency. This latter condition corresponds to the 2:1 internal resonance. In fact, it is seen that the non-linear Coriolis term $2\dot{x}\dot{\theta}$ in the θ EOM provides the source of the parametric drive on θ, and the θ-dependent quadratic terms in the x equation have frequency 2ω, thus resonantly driving x.

To investigate the dynamic response of this system, we carry out averaging using a transformation that assumes a response at frequency 2ω for x, since it responds at the drive frequency, and ω for θ, since it is driven parametrically. To this end we employ a transformation

$$\begin{aligned}
\theta(t) &= \epsilon_1 \left(u_1(\tau)\cos(\omega t) + v_1(\tau)\sin(\omega t)\right) \\
\dot{\theta}(t) &= \epsilon_1 \left(-u_1(\tau)\omega\sin(\omega t) + v_1(\tau)\omega\cos(\omega t)\right) \\
x(t) &= \epsilon_2 \left(u_2(\tau)\cos(2\omega t) + v_2(\tau)\sin(2\omega t)\right) \\
\dot{x}(t) &= \epsilon_2 \left(-u_2(\tau)2\omega\sin(2\omega t) + v_2(\tau)2\omega\cos(2\omega t)\right)
\end{aligned} \tag{36}$$

in the equations of motion, where $\epsilon_1 = 8/3$ and $\epsilon_2 = 4/3$ are introduced for simplifying the system parameters in the averaged equations given below. The result, and the constraint equations described previously, lead to four equations that govern the slowly varying coordinates $(u_1,\ v_1,\ u_2,\ v_2)$, which are averaged over two periods of the drive, $2\pi/\omega$ (as in the case of parametric resonance), resulting in

$$\frac{du_1}{d\tau} = -\alpha_1 u_1 + \sigma_1 v_1 + u_1 v_2 - v_1 u_2 \tag{37}$$

$$\frac{dv_1}{d\tau} = -\alpha_1 v_1 - \sigma_1 u_1 - u_1 u_2 - v_1 v_2 \tag{38}$$

$$\frac{du_2}{d\tau} = -\alpha_2 u_2 + \sigma_2 v_2 + 2u_1 v_1 \tag{39}$$

$$\frac{dv_2}{d\tau} = -\alpha_2 v_2 - \sigma_2 u_2 - u_1^2 + v_1^2 + 1 \tag{40}$$

where $\alpha_1 = \Gamma_\theta/(2\omega_\theta)$ and $\alpha_2 = \Gamma_x/(2\omega_\theta)$ are non-dimensional damping coefficients, $\sigma_1 = (\omega^2 - \omega_\theta^2)/2\omega_\theta^2$ and $\sigma_2 = (4\omega^2 - \omega_x^2)/4\omega_\theta^2$ are non-dimensional detuning coefficients, and $\tau = \omega t$ is the non-dimensional time scale. These, along with the defined values of $\epsilon_{1,2}$ yield this convenient form of Eqs. (37)-(40) (for details, see Sethna (1965); Sethna and Bajaj (1978); Miles (1984b)). Note that these simplified non-dimensional equations for this resonance are valid for models with any type of quadratic non-linearities, and hence they are the normal from for 2:1 internal resonance.

In what follows we take a special case that demonstrates the behavior of interest, by setting $\sigma_1 = \sigma_2/2 \equiv \sigma$, which corresponds to $\omega_x = 2\omega_\theta$, that is, perfect internal tuning, and $\alpha_1 = \alpha_2 \equiv \alpha$, which corresponds to equal normalized damping in the two modes. For a trivial first mode solution (i.e., $\theta = 0$ or equivalently $u_1 = v_1 = 0$) both the EOM and Eqs. (37)-(40) are linear in the second mode (x or equivalently u_2 and v_2), and the unique

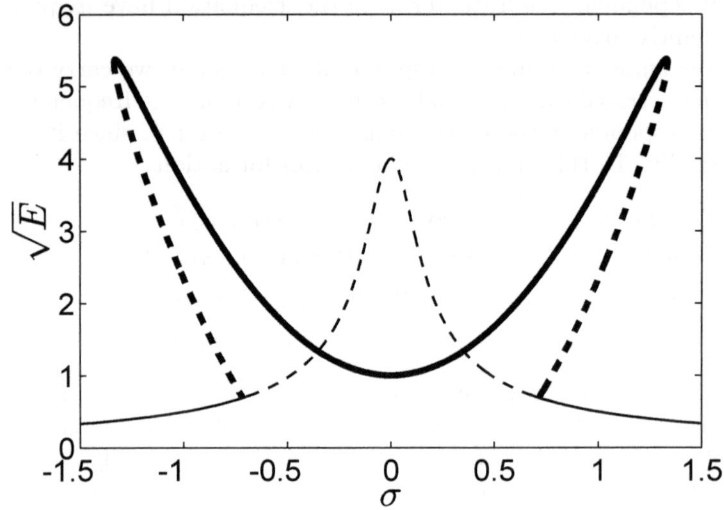

Figure 13. System frequency response curves for $\alpha = 0.25$. Trivial/nontrivial first mode solutions are denoted by thin/thick lines, and stable/unstable solutions denoted by solid/dashed lines.

fixed point in Eqs. (37)-(40) corresponds to a steady-state solution in the EOM given by

$$u_2 = \frac{2\sigma}{\alpha^2 + 4\sigma^2}, \qquad v_2 = \frac{\alpha}{\alpha^2 + 4\sigma^2}$$

and hence $E_2 = u_2^2 + v_2^2 = 1/(\alpha^2 + 4\sigma^2)$, where $E = u_1^2 + v_1^2 + u_2^2 + v_2^2$ is a measure of the system energy in analogy with the spherical pendulum example. For a nontrivial first mode, we obtain the following quadratic equation in E

$$E^2 - 6\sigma^2 E - 1 + 9\sigma^2(\alpha^2 + \sigma^2) = 0 \tag{41}$$

which yields $E = 3\sigma^2 \pm \sqrt{1 - 9\alpha^2\sigma^2}$. Consequently, there are two saddle-node bifurcations at detuning values of $\sigma_{sn} = \pm\frac{1}{3\alpha}$ (given by a zero discriminant of Eq. (41)) and two pitchfork bifurcations at detuning values of $\sigma_{pf} = \pm\sqrt{\frac{1}{8}\left(\sqrt{16 + 9\alpha^4} - 5\alpha^2\right)}$ (the intersections of E_2 with the lower branch of E).

Figure 13 shows the system frequency response curve for $\alpha = 0.25$, which demonstrates the behavior of interest. In this case of perfect internal tuning ($\omega_x = 2\omega_\theta$), the response is symmetric with respect to σ. We note that far from resonance (large values of $|\sigma|$) only the linear response exists (blue curve), in which the first mode (θ) remains at zero, while the second mode (x) responds in a linear manner to the drive. However, when the excitation frequency (2ω) is in the vicinity of the second mode frequency (ω_x), i.e., $|\sigma| < \sigma_{pf}$, the linear x response loses stability and the θ response becomes active. Note that this instability is due to a sub-critical pitchfork bifurcation and hence there is a region of bi-stability which is bounded by saddle-node and pitchfork bifurcations on either side of the resonance, $\sigma_{pf} < |\sigma| < \sigma_{sn}$. In this region the response can be single mode, purely x, or coupled mode, depending on initial conditions. It can be shown that the Hopf bifurcation criterion is never satisfied for zero detuning between the two modes ($\omega_x = 2\omega_\theta$), and therefore this ideal system exhibits only periodic responses, and no secondary bifurcations of the coupled mode response occur. Nevertheless, for detuned modes Hopf bifurcations can occur, as can further sequences of bifurcations which lead to modulated responses, including those of a chaotic nature (Nayfeh, 2000).

This resonance has been thoroughly investigated, including experimental verification as described in detail in Haddow et al. (1984), and it has a number of applications. For example, a 2:1 internal resonance between pitch and roll motions in ships was noted by Froude (1863) to induce large roll motions in heading seas that excited pitch motions. A vibration absorber based on this resonance was proposed in Haxton and Barr (1972), although its features were not favorable when compare to the usual tuned mass damper. More recently, MEMS devices with 2:1 resonances have been proposed for signal processing applications (Qalandar et al., 2014; Vyas et al., 2008, 2009). Other examples and more detailed analysis are provided in Nayfeh (2000); Tondl (2000).

4 Closing

After many decades of study in this area by hundreds of researchers, the question remains: "under what circumstances does one need a non-linear model?" The answer is not, of course, an algorithm, but rather a process that depends on experience, intuition, and the expected requirements of the model, and that is why the title of this volume, and the attendant CISM course, is "The *Art* of Modeling Mechanical Systems." To address the question we have provided a brief introduction to the non-linear vibratory behavior of mechanical systems, focusing on free vibrations and resonant dynamics. While our primary goal was to emphasize modeling, the purpose of such modeling is to describe dynamic responses that are not captured by linear models, and the related issue of determining when linear models break down. To that end, we outlined features of non-linear resonant responses using analytical results based on the method of averaging, which is useful for these cases. However, in some cases other methods, such as multiple scales (Nayfeh and Mook, 2008) or harmonic balance may be more convenient, and/or more intuitive. Nevertheless, the methods serve only as a means for one to predict and understand the non-linear response of a given model. We hope that this brief introduction provides a flavor of the issues involved in answering these questions.

An important issue for future research is the systematic accounting for non-linear behavior in the process of designing a dynamical system. There are at least two aspects to this topic: (i) Can the system be designed to avoid non-linear behavior? (ii) If not, can the non-linear behavior be "tuned" to avoid unfavorable responses? The answer to both questions relies on knowledge of the system behavior that accounts for non-linearities. Hence, even the design of a system for linear response requires an understanding of the limits of the assumed linear model, which in turn requires knowledge about the non-linear effects. A next step in the development of design tools for dynamic systems is to embrace non-linearity in system models, and to develop systematic methods for addressing non-linear behavior so that it can be taken into account in the design phase. The controls community has been doing this for some time, as evidenced by the many books on non-linear control (see, for example, Krstic et al. (1995); Slotine et al. (1991); Isidori (1995); Khalil (1996)). The systematic design for desired open-loop response is in some ways more challenging, due to the lack of feedback and actuation, but the subject is sufficiently mature for the community to consider taking on this task.

Acknowledgements

The lead author is grateful to the organizers of the CISM course on *The Art of Modeling Mechanical Systems*, in particular, Professor Freidrich Pfieffer, for inviting him to contribute to the lectures and to this volume. The authors are grateful to Professor Mark Dykman of Michigan State University for discussions that were particularly helpful in the preparation of this chapter. The authors' current related work on non-linear vibrations is supported by the US National Science Foundation under grants 1234067 and 1100260, by US ARO grant W911NF-12-1-0235, managed by Dr. Samuel Stanton, and by DARPA grant FA8650-13-1- 7301, Mesodynamic Architectures (MESO), managed by Dr. Daniel Green.

Bibliography

Fadi M Alsaleem, Mohammad Younis, Laura Ruzziconi, et al. An experimental and theoretical investigation of dynamic pull-in in MEMS resonators actuated electrostatically. *Microelectromechanical Systems, Journal of*, 19(4):794–806, 2010.

Gregory L Baker and James A Blackburn. The pendulum. A case study in physics. *AMC*, 10:12, 2005.

RC Batra, M Porfiri, and D Spinello. Review of modeling electrostatically actuated microelectromechanical systems. *Smart Materials and Structures*, 16(6):R23, 2007.

NN Bogolyubov and Yu A Mitropolskii. Asymptotic methods in the theory of non-linear oscillations. Technical report, DTIC Document, 1955.

David K Campbell. Nonlinear science. *Los Alamos Science*, 15:218–262, 1987.

TK Caughey and MEJ Okelly. Classical normal modes in damped linear dynamic systems. *Journal of Applied Mechanics*, 32(3):583–588, 1965.

Lamberto Cesari. *Asymptotic behavior and stability problems in ordinary differential equations*, volume 16. Springer Science & Business Media, 2012.

Per Danzl and Jeff Moehlis. Weakly coupled parametrically forced oscillator networks: existence, stability, and symmetry of solutions. *Nonlinear Dynamics*, 59(4):661–680, 2010.

Freeman Dyson. A meeting with Enrico Fermi. *Nature*, 427(6972):297–297, 2004.

Nikolay P Erugin, Richard Bellman, and Scripta Technica. *Linear systems of ordinary differential equations: with periodic and quasi-periodic coefficients*, volume 28. Academic Press New York & London, 1966.

Alexander Fidlin. *Nonlinear oscillations in mechanical engineering*. Springer Science & Business Media, 2005.

W Froude. Remarks on mr. scott russells paper on rolling. *Transactions of the Institute of Naval Research*, 4:232–275, 1863.

O Gottlieb and G Habib. Nonlinear model-based estimation of quadratic and cubic damping mechanisms governing the dynamics of a chaotic spherical pendulum. *Journal of Vibration and Control*, 18(4):536–547, 2012.

John Guckenheimer and Philip Holmes. *Nonlinear oscillations, dynamical systems, and bifurcations of vector fields*, volume 42. Springer Science & Business Media, 1983.

AG Haddow, ADS Barr, and DT Mook. Theoretical and experimental study of modal interaction in a two-degree-of-freedom structure. *Journal of Sound and Vibration*, 97(3):451–473, 1984.

Peter Hagedorn. Nonlinear oscillations. *Oxford and New York, Clarendon Press, 1981. 298 p. Translatio*, 1, 1981.

Robert Steedman Haxton and Allan David Stephen Barr. The autoparametric vibration absorber. *Journal of Manufacturing Science and Engineering*, 94(1):119–125, 1972.

Vladimir Andreevich Iakubovich and Viacheslav Mikhaĭlovich Starzhinskiĭ. *Linear differential equations with periodic coefficients*, volume 2. Wiley, 1975.

Alberto Isidori. *Nonlinear control systems.* Springer Science & Business Media, 1995.

JM Johnson and Anil K Bajaj. Amplitude modulated and chaotic dynamics in resonant motion of strings. *Journal of Sound and Vibration*, 128(1): 87–107, 1989.

William J Kaiser, Thomas W Kenny, Joseph K Reynolds, Thomas R Van Zandt, and Steven B Waltman. Methods and apparatus for improving sensor performance, May 18 1993. US Patent 5,211,051.

Eyal Kenig, MC Cross, LG Villanueva, RB Karabalin, MH Matheny, Ron Lifshitz, and ML Roukes. Optimal operating points of oscillators using non-linear resonators. *Physical Review E*, 86(5):056207, 2012.

Hassan K Khalil. *Nonlinear systems*, volume 3. Prentice hall New Jersey, 1996.

Miroslav Krstic, Petar V Kokotovic, and Ioannis Kanellakopoulos. *Nonlinear and adaptive control design.* John Wiley & Sons, Inc., 1995.

Vijay Kumar, Jacob K Miller, and Jeffrey F Rhoads. Non-linear parametric amplification and attenuation in a base-excited cantilever beam. *Journal of Sound and Vibration*, 330(22):5401–5409, 2011.

Mark Lemkin. MEMS comb-finger actuator, August 31 2001. US Patent App. 09/944,395.

Ron Lifshitz and MC Cross. Nonlinear dynamics of nanomechanical and micromechanical resonators. *Review of non-linear dynamics and complexity*, 1:1–52, 2008.

Michael S-C Lu and Gary K Fedder. Position control of parallel-plate microactuators for probe-based data storage. *Microelectromechanical Systems, Journal of*, 13(5):759–769, 2004.

Leonard Meirovitch. *Principles and techniques of vibrations*, volume 1. Prentice Hall New Jersey, 1997.

Leonard Meirovitch. *Methods of analytical dynamics.* Courier Corporation, 2010.

John Miles. Resonant motion of a spherical pendulum. *Physica D: Nonlinear Phenomena*, 11(3):309–323, 1984a.

John Miles. Resonantly forced motion of two quadratically coupled oscillators. *Physica D: Nonlinear Phenomena*, 13(1):247–260, 1984b.

John W Miles. Stability of forced oscillations of a spherical pendulum. *Quarterly of Applied Mathematics*, 20(1):21–32, 1962.

Francis C Moon. *Chaotic and Fractal Dynamics: Introduction for Applied Scientists and Engineers*. John Wiley & Sons, 2008.

Ali H Nayfeh. *Perturbation methods*. John Wiley & Sons, 2008.

Ali H Nayfeh and Dean T Mook. *Nonlinear oscillations*. John Wiley & Sons, 2008.

Ali H Nayfeh, Mohammad I Younis, and Eihab M Abdel-Rahman. Dynamic pull-in phenomenon in MEMS resonators. *Nonlinear dynamics*, 48(1-2): 153–163, 2007.

Ali Hasan Nayfeh. *Nonlinear interactions*. Wiley, 2000.

Sarah H Nitzan, Valentina Zega, Mo Li, Chae H Ahn, Alberto Corigliano, Thomas W Kenny, and David A Horsley. Self-induced parametric amplification arising from non-linear elastic coupling in a micromechanical resonating disk gyroscope. *Scientific reports*, 5, 2015.

Oliver O'Reilly and Phillip J Holmes. Non-linear, non-planar and non-periodic vibrations of a string. *Journal of Sound and Vibration*, 153(3): 413–435, 1992.

PM Polunin, Y Yang, MI Dykman, TW Kenny, and SW Shaw. Characterization of MEMS resonator nonlinearities using the ringdown response. *revised version to appear in Journal of Microelectromechanical Systems*, 2016b.

PM Polunin, Y Yang, MI Dykman, TW Kenny, and SW Shaw. Independent characterization of mechanical and electrostatic nonlinearities in capacitive MEMS resonators. *near-final version complete*, 2016a.

KR Qalandar, BS Strachan, B Gibson, M Sharma, A Ma, SW Shaw, and KL Turner. Frequency division using a micromechanical resonance cascade. *Applied Physics Letters*, 105(24):244103, 2014.

Jeffrey F Rhoads and Steven W Shaw. The impact of non-linearity on degenerate parametric amplifiers. *Applied Physics Letters*, 96(23):234101, 2010.

Jeffrey F Rhoads, Steven W Shaw, and Kimberly L Turner. The nonlinear response of resonant microbeam systems with purely-parametric electrostatic actuation. *Journal of Micromechanics and Microengineering*, 16 (5):890, 2006a.

Jeffrey F Rhoads, Steven W Shaw, Kimberly L Turner, Jeff Moehlis, Barry E DeMartini, and Wenhua Zhang. Generalized parametric resonance in electrostatically actuated microelectromechanical oscillators. *Journal of Sound and Vibration*, 296(4):797–829, 2006b.

Jeffrey F Rhoads, Nicholas J Miller, Steven W Shaw, and Brian F Feeny. Mechanical domain parametric amplification. *Journal of Vibration and Acoustics*, 130(6):061006, 2008.

Stav Zaitsev, Oleg Shtempluck, Eyal Buks, and Oded Gottlieb. Nonlinear damping in a micromechanical oscillator. *Nonlinear Dynamics*, 67(1): 859–883, 2012.

The Art of Modeling in Solid Mechanics

Peter Wriggers‡,
‡ Institute of Continuum Mechanics,
Gottfried Wilhelm Leibniz Universität, Hannover, Germany

Abstract Modeling is one of the main tasks in engineering in order to predict the behaviour and response of assemblies, structures or vehicles. This contribution is aimed at modeling in solid mechanics. Due to the necessity to use numerical methods for the solution of most theoretical models it will focuss on theoretical models as well as on numerical simulation models associated with engineering applications in solid mechanics.

1 Models

Models can be derived for many tasks in science, buisiness and engineering. The evolution of all these different areas of application shows that without models mankind would not be were it is today. In order to understand this development one has to ask at least one question.

1.1 What is a model?

Modeling of engineering structures is a task that has a long history. There exist many models that can be applied to predict the behaviour of different processes. Since today often numerical simulation tools are used that have a certain model inside, it is more and more necessary to have a deep knowledge of modeling in order to use the correct tools for a proper prediction. Here not all models can be discussed, however the variety of mechanical models will be investigated and ordered such that engineers can pick the ones necessary to describe the behaviour of solids and structures adequately.

In engineering it is generally not possible to apply a full scale model when designing a structure or machine. For example a bridge of a length of 1000 meters willl not be modeled as one piece in a giant computer. This is due to the fact that different length scales have to be resolved when looking at different structural elements that form the bridge. This is e.g. on one

F. Pfeiffer and H. Bremer (eds.), *The Art of Modeling Mechanical Systems*,
CISM International Centre for Mechanical Sciences 570,
DOI 10.1007/978-3-319-40256-7_6

hand the basic structure and on the other side connections between different structural elements. Here the length scale difference can easily be of the order of 1000.

Models can be derived for different length scales. In engineering one could start from an atomistic description of materials. However due to complexity and problems size this is not an adequate modeling approach. Also the molecular dynamics approach that is related to nano and micrometer scales is not feasible to describe real engineering structures at meter scales. Thus we need a continuum approach that is based on the theory of continuum mechanics. Even with these in general three-dimensional models it is not possible to design real structures, like bridges or skyscrapers since three-dimensional models will lead to a number of unknowns that cannot be solved in a design office. Thus reduced models – like trusses, beam, plates and shells – have to be applied that are able to reproduce the main structural behaviour of the building, are consistent with the model assumptions and have to be validated by experimental data.

1.2 Definition of Models

When defining a model in general, the following quote can be made:

A model can be seen as simplified and artificial version of a process that is real or in other words:

A model is a projection of reality onto a set of mathematical equations or onto a set of experiments.

The characteristics of a model can be quite different and depend on the goal and the functionality. Thus models can vary with respect to their application spanning an arc from most simplistic models to very refined models.

The definition of a model starts with the real-world object or process that has to be considered. These objects and processes are projected onto a simplified engineering model, e.g. a real structure like a three-dimensional object in space is reduced to a beam model for a prediction of forces and bending moments. Thus the real world object is transfered to a specific structural model that is reduced to a point and contains only those responses and aspects which are important and necessary for a good prediction. This implies that a model only reproduces parts of the real world phenomena. Hence it cannot be used for general predictions and is restricted in its application. For the choice of a good model we can quote Einstein: "A model should be as simple as possible but not too simple".

The functionality of models depends on their usage. In general one can distinguish different categories of applications which are described below:

Models that can be used for predictions. Once models are constructed they can be applied to predict the development of phenomena in the real world. A bridge for example is exposed to wind loads and subjected to the weight and movement of cars and lorries or trains. A model of the bridge is often used to compute the response of the structure under these loading conditions which can be used to assess the safety of the bridge. Note that in engineering it is often, like in this example, not possible to build a prototype or to do a one-to-one scale experiment since bridges can be several hundred meters long. Other examples are climate models that predict weather or climate change, chemical process models, models for airplanes for prediction of different flight conditions or models for the stock market.

Models that explain physical phenomena. In physics, material science and other areas of science models can be used to explain phenomena that are observed or even only hypothesized. Models that are known for these puerposes are with respect to physics classical Newton's mechanics, thermodynamics or Einsteins theory of relativity. Based on these models there a formal ways on how to derive submodels for specific phenomena. E.g. in material science microscopic models can be introduced to explain behaviour of materials at macro level. Other models can be used to describe the biochemical cycle of blood flow, economic growth or traffic jams.

Models that allow decision making. Hazards like floods can lead to a lot of damage. Stakeholders like adminstrations, people living in the area have to decide whether to evacuate their houses, what kind of additional hazards can occur, like discharge of poisonous chemicals from the ground or damage of bridges, roads etc. Models of the real world have to be reduced to experienced based ones with a lot of uncertainties that still can be reasonably applied. Also in a chemical plant models can be used to make decisions related to production processes. These models contain many decision variables and a lot of a constraints. They can be formulated in ordinary differential equations with inequalities.

Other Models. For communication sketch drawings can be used to model a complex context. The model depends on the person who draws and there is no formal method that will lead to a unique model. Another example are standards that are used to regulate civil engineering constructions, like EU Norms. These norms are often based on predictive models but inside are also measures for uncertainties and random processes and additionally based on the experience of the members of the committees that develop the rules within the norms.

As can be seen, models are not unique. They have to be approved for their application area and range which can be obtained by a verification and validation process. Here verification means that the equations of a model are solved correctly while validation means that the model is checked by e.g. experimental data.

1.3 Mathematical Models

In engineering models of different kind have been developed during the ages. There where models that needed an experimental approach, models that used just observations and the rule of thumbs, but mainly models project the real physics onto a set of mathematical equations. They can be simple analytical formulae, algebraic equation systems, ordinary and partial differential equations. Thus a real world problem or process is represented in formalized manner which is the power of mathematics.

The great advantage of mathematical models lies in the fact that a solution procedure can be devised in general that yields a reliable solution based on algorithms and theory. Such models have a long history, see Section 1.4, starting in the renaissance. At that time the modeling actually drove the mathematics leading to more complex and accurate description of physical objects. Here mathematics developed from simple equations via ordinary differential equations to partial differential eqations. In this line complexity was increased but also the capability of solving partial differential equations analytically is very limited. This lead to numerical approximation methods that have growing application range still today. The latter is due to the enormous increase of computing power since the beginning of the use of computers. One of the pioniers, Conrad Zuse, actually was a civil engineer who did not like to solve large equation systems and thus developed a digital computer. This development in computing power was in its early stages recognized by Civil and Aeronautical engineers. The emerging discretization methods like the finite element, the finite difference and finite volume methods are today used for the solution of real world problems not only in Civil and Aeronautical Engineering but also applied in Mechanical and Electrical engineering.

Based on methematical theory many of the numerical methods can be analysed and it can be shown that numerical schemes and algorithms converge to the solution of the mathematical model. This means that the solution is as good as an analytical solution, but it also means that the accurate solution when using a numerical scheme is not better than the underlying mathematical model.

The knowledge about the mathematical properties of a model is based today on classifications of the models, functional and numerical analysis of mathematical models. These theoretical investigations lead to a great inside. However engineers use today mathematical decscription of real world problems and processes that cannot be sufficiently analysed by mathematicians. Thus existence or uniquness of solutions cannot be guaranteed for many advanced mathematical models like finite plasticity or complex coupled processes. Nevertheless the engineering community uses thes models based on validations by experiments. However this requires engineers that have to understand the mathematical background of a model, the numerical discretizations and algorithmic schemes as well as the real world problem.

The main concept of a mathematical model is based on the introduction of "unknown" variables that have to be computed from a set of equations. These unknowns can be deformations in beams, plates and shells or velocities in rigid body systems. Mostly the unknowns are only partly used to design real world structure. Often derivatives of theses unknowns functions like stresses are of interest in the design process. But depending on the design task also other variables like length and height of a structural member can be unknowns that have to be computed e.g. using optimization algorithms. Parts of structures can be connected which leads to the introduction of constraint equations. Based on the connection these can be equality as well as inequality constraints. These constraints add algebraic equations that contain unknowns to a system of ordinary or partial differential equations and thus result in additional complexity.

1.4 History of Model Development

In order to understand the modeling process in engineering and other fields it is always good to look back in time to see how the field evolved over the years. Modeling is closely connected with the development of mankind. The word modeling stems from the Latin word modellus. It describes a human approach to reality. One can say, the ability to build abstract models gave mankind a competitive edge over animals.

Within the history of model development the first models were numbers and the ability to count. It is documented that numers and even writing of numbers was invented 30.000 BC. Civil Engineering was one of the next areas where models were necessary to construct safe buildings. This was already in place at about 4.000 BC. It is established knowledge that by 2.000 BC the cultures of Babylon, Egypt and India had a decent knowledge of mathematics. They used mathematical models to improve their day-to-day life. Mathematics was not related to a general theory but designed for solving specific problems in many application areas, like e.g. finances, engineering and agriculture.

The combination of philosophy and mathematics in the Hellenic Age led to a more general view on the mathematical theory. With this geometry became a useful tool in analyzing reality in 600BC. The analysis of geometry then inspired a development of mathematics that was independent of its application. An example is the semicircle of Thales. Starting with a method, that could be used to compute the heights of buildings by measuring the lengths of their shadows, a general result was deduced by Thales in that time period which concludes that an angle in a semicircle is a right angle.

The great mathematicians Euclid of Alexandria wrote around 300 BC the book "The Elements", which is a collection containing most of the mathematical knowledge available at that time. This book provides a first concise axiomatic description of geometry and a treatise on number theory. From then on many scientists developed algebra, algorithms and, starting with Newton and Leibniz, calculus. At that time most of the development was application driven. It is apparent that models for real-world problems, especially mathematical models of physical processes, were extremely important for the development of humanity. Thus it is not really astonishing that models were developed independently in Europe, China, India, and the Islamic countries.

In the field of mechanics the ancient Greek philosophers, like Aristoteles 350 BC, started to propose abstract principles that can be used to describe nature. One example is the claim that an object twice as heavy as some other would fall to the ground from the same height in half the time. This of course was not correct but one of the first attempts to describe motion in more general terms. Other results, like the buoyant force exerted on a body that is immersed in a fluid are still used today. The philosophers believed in logic and observation but due to the limitations of that time many of the results were inaccurate.

It took several hundred years before the artist and scientist Leonardo da Vinci, 1452-1519, started to do experiments in the area of mechanics that gave rise to better understanding. Among them are investigations on statics, dynamics and friction. From the latter experiments he concluded "Coulomb's Law" however did not phrase it in mathematical terms. It was the time of renaissance in which the scientists revived knowledge gained in the ancient times of Greek and Roman empires. Based on da Vinci's and own observations scientists like Galieo Galilei, Leibniz, Mariotte, Newton and many others developed mechanics to a high level that allowed predictions. Especially the groundbreaking work of Newton with his axiomatic formulation of mechanics laid the possibility to model rigid body systems for astronomic and engineering applications.

To see that the generation of predictive models for engineering structures is not a straightforward process and not at all intuitive, it is worthwhile to look at the evolution of the theories over time. One example is the model for a beam. Here it took more than 100 years to develop a correct model, starting with Galilei, Leibniz and then Mariotte, see Fig. 1. As can be seen as well Galileo in 1638 as Leibniz in 1684 assumed the wrong point (B) as centre of rotation (green point), leading to a wrong equilibrium statement. Only Mariotte in 1686 had the correct idea to place the centre of rotation in the middle of point (A) and (B) which resulted in the classical beam theory, which was mathematically formulated by Bernoulli in 1691 and named after him. However the complete formulation that related the bending of the elastic line of the beam to the stresses across the cross section was finally deduced by Coulomb (1736-1806). It lead to the classical formula

$$\sigma_x = \frac{M}{I}\, z$$

where the stress σ_x in the direction of the beam axis x is computed for a point (distance z of that point from the beam axis) as a function of the bending moment M and the moment of inertia I. This model is still used today, see e.g. Szabó (1987). The long time for the model development,

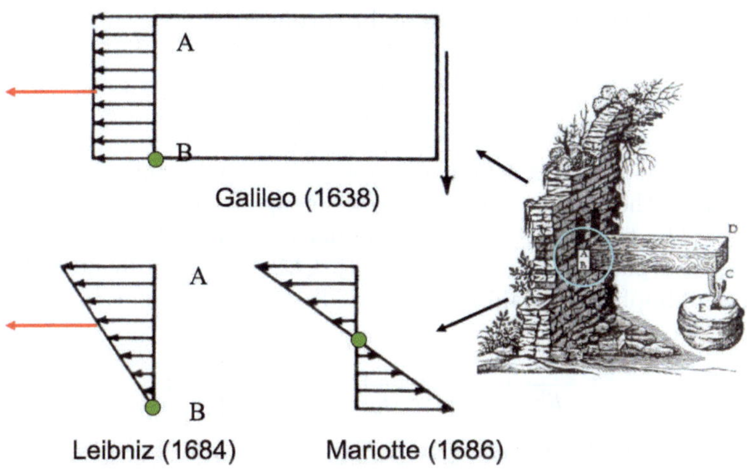

Figure 1. Different kinematical models for the beam.

involving the best scientists at that time shows how complex it is to develop a mathematical model for a physical process from observations.

The first mathematical structural analysis was performed in 1743 by three mathematicians to design measures for the repair of St. Peter's Basilica in Rome. This analysis was based on the scientific knowledge in mathematics and mechanics at that time and thus marks the transition from craftmanship to a modern scientific oriented approach in Civil Engineering. It thus is the first document that is related to the modeling process in Civil Engineering that we still use today.

Another milestone in solid mechanics was the develpoment of plate theories that started from the dynamics. Euler, after having solved the dynamics of strings and beams, wanted to compute around 1770 the sound generation of bells. With this he jumped over static plate theory that assumed plane geometries to dynamic shell theory. His assumptions lead to a wrong model, the final correct version was finally developed by Love in 1888. Regarding plates, it took some years until Chladnis work on experimental observations (one of the first based on a scientific experimental approach) regarding the dynamics of plates lead to great interest in plate models. He obtained in 1787 what is called Chladnis figures showing basically the eigenforms of plate vibrations, see Fig. 2.

This gave rise to effort to derive a mathematical theory that was able to model these figures. Jakob Bernouli II was the first who tried to set up a differential equation to obtain the figures in 1788. Bernoulli failed with the equations (w being the deflection and c a constant)

$$\frac{\partial^4 w}{\partial x^4} + \frac{\partial^4 w}{\partial y^4} = \frac{w}{c^4},$$

but noted that this was a first attempt. The experiments gained such an interest that they led to a competition set up by the French Academy of Sciences. The task was to develop the mathematical theory for elastic vibrations of a plate and to compare with Chladnis experiments. Based on the work of Sophie Germain who obtained in 1813 the equation (t being the time)

$$\frac{\partial^2 w}{\partial t^2} + \lambda^2 \left[\frac{\partial^6 w}{\partial x^4 \partial y^2} + \frac{\partial^6 w}{\partial y^4 \partial x^2} \right] = 0$$

the great scientist Lagrange suggested corrections which resulted in (here k is an, at that time, unknown constant)

$$\frac{\partial^2 w}{\partial t^2} + k^2 \left[\frac{\partial^4 w}{\partial x^4} + 2 \frac{\partial^4 w}{\partial y^2 \partial x^2} + \frac{\partial^4 w}{\partial y^4} \right] = 0\,.$$

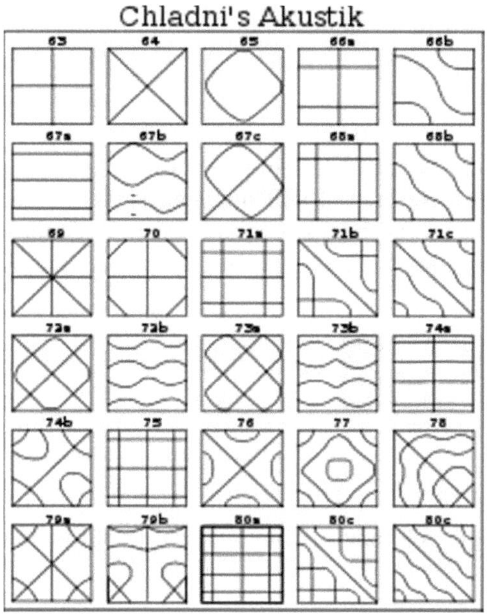

Figure 2. Chladnis figures of eigenforms of a quadratic plate, see Szabó (1987).

This form of the partial differential equation is exaclty the plate equation known today, Sophie Germain obtained in 1816 the prize for her first contribution, even having had help from Lagrange. Again it took more than 25 years and eminent scientists to obtain the correct mathematical model for the physical behaviour shown in the experimental work by Chladni.

There are many more examples in the history of mechanics (elasticity for three-dimensional solids, correct equations for virtual work, definition of energy ....) that show how complicated and time consuming it is to obtain a sound mathematical model for a problem in physics or engineering. The historical interested reader can consult e.g. Szabó (1987), Timoshenko (1983) and Rice (2010).

Since there is no direct and straightforward approach to derive new mathematical models in mechanics one can speak of modeling being an art.

1.5 The Modeling Process

Engineers have to make prediction for objects, like structures, machines, cars and airplanes. These start with a design goal and then the real physical behaviour of the object has to be projected onto a model that can accurately predict the response of the object in the application range that is set by the design goals. Often the model will not lead to the expected results and a different or refined model has to be generated. A generic cycle is depicted in Figure 3. It includes the steps leading to a proper model.

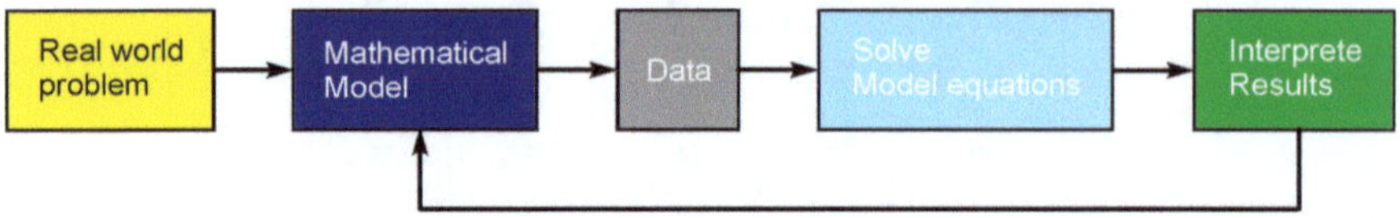

Figure 3. Model development from the real world to mathematics.

One main step in this process is the onstruction of the model. Most of the time this is done by using an existing model that is amended in such a way that it reproduces the behaviour of the real world object. However also the development of a completely new model is possible, when the real world problem demands a new approach.

Another main step is the collection of proper data. In engineering these include the action of a surrounding medium, dead loads and other loads, Additionally initial conditions of states and boundary conditions have to be considered. These data often have random character and thus bring uncertainties into the model. This step actually can also invoke a refined process when it turns out that e.g. actions of the surrounding medium - like air or water - have to be predicted by a model itself that has then to be coupled with the structure leading to a so called fluid-structure-interaction problem.

Once the mathematical formulation of the model is finished, the set of algebraic or differential equations has to be solved. For this a proper method has to be selected. While in early times the mathematical models where selected and tailored such that analytical solutions were possible, it is nowadays standard to use more refined models that are solved by a numerical scheme. Here the engineer needs sufficient mathematical knowledge and knowledge in engineering theory, like continnum mechanics, in order to select a solution scheme that provides reliable results in the most efficient way. This is especially true when nonlinear problems are addressed. In this step a verification of the solution procedure is necessary, answering the question: "Are the equations solved correctly ?". However this is not always done in engineering practise and can lead to wrong interpretations of the results.

A tedious task is the last step where the model results have to be interpreted and validated. This step is of utmost importance since the reliability of a model depends on a good validation. The latter word means: "Are the model equations correct ?". This question can be answered by experiments if possible or by the knowledge of the engineer who can judge the results by experience. Furthermore simplified models can be used to access the general correctness.

On the first view it seems that all steps can be carried out one after the other. However it is obvious that the last and the first step are connected since a change in the model can be caused by the validation. However also the mathematical model might have to be changed if the interpretation of results suggests a new mathematical model. Thus instead of going back to the first step the loop is now just from the fourth to the third step. Again the a loop can also be formed from the validation step to the data collection and thus different paths might have to be followed in the process of projecting real world problems onto a model.

In short, these are the essential questions one has to ask when designing a new model:

Validation: Do we solve the right equations?
Verification: Do we solve the equations right?

For a good model of an engineering problem one may extend the process defined in Fig. 3 by a few steps. The associated methodology can be scetched as follows

1. The starting point is a real physical object, like an engine, a building, a car an airplane etc.
2. Set up an engineering model that includes topology of the system, loading, constraints and eventually substructures like columns and floors of buildings.
3. Select a mechanical model in which the engineering model or its substructures is mapped onto known models like beams, plates, shells or solids. Additionally a constitutive model has to be chosen that describes the material behaviour of the real object.
4. Use a mathematical model in which the underlying ordinary or partial differential equation is formulated together with the boundary conditions. Depending on the type of analysis including e.g. temperature

effects besides the mechanical behaviour in statics or dynamics this can lead to elliptical, parabolic or hyperbolic differential equations.

5. Solve the differential equation with an adequate method. The solution can be performed analytically when the geometry is simple or a one-dimensional model has been selected in point 3. Most of the time a numerical scheme will be used. This should lead to a convergent approximate solution. Classically the boundary or finite element method are applied in solid mechanics.
6. The engineering model has to be dimensioned based on the solution of the mathematical model using engineering codes. In this way the design engineer can provide the correct dimensions for the real object such that it can be safely used.

This process is usually an iterative one in which the dimensioning has an impact on the mechanical model or the solution. The detailed steps that have to be undertaken within each of the points will be described in detail in the next sections.

2 What kind of models exist in solid mechanics?

In solid mechanics one has the choice between models which have different accuracy with respect to the behaviour of the structure in the real world. These models can be grouped in different ways. Here we start with three-dimensional continuum models that however can have different depths in modeling with respect to kinematics or material behaviour. Then one- and two-dimensional structural models are discussed that reduce the mathematical complexity (e.g. from partial to ordinary differential equations). Depending on the application range or the necessary accuracy level an engineer can apply models that provide a safe design. Figure 4 depicts different stages of modeling going from 3d solids to a mass-spring model for the same structure. A the three-dimensional model is necessary to compute stress states at specifiy locations which are assumed to be critical, like e.g. the transition zone between shaft and foundation. The two-dimensional model could be useful for local buckling investigations in the shell-like shaft structure. The one-dimensional beam model is accurate enough to compute stress resultants due to e.g. wind loads. It also can be applied for a vibration analysis of the tower while the mass-spring model might be good enough for a initial layout of the tower with respect to its dynamic response. Often these models are coupled in a structural analysis. In such case the beam model provides stress resultants which show were high stress concentrations might occur. These are then accurately computed by the three-dimensional solid model. However the models with low dimensions, like beams, simplify

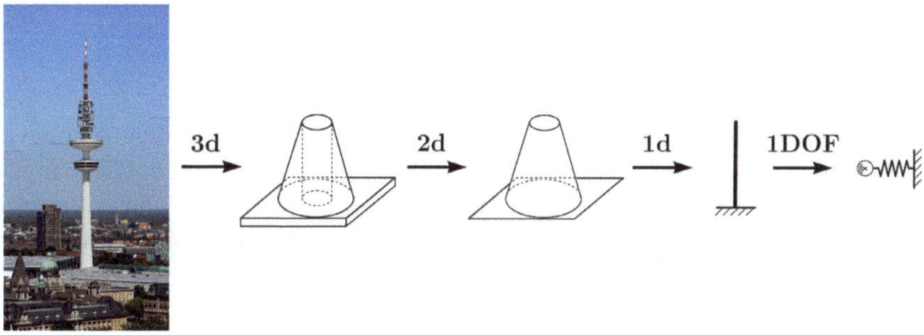

Figure 4. Different models for an I-beam.

responses and thus have reduced accuracy. Hence they have to be used with care.

In Figure 5 another example is depicted which shows an I-beam that is modeled as a 3d-, 2d- and 1d-model. The choice of a specific model then depends on the application range. Here the computation of stress concentrations at the intersection of flange and web need a 3d-model, while local buckling effects can be modeled using a 2d-shell formulation. Finally the 1d-model is sufficient for the computation of normal and shear forces as well as bending and torsional moments.

The engineer who wants to use or amend one of the general models, described e.g. in Fig 5, has to make several decisions. These relate to the model characteristics, boundary and possible constraint conditions and the application range that the design engineer wants to be covered by the model and its mathematical description. Some of the issues are discussed below.

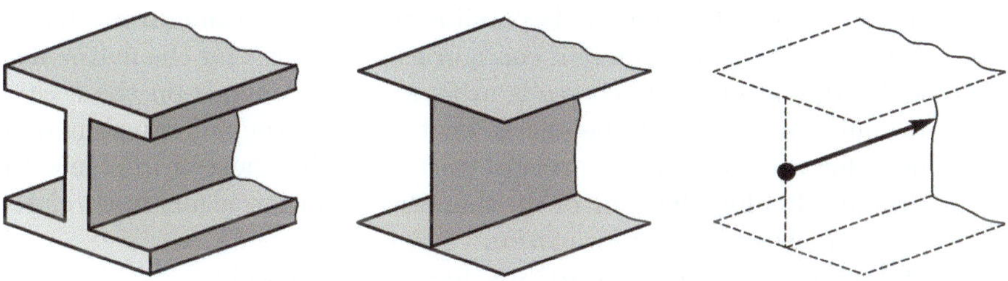

Figure 5. Different models for an I-beam.

What is the best theoretical formulation?

- **Is the small strain assumption sufficient?**
Small strains are in many engineering applications sufficient since engineering structures can only function or only be used safely when no large deformations occur. Of course, there are cases when a structure needs a nonlinear description. This is needed when stability problems can occur or the deformations are very large, like for rubber materials or in forming processes.
- **Which are the optimal model equations?**
Do we need for the mathematical formulation the strong form, e.g. partial differential equations (PDEs) or is our solution techniques based on a variational formulation? The strong form is basically used when analytical solutions are possible or when special numerical techniques are used, like finite difference methods or the isogeometric analysis that enables a solution based on a collocation of the PDEs. The variational approach is usually applied when numerical schemes like the finite or boundary element method are used to solve the underlying mathematical equations that define the model.
- **Which are the correct boundary and initial conditions?**
Is it possible to use classical Neumann boundary conditions or are e.g. the applied forces dependent on the configuration. In case of Dirichlet boundary conditions, isit possible to fix the displacements at the boundary or are the conditions more complex and involve e.g. an elastic foundation. To make it even more complex one can formulate unilaterial boundary conditions involving inequalities that have to be used within contact and frictional contact problems. Usually the latter formulation occur when the model is more refined and then these conditions have to be taken into account.
- **Which continuum description is optimal?**
In case of large strains, it is possible from the continuum mechanics point of view to use different configurations as basis for the mathematical formulation. The choice is arbitrary and depends on the ease of formulation and has to be selected with view to the numerical method used for the solution in terms of ease and robustness of implementation and finally efficiency of the numerical solution algorithms related to the mathematical formulation.
- **What constitutive equations have to be used?**
A complex task is the selection of the adequate constitutive equation. In solid mechanics this is actually the most difficult task. The choice

depend on the expected response which can be quite different, like elastic, visco-elastic, elasto-plastic behaviour. On top of that fracture due to damage may occur, or fatigue has to beconsidered which make the selection even more complicated. The art of selecting a constitutive equation lies additionally in the fact that one has to rely on experimental results and thus a model that needs only few parameters is a lot more accepted in industrial applications than a very refined model with a lot of constiutive parameters.

- **Are there constraint equations that have to be included in the formulation?**
  Constriants within the formulation can have a certain impact on the mathematical formulation. This is well known for the incompressibility constraint within the elastic responce of rubber materials or the plastic response of metals. However also other constraints like directional constitutive behaviour or contact may lead to different mathematical models.
- **Multiscale or macroscale modeling?**
  In case that more insight in the real physical behaviour of a material is needed, it is appropriate to apply micromechanical models that are valid at small length scales. These models can often not be applied at structural level and hence a process, called homogenization, is needed to transfer the results from microscale solutions to the structural level. Modern approaches also use concurrent multiscale methods that combine the different scales in one model. However this is computationally not very efficient and thus only applicable for studies at research level.
- **Do we need to couple different fields?**
  In some processes it is neccesary to couple different field equations which allow the prediction of interaction of fields. Thus coupled formulations combine the action of different fields, like temperature and mechanics. Also electrical and magnetical fields can be coupled to the mechanical response of a solid or structure. These multifield approaches add complexity to the theoretical and numerical formulation since the different fields usually relate to specific partial differential equations that do not always have the same structure.
- **Coupling in the solid or at interfaces?**
  Besides the coupling of different fields acting in the same solid, it is also possible to couple different partial diffential equations at their boundary. This is e.g. the case for fluid-structure interaction were the influence of a flow field on a structure and vice versa is taken into account. Such coupling is also possible between particle and continuum methods, see e.g. Wellmann and Wriggers (2012).

Once the set of partial differential equations or the respective variational formulation is set up, the next question arises:

What is the best numerical solution technique?

- During the last sixty years methods like the finite differentce method, the finite element method, the boundary element method were developed and are now quite mature for application in engineeing design problems. Many of these methods cover the problem ranges that are described above (e.g. small/large strains, different constitutive equations or coupled field equations). Additionally new approaches like meshless methods for arbirary deformations and the extended finite element method (XFEM) for fracture mechanics problems evolved and can be efficiently used within a specific problem range. Thus the art of modelling means here to pick the right numerical solution method that provides accurate results in the most time efficient way.
- Each of the methods described above has its own specifications and thus needs experts for a correct and efficient application. As an example: within the finite element method there are hundreds of different finite element formulations for different applications that even can fail when applied to an inadequate problem. The same is true also for all the other methods.

Classically all models in solid mechanics are based on three different sets of equations. These are the kinematical relations which connect the deformation to strains, the balance equations and the constitutive equations. While balance equations and kinematical relations in three-dimensional models are settled from the theoretical point of view, constitutive modeling is still an open question for complex nonlinear behaviour for materials such like concrete, soil or even rubber and metals.

2.1 Mathematical model for the three-dimensional continuum

A continuum model is formulated when the engineer has to compute accurate three-dimensional stress fields in geometrically complex structures. Here the design engineer has to distinguish between loading conditions that produce a response with small strains, e.g. less than 3% or action of forces that lead to large or finite strains in the solid, like in forming or cutting simulations. This differentiation leads to the use of different kinematical relations which enhance the complexity of the model equations and the solution[1]. But also the constitutive behaviour can be nonlinear. In the following we will give a very short introduction into the continuum mechanics background of solid models. For detailed derivations and additional information the books of e.g. Truesdell (1985), Chadwick (1999), Holzapfel (2000) and Wriggers (2008) can be consulted.

Kinematics

The deformation of a solid body can be geometrically described by introducing two configurations. The initial configuration B and the current or deformed configuration $\varphi(B)$, see Fig. 6 Here the position vector $\boldsymbol{X}$

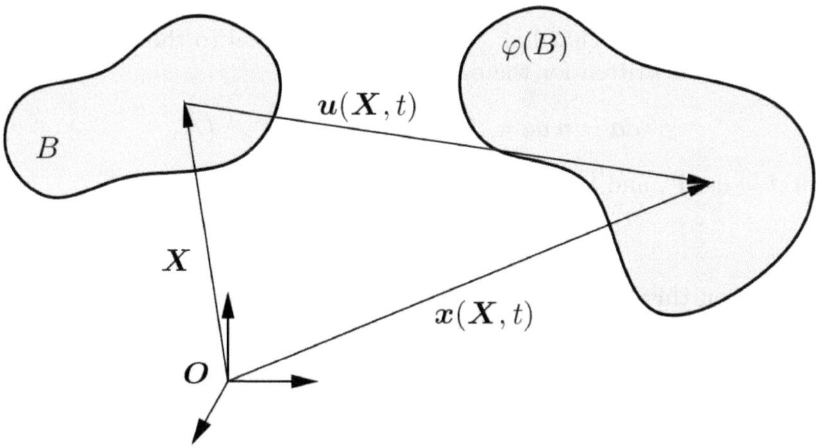

Figure 6. Deformation of a solid.

denotes a position of a point within the initial configuration whereas the vector $\boldsymbol{x}(\boldsymbol{X}\,,t)$ describes the current position of the same point. Both configurations are connected via the displacement vector $\boldsymbol{u}(\boldsymbol{X}\,,t)$. Thus the equation

$$\boldsymbol{x}(\boldsymbol{X}\,,t) = \boldsymbol{X} + \boldsymbol{u}(\boldsymbol{X}\,,t) \tag{1}$$

[1]However also other phenomena like stability can lead to the necessity to use a nonlinear kinematical relation in order to be able to model the effects of snap-through or bifurcations of solutions

holds. To obtain measures that enable the computation of changes of length and angles, the strains, it is necessary to introduce infinitesimal changes of the position vectors, the so called line elements $d\boldsymbol{X}$ and $d\boldsymbol{x}$. These are connected via the deformation gradient $\boldsymbol{F}$

$$d\boldsymbol{x}(\boldsymbol{X}\,,t) = \boldsymbol{F}(\boldsymbol{X}\,,t)\,d\boldsymbol{X} \quad \text{with} \;\; \boldsymbol{F}(\boldsymbol{X}\,,t) = \operatorname{Grad}\boldsymbol{x}(\boldsymbol{X}\,,t)\,. \tag{2}$$

Here 'Grad' means that the gradient of the position vector $\boldsymbol{x}(\boldsymbol{X}\,,t)$ has to be differentiated with respect to $\boldsymbol{X}$. Note that $\boldsymbol{F}$ is a two-point tensor that connects initial and current configuration. Using (1) we can write

$$\boldsymbol{F}(\boldsymbol{X}\,,t) = \mathbf{1} + \operatorname{Grad}\boldsymbol{u}(\boldsymbol{X}\,,t) = \mathbf{1} + \boldsymbol{H}(\boldsymbol{X}\,,t)\,. \tag{3}$$

To shorten notation we will omit the arguments from now on, hence e.g. $\boldsymbol{F}(\boldsymbol{X}\,,t)$ will be $\boldsymbol{F}$.

Further relations that link elements of the initial to the current configuration can be written for the area elements

$$d\boldsymbol{a} = \boldsymbol{n}\,da = J\,\boldsymbol{F}^{-T}\,\boldsymbol{N}\,dA = J\,\boldsymbol{F}^{-T}d\boldsymbol{A}\,, \tag{4}$$

with $J = \det\boldsymbol{F}$, and for the volume elements

$$dv = J\,dV \tag{5}$$

Based on this basic kinematical relation the strain measures can be defined. Among many options the Green-Lagrange strain tensor

$$\boldsymbol{E} = \frac{1}{2}\left[\boldsymbol{F}^T\boldsymbol{F} - \mathbf{1}\right]\,. \tag{6}$$

is often used. The first term in the bracket is called Cauchy-Green tensor $\boldsymbol{C} = \boldsymbol{F}^T\boldsymbol{F}$. When using (3) to compute the Green Lagrangian strain then the nonlinearity of this strain measure appears

$$\boldsymbol{E} = \frac{1}{2}\left[\boldsymbol{H} + \boldsymbol{H}^T + \boldsymbol{H}^T\boldsymbol{H}\right]\,. \tag{7}$$

In the linear theory the last term is neglegted leading to the linear strain measure

$$\boldsymbol{\epsilon} = \frac{1}{2}\left[\boldsymbol{H} + \boldsymbol{H}^T\right]\,. \tag{8}$$

Balance laws, weak forms

Balance laws exist for different physical quantities. Here we will only introduce the balance of mass, linear momentum and angular momentum.

Balance of mass. The balance of mass has to be fulfilled for classical engineering applications. It can be cast in the form $m = \int_{\varphi(B)} \rho\, dv = \int_B \rho_0 dV = const.$ or the change of the mass has to be zero: $dm = 0$, which leads to, see also (5),

$$\rho\, dv = \rho_0\, dV \Leftrightarrow \rho\, J = \rho_0 \tag{9}$$

where ρ is the densitiy and dv the volume element in the current and ρ_0 the density and dV the volume element in the initial configuration.

Balance of linear and angular momentum. The linear momentum or the translational momentum can be stated in the current and initial configuration with

$$\boldsymbol{L} = \int\limits_{\varphi(B)} \rho\, \boldsymbol{v}\, dv = \int\limits_{B} \rho_0\, \boldsymbol{v}\, dV \tag{10}$$

for the continuous case where $\boldsymbol{v}$ denotes the velocity. The balance of linear momentum reads: *The change of linear momentum* $\boldsymbol{L}$ *in time is equal to the sum of all external forces (volume and surface forces) acting on body* B. Mathematically this statement can be expressed by

$$\dot{\boldsymbol{L}} = \int\limits_{\varphi(B)} \rho\, \bar{\boldsymbol{b}}\, dv + \int\limits_{\varphi(\partial B)} \boldsymbol{t}\, da\,. \tag{11}$$

$\rho\, \bar{\boldsymbol{b}}$ defines the volume force (e.g. gravitational force). $\boldsymbol{t}$ is the stress vector acting on the surface of the body.

With Cauchy's theorem which relates the stress vector $\boldsymbol{t}$ to the surface normal $\mathbf{n}$ via the linear mapping

$$\boldsymbol{t} = \boldsymbol{\sigma}\, \boldsymbol{n} \tag{12}$$

the stress vector can be expressed in terms of the Cauchy stress tensor $\boldsymbol{\sigma}$ that is defined in the current confguration $\varphi(B)$. All these equations can be combined and lead to the local form of the linear momentum

$$\operatorname{div} \boldsymbol{\sigma} + \rho\, \bar{\boldsymbol{b}} = \rho \dot{\boldsymbol{v}} \tag{13}$$

The Cauchy theorem can also be formulated with respect to the initial configuration. It then reads

$$\boldsymbol{t}_0 = \boldsymbol{P}\, \boldsymbol{N} \tag{14}$$

Here $\boldsymbol{P} = J\, \boldsymbol{\sigma}\, \boldsymbol{F}^{-T}$ is the first Piola-Kirchhoff stress tensor and ρ_0 the density related to the initial configuration.

Using now the divergence theorem the local balance equation of linear momentum is derived from (11). With reference to the initial configuration B relation

$$\mathrm{Div}\,\boldsymbol{P} + \rho_0\,\bar{\boldsymbol{b}} = \rho_0 \dot{\boldsymbol{v}} \tag{15}$$

is obtained. $\rho_0 \dot{\boldsymbol{v}}$ describes the inertial forces which can be neglected in case of purely static investigations.

The angular momentum with reference to a point O yields after some manipulations the local balance of angular momentum which simply demands the symmetry of the Cauchy stress tensor

$$\boldsymbol{\sigma} = \boldsymbol{\sigma}^T\,. \tag{16}$$

The first Piola-Kirchhoff tensor is not symmetric, one obtains

$$\boldsymbol{F}^T\,\boldsymbol{P} = \boldsymbol{P}^T\,\boldsymbol{F}\,. \tag{17}$$

Boundary and initial conditions. Boundary and initial conditions are of utmost importance for a correct model. This is due to the fact that some of the models are tolerant to incorrect choices while others are extremely sensitive and thus even a small mistake can lead eventually to totally different results. DIfferent boundary conditions have to be formulated in solid mechanics. These are related to the displacements and the stresses. Furthermore inequality constraints have sometimes to be formulated for certain problem classes. The following list describes different boundary and initial conditions.

- Von Neumann conditions prescribe forces in form of stresses at the boundary ∂B_σ of the body. Mathematically this leads to

$$\boldsymbol{t}_0 = \boldsymbol{P}\,\boldsymbol{N} = \widehat{\boldsymbol{t}}_0 \tag{18}$$

  where $\widehat{\boldsymbol{t}}_0$ is the applied traction at ∂B_σ.

  The applied traction forces can be deformation dependent as e.g. pressure loads steming from a fluid. In this case the boundary condition can be written at the deformaed boundary $\varphi(\partial B)_\sigma$ as

$$\boldsymbol{t} = \widehat{p}\,\boldsymbol{n} \tag{19}$$

  where $\widehat{p}$ is the applied pressure that acts always in normal direction.
- Dirichlet conditions are formulated in terms of the displacement field at the boundary ∂B_u. Usually these boundary conditions can be formulated as

$$\boldsymbol{u} = \widehat{\boldsymbol{u}} \tag{20}$$

  where $\widehat{\boldsymbol{u}}$ is the prescribed displacment at the boundary ∂B_u.

- Some conditions are more complex and can even change during deformation process. These are e.g. constraint equations that stem from the non-penetration condition in contact analysis. These have to be modeled as inequalities and need a special treatment. For details see e.g. Johnson (1985) for classical contact analysis and Wriggers (2006) and Laursen (2002) for computational contact mechanics.

Weak forms. The momentum balance equation can be recast in a variational or weak form by multiplying (15) by a test function $\boldsymbol{\eta}$ and then integrating this expression over the initial volume. After several mathematical manipulations, see e.g. Wriggers (2008), the weak form is derived

$$\int_B \boldsymbol{P} \cdot \operatorname{Grad} \boldsymbol{\eta} \, dV - \int_B \rho_0 \, (\bar{\boldsymbol{b}} - \dot{\boldsymbol{v}}) \cdot \boldsymbol{\eta} \, dV - \int_{\partial B_\sigma} \bar{\boldsymbol{t}} \cdot \boldsymbol{\eta} \, dA = 0 \tag{21}$$

where $\bar{\boldsymbol{t}}$ is now the prescribed traction vector that acts at the boundary ∂B_σ. This equation is basis for the derivation of numerical simulation methods, like the finite element method. It is often written in a slightly different way, using a symmetric stress tensor, the second Piola-Kirchhoff stress,

$$\int_B \boldsymbol{S} \cdot \delta \boldsymbol{E} \, dV - \int_B \rho_0 \, (\bar{\boldsymbol{b}} - \dot{\boldsymbol{v}}) \cdot \boldsymbol{\eta} \, dV - \int_{\partial B_\sigma} \bar{\boldsymbol{t}} \cdot \boldsymbol{\eta} \, dA = 0 \tag{22}$$

with $\boldsymbol{S} = \boldsymbol{F}^{-1} \boldsymbol{P}$ and the variation of the Green-Lagrange strain tensor (6)

$$\delta \boldsymbol{E} = \frac{1}{2} \left[(\operatorname{Grad} \boldsymbol{\eta})^T \boldsymbol{F} + \boldsymbol{F}^T \operatorname{Grad} \boldsymbol{\eta} \right] . \tag{23}$$

These equations are also referred to as virtual work. Then the test function is interpreted as a virtual displacement.

Equations (21) and (22) are valid for finite deformations. In case of linear deformations the variational equation has actually the same form as (22)

$$\int_B \boldsymbol{\sigma} \cdot \delta \boldsymbol{\epsilon} \, dV - \int_B \rho_0 \, (\bar{\boldsymbol{b}} - \dot{\boldsymbol{v}}) \cdot \boldsymbol{\eta} \, dV - \int_{\partial B_\sigma} \bar{\boldsymbol{t}} \cdot \boldsymbol{\eta} \, dA = 0 \tag{24}$$

with

$$\delta \boldsymbol{\epsilon} = \frac{1}{2} \left[(\operatorname{Grad} \boldsymbol{\eta})^T + \operatorname{Grad} \boldsymbol{\eta} \right] . \tag{25}$$

Interim conclusions related to continuum modeling

The balance law and the kinematical relations are well established and since they can be deduced from purely geometrical relations and from the classical laws of Newton are not questioned and hence have to be the basis for any mathematical model of a three-dimensional solid. The only modeling question that remains at this stage is the decision whether engineering problem is subjected to small or a large deformations and based on that the right equations have to be selected.

Other models that are reduced in dimension, like plate, shells or beams, have to obey also these kinematical relations and the balance laws. Ideally one derives the reduced equations as limiting cases of the three-dimensional equations. Such that the model assumptions are in unison with the general kinematical relations, e.g. (6), and the balance laws, e.g. (15), (17) and (21).

The real art of modeling starts for the design engineer with the correct selection of the constitutive equations for a specific application. Here many different models are possible that will be mentioned in the next section. Also the choice of boundary conditions for the displacements but often more important for the forces will influence the solution. There are engineering problems that can be very sensitive to displacement boundary conditions but also to force and traction boundary conditions.

Constitutive equations

Elastic response. The mostly used constitutive model is the classical Hooke's law which relates in the linear theory the stresses via a linear relation to the strains

$$\boldsymbol{\sigma} = I\!E\,\boldsymbol{\epsilon} \quad \text{with} \quad I\!E = \lambda \mathbf{1} \otimes \mathbf{1} + 2\,\mu\,I\!I \tag{26}$$

where $I\!E$ is a fourth-order tensor that has two material constants. Its special form includes the tensor product of the second order unit tensor $\mathbf{1}$ and the fourth order unit tensor $I\!I$. As material constants, the Lame constants λ and μ where used. These can be linked to other elastic constants, like Young's modulus $E = \frac{\mu(3\lambda+2\mu)}{\lambda+\mu}$ and the Poisson ratio $\nu = \frac{\lambda}{2(\lambda+\mu)}$. These constants have to be deducted from experiments. The evaluation of (26) yields

$$\boldsymbol{\sigma} = \lambda\,\text{tr}\boldsymbol{\epsilon}\,\mathbf{1} + 2\,\mu\,\boldsymbol{\epsilon} \tag{27}$$

where $\text{tr}\boldsymbol{\epsilon} = \boldsymbol{\epsilon} \cdot \mathbf{1}$ is the trace of the linear strain tensor $\boldsymbol{\epsilon}$, see (8).

This material behaviour is called elastic since no permanent deformation occurs when the structure is unloaded. This elastic behaviour can be seen in Fig. 7 for a one-dimensional tension loading where the black line approximates the real material behaviour (in red) that is observed in an experiment.

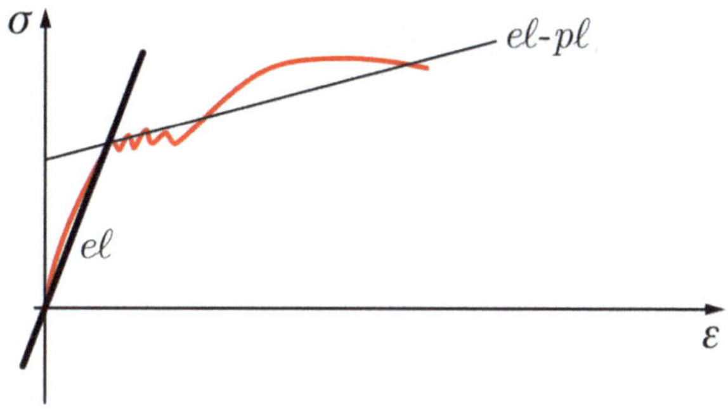

Figure 7. Elasto-plastic approximations (in black) of the real material behaviour (in red).

Since the elastic material behaviour is an approximation, the design engineer has to make the decision whether this model describes the real behaviour appropriately for the range of loading that is applied to the structure to be designed. Sometimes it is of interest to approximate nonlinear material behaviour by a linear relation like in (27). This is for example the case for rubber bearings of engines or buildings that undergo large deformations in the static case of dead load. Then a linear approximation of the material behaviour in a certain range a as shown in Fig. 8 can be used to model e.g. the dynamic response around that specific deformation state. Again the design engineer has to make sure that the model assumptions – linear material and rang a – are not violated by the response of the model under dynamic exitations.

Models that predict nonlinear response of elastic materials are usually called hyperelastic constitutive equations. They have to be formulated for e.g. rubber materials or specific foams. Usually these equations are deduced from strain energies W that depend on material parameters and nonlinear strain measures, for details see e.g. Truesdell and Noll (1965), Holzapfel (2000) and Wriggers (2008). Special strain energies are also formulated using the principal strains, like in the Ogden material, see e.g. Ogden (1984). A typical and relatively simple response function the strain energy $W(I_C\,, J) = \frac{\lambda}{4}\,[(J^2-1) - 2\ln J] - \mu \ln J + \frac{\mu}{2}\,(I_C - 3)$ with $I_C = \operatorname{tr} \boldsymbol{C}$ is

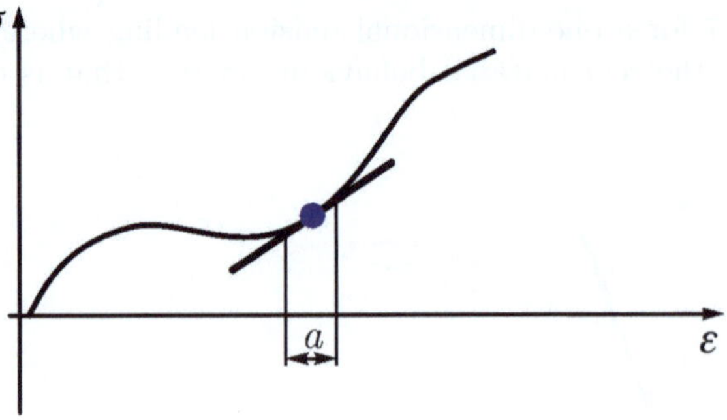

Figure 8. Linearized constitutive behaviour at a specific configuration and application rang a.

known as Neo Hooke material. It has the form

$$\boldsymbol{S} = 2\,\frac{\partial W}{\partial \boldsymbol{C}} = \frac{\lambda}{2}\,(\,J^2 - 1\,)\,\boldsymbol{C}^{-1} + \mu\,(\boldsymbol{1} - \boldsymbol{C}^{-1}) \tag{28}$$

where λ and μ are the Lame constants, J is the determinant of the deformation gradient, see (4), and $\boldsymbol{C}^{-1}$ is the inverse of the Cauchy-Green tensor, see (6) for the definition. This model is highly nonlinear, but only has two material parameters and hence parameter identification using experimental results is simple.

There are extensions to other constitutive equations which can be used to model composite material. These have to take anisotropic behaviour into account, see e.g. Hashin (1983) and Aboudi (1991). For large strain such formulations can be found in Schröder (2009). These constitutive relations are needed to model soft anisotropic materials. Often these are biological materials like blood vessels, skin and other tissues.

In composite materials there are often some directions much stiffer than other ones. This can lead to a formulation that introduces a directional constraint of e.g. an inextendable fiber in a certain direction. The same is true for incompressible material behaviour, like in rubber materials or plastic flow. Here a constraint equation has to be formulated regarding the volume change, leading to $J = 1$ where $J = \det \boldsymbol{F}$[2].

[2]This type of formulation leads often numerically to problems since the underlying variational principles assume mixed character, see Section 3.2.

Inelastic response. The response of materials is however not only in the elastic range. There exist many technical applications where inelasic response has to be taken into account. Examples for this is rubber which at a first glance behaves elastic, but the deformation of rubber generates dissipation (loss of energy) which can be observed in experiments where dissipation leads to temperature increase in the material and damping of vibrations. To model this behaviour one can introduce visco-elastic material constitutive equations, see Lubliner (1985), Reese and Govindjee (1998) and Christensen (1980).

Another class of material models that is essential for the predictive design in many engineering application relates to elasto-plastic behaviour. This behaviour is depicted for a tension test in Fig. 7, see red curve for experimental results. Such behaviour needs to be considered when the material in a structures leaves the range of elastic behaviour like in forming, cutting or crash problems. Generally an elasto-plastc model has to take into account the change from the elastic to the plastic part. Thus a model, the yield surface, is needed to distinguish between the elastic and plastic deformation. Furthermore the direction of the plastic deformation (flow) has to be determined. For that purpose one introduces a flow rule, also named evolution equation of the plastic flow and some rules that determine the hardening behaviour that often is nonlinear, see the red experimental curve in Fig. 7.

Next we will formulate elasto-plastic constitutive models in a standard framework which is valid for a large class of elasto-plastic materials. This reflects the fact that there exist many materials of technical interest which can be either described by associative or non-associative flow rules or need multi-surface flow conditions for a proper modeling. This includes most plasticity models e.g. metals, concrete or geo-materials like sand, clay or rock. For an indepth treatment of such material laws see e.g. Desai and Siriwardane (1984), Khan and Huang (1995), Wriggers (2008) and de Souza Neto et al. (2008).

The linear strain tensor $\boldsymbol{\varepsilon} = \frac{1}{2}(\text{Grad}\mathbf{u} + \text{Grad}\mathbf{u}^T)$ can be split additively into an elastic and a plastic part when only small strains are present

$$\boldsymbol{\varepsilon} = \boldsymbol{\varepsilon}^e + \boldsymbol{\varepsilon}^p \,. \tag{29}$$

For the general case of an elasto-plastic material with m independent flow surfaces the subsequent equations and evolution laws are obtained based on the introduced notation and $\boldsymbol{\alpha}$ as a set of hardening variables. Associative and non-associative plasticity is distinguished in the following

- Stress $\boldsymbol{\sigma}$ and back stress $\mathbf{q}$ follows from the strain energy $W(\boldsymbol{\varepsilon}^e, \boldsymbol{\alpha})$

$$\boldsymbol{\sigma} = \frac{\partial W(\boldsymbol{\varepsilon}^e, \boldsymbol{\alpha})}{\partial \boldsymbol{\varepsilon}^e} \quad \text{and} \quad \boldsymbol{q} = -\frac{\partial W(\boldsymbol{\varepsilon}^e, \boldsymbol{\alpha})}{\partial \boldsymbol{\alpha}} \tag{30}$$

- m flow conditions / yield criteria[3] (restrictions for elastic domain), $1 \leq s \leq m$

$$f_s(\boldsymbol{\sigma}, \boldsymbol{q}) \leq 0 \tag{31}$$

- Flow rule and evolution equation for hardening
  1. associative plasticity, the flow direction is normal to the yield surface,

$$\dot{\boldsymbol{\varepsilon}}^p = \sum_{s=1}^{m} \lambda_s \frac{\partial f_s(\boldsymbol{\sigma}, \boldsymbol{q})}{\partial \boldsymbol{\sigma}} \quad \text{and} \quad \dot{\boldsymbol{\alpha}} = \sum_{s=1}^{m} \lambda_s \frac{\partial f_s(\boldsymbol{\sigma}, \boldsymbol{q})}{\partial \mathbf{q}} \tag{32}$$

  2. non-associative plasticity

$$\dot{\boldsymbol{\varepsilon}}^p = \sum_{s=1}^{m} \lambda_s \boldsymbol{r}_s(\boldsymbol{\sigma}, \boldsymbol{q}) \quad \text{and} \quad \dot{\boldsymbol{\alpha}} = \sum_{s=1}^{m} \lambda_s \boldsymbol{h}_s(\boldsymbol{\sigma}, \boldsymbol{q}) \tag{33}$$

- Loading-/unloading conditions in Kuhn–Tucker form

$$\lambda_s \geq 0, \qquad f_s(\boldsymbol{\sigma}, \boldsymbol{q}) \leq 0, \qquad \lambda_s f_s(\boldsymbol{\sigma}, \boldsymbol{q}) = 0 \tag{34}$$

All relations hold for models of plasticity with m flow surfaces. The special case of only one flow surface is naturally included by setting $m = 1$. Tensors $\mathbf{r}_s$ and $\mathbf{h}_s$ describe the flow direction and the change of hardening for non-associative plasticity in (33).

As can be seen, equations (30) to (34) can depend on many different constitutive parameters[4] that describe the elastic and the hardening as well as the yield criteria. In the most general case of multiple yield surfaces and evolution equations this is immediately evident.

There exist many specific sets of yield functions and evolution equations to describe material behaviour of different materials – like concrete, steel, soil, wood and polymers – that exhibit elasto-plastic response. The general model, shown above, is formulated for problems with small strains that

[3] Here the flow conditions depend upon the stress $\boldsymbol{\sigma}$ and not solely on the deviatoric stresses $\mathbf{s}$ as in classical J_2-plasticity. The reason for this is that inelastic processes of general non metallic materials or of metals in which damage has to be considered are pressure sensitive and hence the flow condition has to depend upon the full stress tensor.

[4] At the moment the constitutive parameters are hidden due to the general formulation.

occur in elasto-plastic deformations of structural members like plates and shells and in solids that are mainly designed for elastic response. Models for specific applications are listed below. Here only references to the literature can be provided.

- The most classical flow condition is the so-called J_2 flow criterion which is related to the Huber-von Mises model for inelastic response of metals. The flow rule depends only upon the second invariant J_2 of the stress deviator. Since it was experimentally observed that metals do not yield under hydrostatic pressure the J_2 criterion is the correct choice. Additionally it was shown that an associative plasticity formulation where the flow direction is normal to the yield surface, see (32) is sufficient to describe the elasto-plastic response of metals.
- A flow condition which is often used in soil mechanics is the Mohr-Coulomb criterion or the Drucker-Prager flow condition, see e.g. Khan and Huang (1995). These depend not only upon the second invariant J_2 of the stress deviatoras in the classical Huber - von Mises theory of metal plasticity but also on a hydrostatic stress term that can be represented by the first invariant I_1 of the stress tensor.
- Experimentally one observes the development of cracks and cavities at micro scales that start to deteriorate the material properties and can lead to failure of structural parts. Thus models are needed to predict the material behaviour in such cases and a lot of work was devoted to model the response based on damage behaviour at micro scales, see e.g. Lemaitre (1996) and Bazant and Cedolin (2003).
- During elasto-plastic deformations the microstructure of a material can change. There is the possibility of void nucleation, void growth and ductile fracture in the micro structure of a metal .One model that describes this type of damage behaviour is the Gurson model, see e.g. Gurson (1977). The original model from Gurson was corrected by Tvergaard and Needleman (1984).
- Different loading conditions like pulsating or dynamic loads can lead to so called ratcheting strains. Special constitutive descriptions are needed to model these phenomena, see e.g. Ekh et al. (2000) and Johansson et al. (2005).
- Hypo-plastic constitutive equations are frequently used to describe complex soil behaviour. These models are especially suited for high-cycle loadings in order to compute the accumulation of strain in sand, see e.g. Kolymbas (1991) and Niemunis et al. (2005).
- In case of impact loading, strain rate dependent characteristics have to be considered. These need elasto-visco-plastic constitutive equation,

see e.g. Perzyna (1966) for very early work and references in Lubliner (1985) and Simo and Hughes (1998).

- Modern approaches like to combine models at micro-scale with models at macro-scale in a so called multi-scale setting, see e.g. Fish and Shek (2000), Liu et al. (2005).
- Micro-scale models can be used for a prediction of the macro-scale material behaviour. This approach is used in order to introduce more insight with respect to the physical phenomena at micro-scale into macro-scale models. Examples for this approach can be found e.g. in Miehe et al. (1999), Lehmann et al. (2012) and Lehmann et al. (2013) for metals and Hain and Wriggers (2008b), Hain and Wriggers (2008a) for cement paste.
- Material models can include softening behaviour such as most damage models, like e.g. the Gurson model. The latter describes softening behaviour of the material due to the growth and nucleation of voids. This type of material behaviour depicts localizations and leads in the numerical formulation and simulation to mesh dependent solutions, see e.g. Oliver (1995). This can be observed in any other elasto-plastic constitutive models used e.g. in engineering analysis that includes softening. Specific engineering materials that incorporate softening behaviour are concrete or soil.

  Localizations represent the step from a continuum to a discontinuum in which a part of the body slides along the other at the localization surface. Based on this observation, methods which directly introduce the discontinuous behaviour were developed. These are based on so-called strong discontinuity approach, see e.g. Simo et al. (1993a), Larsson et al. (1993), Miehe and Schröder (1994) and Oliver (1995). They can be combined with adaptive techniques or special interface elements, see e.g. Leppin and Wriggers (1997).
- Finite strain models for plasticity have the same variety of models as in the geomertically linear case. They will be applied in forming processes, in crash analysis, in large deformations of soil, like pile installation processes and many other applications. Models that are in line with modern computational techniques can be found in books like Simo (1998), Simo and Hughes (1998), de Souza Neto et al. (2008) and references therein.
- In the beginning of modeling finite inelastic strains it was common to use hypo-elastic constitutive equation for the elastic part, see e.g. Khan and Huang (1995). Such material assumption does not represent elasticity in the strict sense, see e.g. Truesdell and Noll (1965) or Simo and Pister (1984). Besides this restriction which can result

in unwanted, physically meaningless effects, see e.g. Atluri (1984), there exist also problems of numerical nature when these constitutive equations are numerically integrated, see e.g. Simo and Hughes (1998).

This listing could go on for a long time. Many more different constitutive models which stem from experimental observations and describe plastic flow can be formulated. These are related to different types of materials, for an overview see e.g. Desai and Siriwardane (1984), Lubliner (1990), Hofstetter and Mang (1995), Khan and Huang (1995) and de Souza Neto et al. (2008) Generally one can observe that many different models were developed to model specific types of inealstic material response. Many of these models are implemented in standard software and thus design engineer has to select one of these models that will be able to accurately predict the behaviour of a structural or solid part under investigation.

Interim conclusions related to material modeling

Material modeling has advanced over the last decades. Many engineering problems can be solved efficiently and with the required accuracy. While the balance laws and the kinematical relations are nowadays not questioned anymore and are accepted to be correct models as well in the linear as in the nonlinear range. This is not true for the constitutive equations. These are still in a state of development, especially in the finite strain application range. Materials like steel, alloys, but also concrete and soil are not modeled by equations that are generally accepted and there is still a lot of room for improvement. This can also be deducted from the list above that includes many different approaches to model the same material.There are still open questions related to different issues.

- The constitutive model is developed by a curve fitting procedure to match experimental results. This can lead to a description that is thermodynamically not consistent and thus the application range is very limited or it even can lead to erroneous responses in multi-dimensional loading states.
- The constitutive model needs too many parameters to describe the physically observed behaviour. It is hard to use this type of models in real engineering applications because on one hand it is too costly and time consuming to determine all parameters from a set of experiments and on the other hand parameter sets might not be uniquely defined and thus might not cover the entire application range.
- The parameters of a constitutive model are determined from a certain set of experiments that is related to specific loading paths. If now this

model is applied to problems with different loading paths it might yield erroneous results.

From a theoretical point of view, the art of modeling in solid mechanics depends an the selection of the best material model for a specific engineering application.

2.2 Structural models

Structural models serve the purpose of providing a solution that is as accurate as necessary in order to do a proper engineering prediction. These models will not yield accurate three-dimensional stress states but they will lead to an understanding of the force distribution in a structure and its deformation state. Different models for the same structure where depicted in Fig. 5 that showed solid, shell and beam models for an I-beam. The selection of the model depends on the depth of modeling needed for an accurate prediction of the mechanical response and a safe design of the structure. Some engineering structures that can be modelled by one- or two-dimensional structural models are summarized in the following list.

- **One-dimensional description.** Engineering structures that can be descibed using one-dimensional models like trusses and beams are: masts, domes, bridges, halls, high-rise buildings, robots and machines.
- **Two-dimensional descitprion.** Engineering structures that can be descibed using two-dimensional models like shells are: cooling towers, bridge decks, vessels, containers, ships, air planes, car bodies and many parts like castings, boxes and others

One-dimensional models

Beam and truss models are the most common one-dimensional models that have been applied over the last 200 years in Civil, Mechanical and Naval Engineering to design structural parts. Beam and truss models have the big advantage that they reduce a three-dimensional solid beam to a one-dimensional object that can be described by an ordinary differential equation. These equations can be solved analytically when the geometrical form of the structure is not too complex.

The reliable mechanical and mathematical description of trusses, beams and shells is of great significance. Still today new formulations for beams are being developed based on classical approaches steming e.g. from Bernoulli, Euler, Timoshenko and Vlasov. These theories are based on different kinematical approximation of the deformation of a cross section with respect to the beam axis. The Euler-Bernoulli beam neglects shear deformations while the Timoshenko beam includes shear deformations. Vlasov extended these kinematical assumptions by assuming arbitrary deformations of the cross section. Details can be found in classical textbooks, see. e.g. Timoshenko and Woinowsky-Krieger (1959), Donnell (1976), Vlasov (1984) and Crisfield (1991).

Linear and approximate nonlinear theories (second order theory, theory of moderate rotations etc.) are known for a long time and have been introduced to the engineering codes. Especially stability problems were solved by different approximate theories and associated numerical methods.

Due to the development of inexpensive computer hardware it is today possible to perform numerical simulations based on completely nonlinear theories. With this the general description of finite deformation within one-dimensional structural members has found its way into modern numerical simulation tools like the finite element method. Due to this development it is not necessary to discuss the validity of approximate theories since no restrictions with respect to deflections and rotations are made in a fully nonlinear approach, see e.g.Simo and Vu-Quoc (1986), Pimenta and Yojo (1993), Gruttmann et al. (1998) and Pimenta et al. (2008). This can lead to a paradigm shift in the application of models in engineering.

The one-dimensional models of trusses and beams are imbedded in the three-dimensional space. All models are characterized by a description of the geometry as a curve in space. Formulations which can be applied to describe the spatial curves or surfaces are provided by introduction of the arc-length of a curve, convective coordinates for curved lines. But there exists also the simple reference to a cartesian coordinate system using an approximation of the initial geometry by polynomial patches. All these formulations can be considered to derive the associated strong or weak forms for trusses and beams. Often the numerical approximation model, like e.g. the finite element or meshless method define the choice of the theoretical description, see e.g. Dvorkin et al. (1988) and Ibrahimbegovic (1995).

E.g. the use of a straight beam axis using Cartesian coordinates leads to a discretization of curved spatial beams by a number of straight finite elements with linear interpolation. Such approximation of the geometry however will create additional errors besides the discretization error of the deformation field. Often this error, due to the approximation of the geometry, vanishes with increasing number of finite elements. In that case it is essential that the additional element coordinates are always related to the exact geometry. These errors diminish for higher order finite element approximations. Additionally, in the last few years different approaches have been proposed which link the geometry directly to the finite element approximation by using the same ansatz, like NURBS, for geometry and finite element interpolation, see e.g. Hughes et al. (2005), Bouclier et al. (2012) and Cazzani et al. (2014).

Further considerations are required in order to formulate constitutive equations for beams, especially when nonlinear behaviour – like elasto-plastic responses – have to be considered. There are basically two different concepts.

The first formulation uses the constitutive equations of the three-dimensional solid directly. Thus the constitutive equations are represented by the stresses as discussed in section 2.1. An associated formulation can be found in e.g. Gruttmann et al. (2000). The advantage of this formulation is that any constitutive equation formulated in section 2.1 can be applied within the model of the beam.

The other formulations uses the stress resultants to derive inelastic constitutive equations for beams. This leads to the notion of plastic hinges. Different approaches for steel can be found in Windels (1970). The approach leads to a nonlinear interaction between the different stress resultants of normal, shear force and bending, torsional moment. Hence the flow condition has to be formulated in terms of the stress resultants. We note, that such formulations are limited in their range of application, but they provide satisfactory results for a number of practical problems, see e.g. Henning (1975), Vogel (1985) and Becker (1985). Formulations of the inelastic constitutive behaviour using the plastic hinge model for beams undergoing large deflections and rotations can be found together with the associated finite element formulation in e.g. Kahn (1987), Simo et al. (1984) and Ehrlich and Armero (2005).

Two-dimensional models

Structural modesl that have two-dimensions but are located in the three-dimensional space are plates, membranes and shells, see Fig. 9. The difference of these models stems from the type of loading. Membranes are usually loaded in the same plane that is defined by the initial geometry of the membrane whereas plates have to sustain loads that act perpendicular to the surface defining the plate. While plates have an initial geometry that describes a flat surface, membranes can have arbirary surface geometries but their model does not include any bending stiffness. Shell models are a combination of plates and membranes for arbitrary surface geometries. They include plate and membrane effects and hence are more complicated but also the application range of shells is more general.

The theoretical model for a plate was finalized by the work of Kirchhoff, see Kirchhoff (1850), who included the correct boundary conditions for the complete mathematical description of the plate model. This model includes

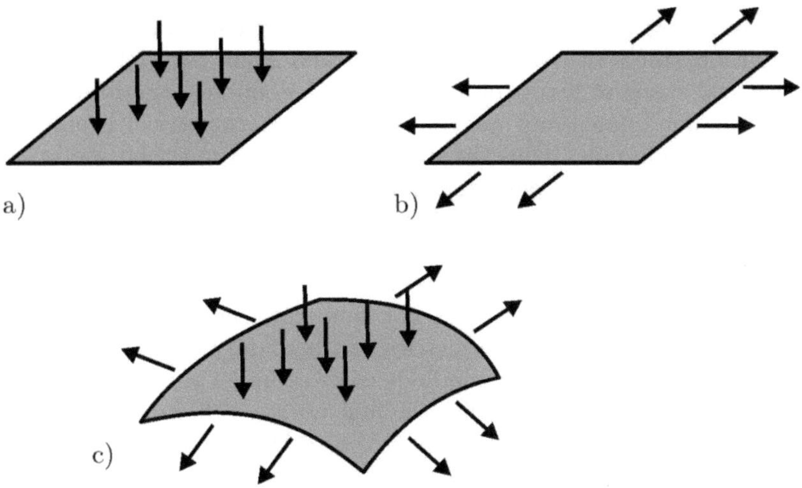

Figure 9. a) Plate- , b) membrane- and c) shell-structures.

an analogous kinematical assumption as the Euler-Bernoulli beam. Thus an advanced theory, including shear deformations, was also developed for the plate. It is analogous to the Timoshenko beam. The model is known under the names of Reissner and Mindlin. Further plate models were developed in order to model composite structures that have many applications in structures related to aviation like airplanes. Here multi-layer approaches were proposed in order to obtain good predictive models, see e.g. Jones (1975), Reddy (1984), Hull and Clyne (1996) and Reddy (2004).

The first nonlinear extension of the plate theory that included nonlinear membrane parts was developed by von Karman. This theory was extensively applied for the solution of stability problems of plates.

The development of shell equations started with the work of Love, see Love (1927). Since then other, more refined shell models were derived in the literature, see e.g. Koiter (1960) who obtained a consistent first order shell model. Further development, using convective coordinates to derive general shell models were introduced by Naghdi (1972) and Flügge (2013). However these results are not of interest for modern shell analysis since modern approximate models are based on isoparametric or isogeometric formulations. Thus they do not need the notion of convective coordinates and the related complex calculus, see e.g. Cirak et al. (2000), Pimenta et al. (2004), Kiendl et al. (2009) and Benson et al. (2011).

Analytical solutions of the partial differential equations describing the deformation states of plates, membranes and shells are only possible for very special cases in terms of geometry and boundary conditions. These are rectangular plates and membranes and axisymmetrical geometries of shells and membranes. Thus the use of plate, membrane and shell theories is combined very much with the development of computational facilities and numerical approaches. The first numerical solution methods used in engineering actually targeted these models. In the early stages of the numerical schemes finite difference methods were used due to simplicity. But it turned out that this method is not well suited for structural engineering problems with complex geometries and boundary conditions. With the development of finite element methods more complex problems in solid mechanics could be solved. Actually the first notion of finite elements can be found in Courant (1943) who derived the method and solved the torsion problem of a quadratic box and - as a mathematician - even provided a convergence study. More than ten years later the first engineers developed the method to solve structural problems in aircraft industry related to plate problems, see e.g. Turner et al. (1956) and Argyris (1960).

Today shell theories are available that model general nonlinear behaviour such as finite strains and include inelastic constitutive models. A first contribution that categorized different shell theories for nonlinear problems in terms of rotations can be found in Pietraszkiewicz (1978).

Interim conclusions related to structural models

Structural models have a wide application range in engineering design analysis. The one-dimensional models lead to ordinary differential equations that can be solved directly in the geometrically linear case. For shells similar simple ordinary differential equations exist for axisymmetric case. Here also analytical solutions are possible and can be found in classical textbooks like Timoshenko and Woinowsky-Krieger (1959) and Pflüger (1981).

Starting with two- or three-dimensional beams it is possible to use a very efficient and robust model to predict the basic force and moment distribution in a structural member. Here the user has to select the model for the type of beam. E.g. the ratio of length over height of the beam $(l\,/\,h)$ defines the slenderness. For a thin beam often this ratio is limited by $8 \leq \frac{l}{h} \leq 100$. Here the theory of Bernoulli-Euler is sufficient. For thick beams one has $\frac{l}{h} < 8$. In this case the shear effects have to be included which leads to the Timoshonko beam model. Finally for very thin beams $\frac{l}{h} > 100$ it is possible that large deformations occur and then a second order or fully nonlinear beam model has to be used.

Plates are generally used in structural analysis for buildings to model floors and walls. The differentce is that plates for floors are loaded perpendicular to the defining surface while plates for walls are loaded in plane. Shells are more used in mechanical engineering since e.g. car bodies are naturally shaped like a complex shell. This is also true for ship hulls in naval architecture. Also modelling forming processes has to be based on shell models. While the car body and the ship hull often can be analyzed using a geometrically linear theory, Forming of parts or crash analysis of cars demands highly nonlinear shell models including large elasto-plastic deformations. Regarding the slenderness of plates and shells, similar relations can be found in order to select a Kirchoff-Love or Reissner-Mindlin model with respect to the slenderness now defined in two spatial directions.

Plate and shell models are very sensitive with respect to the boundary conditions. This is related to the fact that boundary layers of high stresses can occur in some models. Due to that one has to be careful to select the appropriate boundary conditions that are consitent with the model. For details see e.g. Timoshenko and Woinowsky-Krieger (1959), Donnell (1976) and Flügge (2013).

2.3 Coupling of models

There are different possibilities to couple the aforementioned models and others in order to reduce computational cost and effort. The classical approach in engineering is to decouple models. With this substructures are introduced that are analyzed based on the response of the global structure. In such cases the loads, stresses and deformations can transferred from the global structure to the substructures. However in nonlinear cases such decoupling is not possible or only in a complex iterative manner. Thus direct coupling approaches were developed for a number coupling approaches where different dimensional models with the same underlying constitutive behaviour (like beam, shells and solids) but also completely different models are coupled. Some combinations can be found in the following list:

- Shell-beam,
- Solid-shell,
- Solid-particles,
- Fluid-Solid (FSI),
- Fluid-Particle (FPI).

Thus a beam can be coupled with a shell. E.g. the shell model can be used in places where the local stability of a flange is of interest and the beam model then is applied in parts of the structure where no local effects regarding the cross section of the beam have to be considered, see e.g. Fig. 10. These types of models need special interface formulations that allow

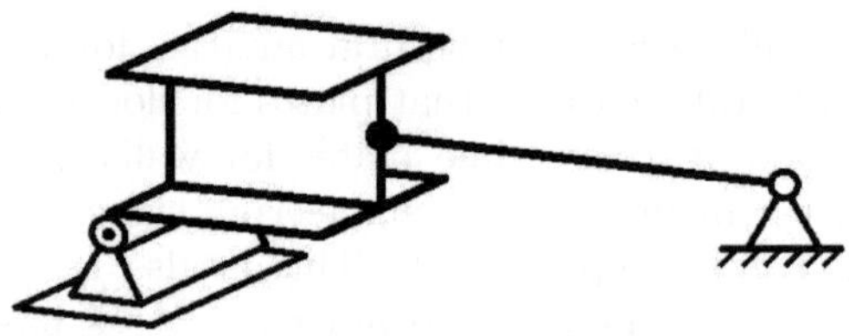

Figure 10. A coupling model for beam and shells.

the consistent transition of the variables. Examples for beam shell coupling can be found in Wagner and Gruttmann (2002) and Chavan and Wriggers (2004).

Besides modeling local buckling in parts of the structure local-global responses can be predicted using coupled models. These are e.g. bolt or screw connections where the local behaviour of the bolt/screw is governed by the global structural response and thus a coupled analysis is necessary[5].

Also problems like pile driving can be treated successfully with coupled solid-particle models, see e.g. Wellmann and Wriggers (2012), where near the pile rigid particles are used to model the penetration process and away form penetration a solid model is sufficient, see Fig.11. In such cases the design engineer has to make sure that the constitutive behaviour of the particles are the same like the constitutive behaviour in the solid. These type of models are more and more used, also together with meshless methods, in order to compute nonlinear problems where large deformations occur that cannot be captured by standard numerical approaches.

Coupled models or a model hierarchy provide sufficient accuracy at different spatial regions. Thus they lead to a more efficient way than the use of one model that provides an "overkill" at certain regions where its accuracy is not necessary or of interest. These models are hence inexpensive for a given accuracy.

The art of modeling with respect to coupling of models is to select the regions for the different models. There exist some mathematical attempts to change models based on error estimation, see e.g. Oden et al. (2001), Stein and Ohnimus (1996), Temizer and Wriggers (2011) and Radermacher and Reese (2014) but they are so far not applied in general engineering applications. Furthermore it is not obvious how these methodologies can

[5]Classically also Fluid-Structure-Interaction (FSI), Fluid-Particle-Interaction (FPI) or any type of multi-physics problem needs the coupling of different mathematical models, but this is not the main topic of this work

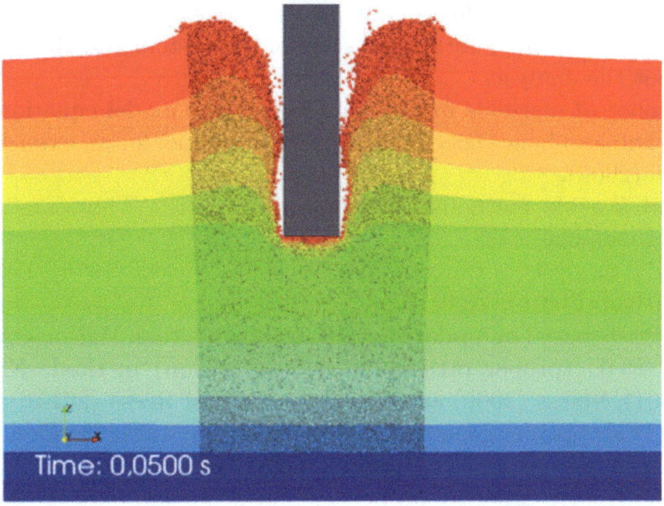

Figure 11. Pile penetration using a coupled Particle-Soil approach.

be used in an efficient and robust way in nonlinear problems. Thus the engineer still has to use intuition to find an accurate but efficient model.

2.4 Mathematically reduced models

It is possible to reduce the space of solutions for complex structural models. This is of interest for fast computations that have to be done in real time and for large systems that cannot be solved in an acceptable time frame. The general idea is that reduction techniques act on the mathematical model and try to simplify the model equations such that with almost the same accuracy a problem can be solved in a fraction of the time that would be needed to the full problem. These models are typically introduced when the dynamic response of an engineering structure has to be computed.

In linear dynamical analysis modal analysis techniques are well established. These are based on the computation of the eigenvalues and vectors that lead to a diagonalization of the algebraic equations and hence reduce the equation systems considerably, see e.g. Bathe (1996). On top of it, it is well established that a large amount of the system's energy can be associated with a few eigenmodes and hence the behaviour of the dynamical system can be described by these modes. This works well for linear systems, however for nonlinear systems the modes that contain most of the energy depend on the deformation state of the system and thus will change throughout the analysis.

A number of methods exist which reduce the model equations or their discrete form and hence speed up computation time, for an extensive review see Spiess (2006).

Several feasible methods are mentioned below. These are projection based methods like

- modal analysis in tangent space, see e.g. Nickell (1976), Leger (1993) and Kirsch et al. (2005),
- methods based on updated eigen- or Ritz vectors, Wilson et al. (1982),
- the POD (proper orthogonal decomposition) method, Krysl et al. (2001), Spiess (2006) and Radermacher and Reese (2014), and
- the PGD (proper generalized decomposition) methods, for an overview see Ladeveze et al. (2010) and Chinesta et al. (2010).

The methods have different properties and hence are optimal for different applications. However there does not exist a unique way of treating nonlinear dynamical equations. Again the intuition of the analyst is needed and the application range of the systems and their degrees of nonlinearity has to be known.

3 Numerical methods: finite element analysis

Once the mathematical model is known which results typically in an ordinary or partial differential equation a solution technique has to be selected to solve the set of equations. Analytical methods that were used in the last centuries are usually not sufficient for complex engineering problems. These have either a very complicated geometrical form and require inelastic constitutive models for the analysis. In this case, analytical methods, that are usually limited to simple geometries and constitutive equations cannot be applied. Hence it is necessary to use numerical schemes that provide efficient and accurate solutions. All numerical schemes approximate the solution space in one or another way. Depending on the nature of the problem different approaches can be selected. Among these are

- finite difference methods (FDM),
- finite volume methods (FVM),
- boundary element methods (BEM),
- finite element methods (FEM),
- meshless methods based on smooth partical hydrodynamics (SPH) or other particle methods and
- discrete element methods (DEM).

This list is by no means complete, but depicts the different possibilities that an engineer has once he wants to solve a problem numerically. Each of these methods has an application range where it is optimal from the point of view of accuracy, effictiveness and manageability. Thus the art of modelling includes in a way also the art of selecting proper solution methods that converge to correct and reliable solutions.

In this chapter we will concentrate on the finite element method (FEM) since it has been proven over the last 70 years, starting from the early work of Courant, see Courant (1943), that this method can be applied to most problems stemming from structural analysis in civil engineering and mechanical engineering. This does not mean that there are methods, like BEM, that have advantages in far field wave propagation problems, but generally the finite element method has become the tool for predictive computations in structural analysis.

However even when an established method like the finite element method is applied to numerically solve an engineering problem, the user has to make decisions regarding the numerical model. There are hundreds of finite elements that are based on a variety of mathematical models and different ansatz spaces and hence again it becomes an art to select the right finite element for the task at hand. However in the case of geometrically linear problems solved by finite elements there exists established mathematical theory that helps the engineer to make his selection. In case of nonlinear problems mathematics can provide ideas stemming from the linear resuts. These help to obtain reliable solutions also in the nonlinear application range. However some methods that work in linear cases might not work when applied to nonlinear problems[6].

3.1 Set of equations: linear finite elements

Linear finite element formulations lead from ordinary and partial differential equations and their related weak forms, as discussed in the previous chapter, to algebraic matrix equations. Here stiffness matrices which are related to a local problem are defined at the element level $\mathbf{k}^e$ (finite element e) and will be assembled to the global stiffness matrix $\mathbf{K}$. In this process the compatibility between the elements has to be fulfilled. The same procedure is applied for the element load vectors $\mathbf{f}^e$ that are assembled to the global load vector $\mathbf{F}$. Additionally the boundary conditions for the primary

[6]This can be the case in constraint problems when the mathematical form of the constraint condition is different in the linear and nonlinear mathematical model. An example is incompressibility where the constraint condition is $\text{div}\boldsymbol{u} = 0$ in the linear and $\det \boldsymbol{F} = 1$ in the nonlinear case.

variables are directly fulfilled during the assembly process. This leads to a reduced size of the global matrix. For static problems which do not depend on time this process yields the global equation

$$\mathbf{K}\,\mathbf{u} = \mathbf{F}\,.$$

In the dynamic case the momentum balance yields together with element matrices $\mathbf{m}^e$ the mass matrix $\mathbf{M}$ an discrete ordinary differential equation in time t

$$\mathbf{M}\,\ddot{\mathbf{u}}(t) + \mathbf{K}\,\mathbf{u}(t) = \mathbf{F}(t)\,.$$

The basic form of both equations is independent on the chosen model (trusses, beams, plates, shells or three-dimensional solids). Of course the size of the system changes as well as the quantities entering the displacement vectors etc.

In case of dynamics often damping effects have to be considered which then leads to

$$\mathbf{M}\,\ddot{\mathbf{u}}(t) + \mathbf{C}\,\dot{\mathbf{u}}(t) + \mathbf{K}\,\mathbf{u}(t) = \mathbf{F}(t)$$

with the velocity dependent damping term $\mathbf{C}\,\dot{\mathbf{u}}$. Often the damping matrix $\mathbf{C}$ is selected as a linear combination of the mass- and stiffness matrix: $\mathbf{C} = \alpha_1\,\mathbf{M} + \alpha_2\,\mathbf{K}$ which is called modal damping. The parameters α_1 and α_2 determine the damping magnitude and depend on the modeled system. The solution of the ordinary differential equation in time can be obtained by modal transformation or by numerical integration procedures.

3.2 Nonlinear finite elements

Nonlinear finite element formulations have to be applied when either large displacements or strains occur (geometrical nonlinearity) or when inelastic deformations (physical nonlinearity) are present. The general formulation (weak form) leads to a matrix equation in vector form

$$\mathbf{R}\,(\,\mathbf{u}\,) - \mathbf{F} = \mathbf{0}\,. \tag{35}$$

Here $\mathbf{R}\,(\,\mathbf{u}\,)$ represents the nonlinear internal virtual work (residual), stemming from the first term in (22), while $\mathbf{F}$ is associated with the external forces (loading) that result from the last two terms in (22). This algebraic equation is generally nonlinear in the displacements. It can be solved by different algorithms, e.g. fix point method, secant method or Newton method. The most appropriate solution scheme is the Newton method which leads to the following algorithm

$$\begin{aligned}
\mathbf{K}\,(\,\mathbf{u}_n\,)\,\Delta\mathbf{u}_{n+1} &= -[\,\mathbf{R}\,(\,\mathbf{u}_n\,) - \mathbf{F}\,] \\
\mathbf{u}_{n+1} &= \mathbf{u}_n + \Delta\mathbf{u}_{n+1}
\end{aligned}$$

Here an initial value must selected for the displacements, often $\mathbf{u}_0 = \mathbf{0}$ is chosen. The algorithm will be executed for $n = 0, 1, 2, \ldots$ steps until convergence $\| \mathbf{R}(\mathbf{u}_n) - \mathbf{F} \| \leq \varepsilon$ with tolerance ε.

In case of strongly nonlinear problems, the load cannot be applied in one step. Then an incremental loading is necessary which leads to $\mathbf{R}(\mathbf{u}) - \lambda\,\mathbf{F} = \mathbf{0}$ with the load parameter λ.

For dynamical problems the inertia terms have to be considered additionally which yields the matrix form

$$\mathbf{M}\,\ddot{\mathbf{u}} + \mathbf{R}(\mathbf{u}) - \mathbf{F} = \mathbf{0}\,. \tag{36}$$

In this form the mass matrix can be computed using the formulations stated for linear finite element methods. The solution of the nonlinear ordinary differential equations is usually obtained by numerical integration.

3.3 Modeling with continuum finite elements

Finite elements can be applied to arbitrary problem classes within solid mechanics. It is clear that some general requirements have to be fulfilled when finite elements are constructed. These are related to the mathematical demands such that the finite element solution is based on a convergent scheme when the number of elements are increased. Here we will not go in details regarding the construction of classical deformation based elements, for details see e.g. Zienkiewicz and Taylor (1989) and Hughes (1987) and for the mathematical analysis, Braess (2007).

There exist many different element formulations that were developed for special applications. Some points that lead to different finite elements are summarized below.

- **Efficiency** is one of the main requirements when large structures in solid mechanics have to be solved since often more than 10^6 finite elements are needed to model the structure accurately. Within this considerations efficiency is not only related to speed of the element formulation but also to the memory requirement. The latter demand is essential when e.g. inelastic problems have to solved within a given time frame. This leads to element formulations with as few as necessary integration points. The speed of the element formulation is essential when iterative solvers are applied since in that case the time for the computation of residuals and tangent matrices is of the same order as the time used by the solver within one iteration.
- **Good bending performance** is of significance when three-dimensional solid elements are used to solve beam- or shell problems. These structural problems are often dominated by bending behaviour and

hence classical low order elements will not provide a good solution. Here either low order mixed elements, see e.g. Wriggers (2008), or higher order classical finite elements have to be employed. It has been shown that a high order of interpolations (so called p-version of finite elements) can be applied successfully to finite deformation problems for rubber like materials, see e.g. Heisserer et al. (2007). Lately isogeometric elements were developed. These elements are of higher order and thus perform well in bending dominated problems, see e.g. Hughes et al. (2005) and Cottrell et al. (2009).

For low order elements usually mixed variational principles, see e.g. Washizu (1975) , are used as basis for the finite element discretization. Here different possibilities exist for the construction of the finite element matrices. Some of them rely on conforming displacement fields together with non-conforming stress or strain fields, others rely on conforming stress fields but allow non-conforming displacement fields. Theoretical background for linear mixed methods can be found in e.g. Braess (2007) and Brenner and Scott (2002). For linear elasticity specific hybrid elements where first described in Pian (1964) which has led to many different finite element formulations. Within this approach Pian and Sumihara (1984) developed a linear finite element which is efficient and accurate, it has good bending performance and overcomes locking, see below. However due to the need to invert the constitutive equations within the formulation in order to obtain the constitutive equations in terms of the stress field, there are only few elements for St. Venant materials known that work for large deformations.

- **Locking** occurs in general when constraint conditions have to be considered in the mathematical model. This is associated with the numerical simulation of special problem classes that include rubber like materials and constitutive formulations of elasto-plastic materials in the framework of J_2-plasticity. But also the use of three-dimensional elements for thin structures like beams, plates and shells leads to locking, however these elements allow a simple implementation of three-dimensional constitutive equations which is not so easily possible when classical beam or shell models are employed.

  As an example a thin clamped rectangular plate is considered that is subjected to uniform loading, see Korelc et al. (2010). Geometry and material data are depicted in Fig. 12 as well as the deformed configuration. This test compares different three-dimensional finite elements. These are linear and quadratic tetraeders (O1 and O2), linear and quadratic hexaeders (H1 and H2), two enhanced elements

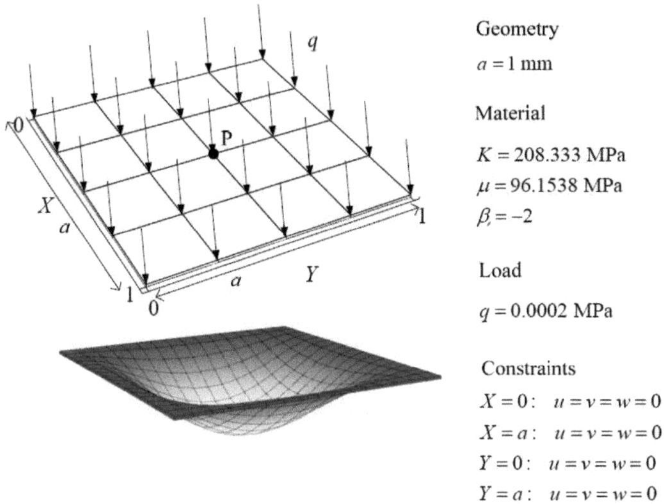

Figure 12. Thin plate under uniform load, geometry and deformed configuration, see Korelc et al. (2010).

(H1E9, see Simo and Rifai (1990), and CG9, see Korelc and Wriggers (1996)) and two higher order enhanced elements (H1E21, see Büchter et al. (1994) and TSCG12, see Korelc et al. (2010)). One can conclude from Fig 13 that only two elements, H1E21 and TSCG12, match the analytical solution of this rather simple problem. This is due to the severe shear locking of all the other, even quadratic elements, at a high ratio between the length and thickness of the plate.

Thus constraints lead, even for geometrical linear elements, to *locking*, see e.g. Braess (2007), Zienkiewicz and Taylor (1989) and Hughes (1987). During the years different special finite elements were developed to overcome locking, The most simple method is the "reduced integration" of the integrals leading to the finite element vectors and matrices. This approach is very efficient and safes memory for history data storage since less integration points are used. Reduced integration uses less Gauss points for the integration of tangent matrices and residual vectors than necessary for the chosen polynomial degree of the ansatz functions, for first applications see e.g. Zienkiewicz et al. (1971). Reduced integration was developed to avoid locking in case of incompressibility. Many variants were developed to make reduced

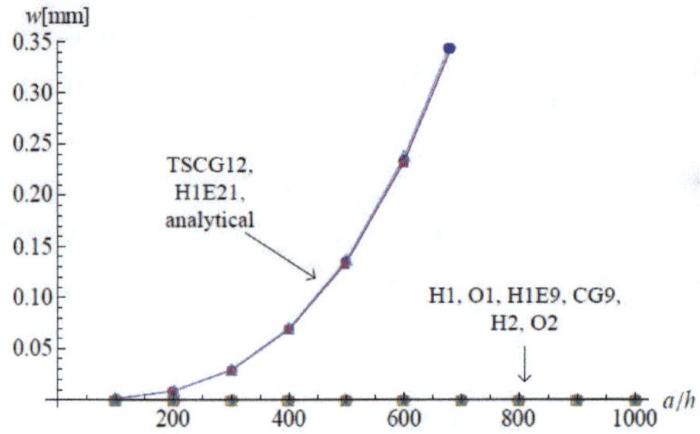

Figure 13. Deflection at the central point of the plate using $4 \times 4 \times 4$ mesh, see Korelc et al. (2010).

integration techniques robust. This stems from the fact that reduced integration is always associated with a rank deficiency of the tangent matrices which has to be cured. The related methods are known as stabilization techniques, see e.g. Belytschko et al. (1984).

Another class of finite elements that overcomes locking is based on the enhanced strain formulations. Here non-conforming strain measures are introduced within the Hu-Washizu principle, see Simo and Rifai (1990) who developed enhanced strain elements for the geometrical linear theory using linear ansatz functions for the displacement field. Then Simo and Armero (1992) and Simo et al. (1993b) derived a family of enhanced elements for large deformations and inelastic constitutive equations. This class of elements is related to the incompatible mode elements which were developed by Wilson et al. (1973) and Taylor et al. (1976) for linear problems. Enhanced strain elements have good bending properties and work for incompressible materials. However these elements have some disadvantage. They need a statical condensation at element level. This reduces the efficiency of the enhanced strain elements. Additionally tese elements depict hourglassing when negative stresses (pressure) occur, this was discovered by Wriggers and Reese (1996). It has led to formulatons in which these elements are stabilized, see e.g. Reese (2005) and Korelc et al. (2010).

A nonlinear example depicts the behaviour of the different elements that were already described, it can be found in Korelc et al.

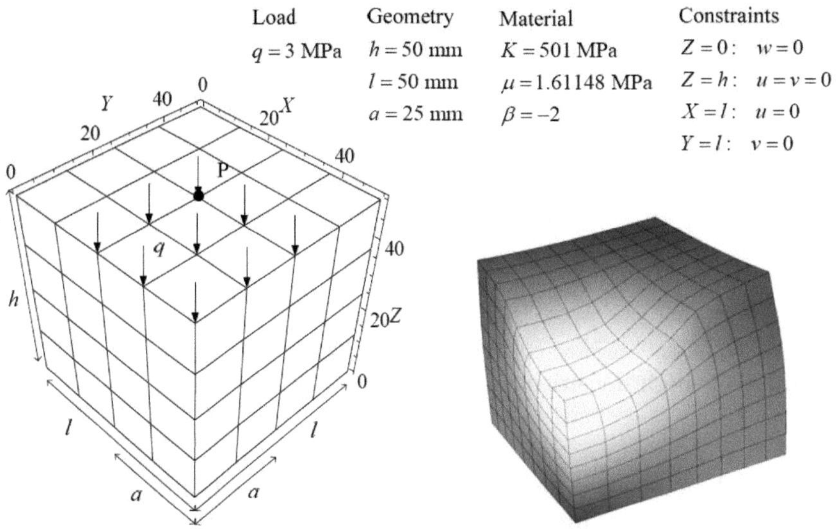

Figure 14. Incompressible block, data and deformed configuration, see Korelc et al. (2010).

(2010). The block consisting of nearly incompressible rubber material is loaded by a constant surface load $q = 3$ MPa at the top central quarter. Only one quarter of the block is discretized due to the symmetry of the system, as shown Fig 14. This figure contains also the geometrial data and boundary conditions as well as a deformed configuration. In Fig 15 the displacement of the point P resulting from the loading is displayed as a function of mesh density for all elements discussed already in the example above. The H1 and O1 elements depict locking behavior, especially for coarse meshes. All other elements converge to the same result with mesh refinement. There are only slight differences, however it can be seen that the TSCG12 element has the best coarse mesh accuracy.

- **Sensitivity against mesh distortions** is essential when modern methods for mesh generation are applied to construct finite element meshes. These methods lead for arbitrary geometries to so called unstructured meshes that consist of finite elements shapes with arbitrary geometry, hence a finite element has to be able to produce good results also for distorted meshes. Another source for the distortion of finite elements is the movement of nodal coordinates during nonlinear

Mesh	DOF	TSCG12	H1	H1E9	CG9	O1	H1E21
4×4×4	260	20.1379	7.61980	19.9773	19.0326	14.4240	19.7107
8×8×8	1800	20.0974	13.0375	20.0260	20.0660	17.8659	20.0582
16×16×16	13328	20.0303	17.4522	19.9500	20.0336	18.9050	20.0297
32×32×32	102432	20.0069	19.4629	20.0107	20.0072	19.5870	20.0069
Mesh	DOF	H2	DOF	O2			
2×2×2	148	18.8387	236	20.2283			
4×4×4	968	20.3535	1608	20.4907			
8×8×8	6928	20.3175	11792	20.1177			
16×16×16	52256	20.1182	90144	19.9730			

Figure 15. Convergence study for the displacement of point P of the block for different mesh densities, see Korelc et al. (2010).

simulations which can lead to severely deformed finite elements.

- **Good coarse mesh accuracy** is related to the fact that in real engineering applications often three-dimensional components have to be analyzed. Often the size of the model is so large and has such a complexity that it cannot be modeled using a converged mesh, especially when the simulation is nonlinear. Hence there is a need for elements that depict a superior accuracy, even when used within a coarse mesh. This is a point which will diminish with the increasing computing power, but at the moment it is still of concern.
- **Simple implementation of nonlinear constitutive equations** is a requirement that stems from the need for accurate mathematical and physical models that have to be used within the simulation of nonlinear engineering structures. Within the modeling process new complex nonlinear constitutive equations have to be used. Here a simple interface to the finite element should support the user in order to efficiently change existing finite elements and to be able to implement new complex constitutive equations.

There exist many more finte element formulations that were developed for special problem classes. An overview can be found in e.g. Zienkiewicz and Taylor (2000) and Wriggers (2008).

3.4 Modeling aspects for solid elements

When modeling the order of the element interpolation has to be selected with care in relation to the properties of the mathematical model. Low continuity, non smooth solutions, like in plasticity, require low order interpolation functions which have been proven to be robust for many nonlinear simulations, while high continuity, like in elasticity, can be treated with high order interpolation schemes.

Generally one has to compare linear and quadratic or other higher order elements with respect to all points mentioned above. It is clear that elements with higher order have a higher order of convergence when the solution has the necessary regularity. In that case less finite elements of higher order can be used which yield results with the same accuracy. However in order to compare the finite element discretizazions of different interpolation orders the total solution time needed to obtain a result with the same accuracy has to be considered.

Memory requirement for history variables plays an essential role when explicit integration schemes are employed to simulate impact or shock problems. Here only the residual has to be stored, which leads to the storage of three values per node. In order to reduce the overall computing time, all quantities have to be retained in the main memory. In such case the storage of the history variables is of major concern for explicit simulations. Hence most of the explicit finite element codes use special finite elements with as few integration points as possible.

It is well known that the pure displacement element with bi- or tri-linear ansatz function has bad convergence behaviour in bending problems, especially if the length in one direction is a lot smaller than in the other ones, e.g. for beam or shell structures. Hence special elements have been developed for such problems to avoid locking. Such elements still base on linear ansatz functions. Hence the convergence order cannot be increased but the constant governing the mathematical inequality which is related to the convergence behaviour is reduced considerably. Thus the required accuracy of a finite element solution can be gained with considerably less elements.

3.5 Finite element methods for beams and shells

Structural elements are modeled in case of trusses and beams by one-dimensional objects that are imbedded in three-dimensional space. The models are characterized by a description of the geometry as a curve in space. Formulations which can be applied to describe the spatial curves are provided by the introduction of the arc-length of a curve or simply by reference to a cartesian coordinate system using an approximation of the initial geometry by straight lines.

The latter leads often to a discretization of curved spatial beams by a number of straight finite elements with linear interpolation. Such approximation of the geometry however will create additional errors besides the discretization error of the deformation field. The error, due to the approximation of the geometry, vanishes with increasing number of finite elements. In that case it is essential that the additional element coordinates are always related to the exact geometry.

Beam element need special interpolations. For the Bernoulli-Euler beam C^1-continuous interpolations have to be introduced. This leads to an interpolation with at least cubic polynomials. Classically Hermitian polynomials are employed. For shear elastic Timoshenko-Reissner beams only C^0-continuous ansatz functions are required. For thin beams the shear deformations are extremely small. This leads to locking since the constraint (no shear deformation) is approximated. Thus special elements have to be developed, similar to the ones discussed in section 3.3.

Several lines for the formulation of beam elements within the finite element methodology were developed. The first is based on the assumption of small strains. This formulation introduces a rigid frame undergoing finite rotations and introduces strains and stresses relative to the rotations of the frame. It is known as co-rotational formulation. The strains have to be small but large deflections and rotations can be investigated. Finite element schemes that base on such formulations can be found in e.g.. in Oran and Kassimali (1976), Wempner (1969), Rankin and Brogan (1984), Lumpe (1982), Crisfield (1991) and Crisfield (1997).

Another approach employs directly the continuum equations and introduces the beam kinematics by special isoparametric finite element interpolations. This approach is known as degenerated continuum approach, see e.g. Bathe and Bolourchi (1979), Dvorkin et al. (1988), Bathe (1996) and Crisfield (1997).

A third formulation is based on the direct us of nonlinear rod and beam theories that have as only restriction the classical Timoshenko assumption of "plane cross sections remain plane". No other approximations are made. Thus strains, deflections and rotations can be finite. The associated theories are called geometrically exact. The development of these beam theories started with the work of Reissner (1972). A generalization for the three-dimensional case can be found in Simo (1985). Based on this theoretical background, several authors developed associated finite element formulations that have different complexity, such as consideration of cross-section warping, see Simo and Vu-Quoc (1986), Pimenta and Yojo (1993), Jelenic and Saje (1995), Gruttmann et al. (1998) and Mäkinen (2007). A nonlinear formulation of curved beam elements can be found in e.g. Ibrahimbegovic (1995). Gruttmann et al. (2000) derived a three-dimensional beam element for elasto-plastic material and in Romero and Armero (2002) dynamics was considered within the geometrically exact framework.

Within all formulations classical interpolation functions were employed. Isogeometric interpolations provide also a good basis for Euler-Bernoulli beams since they provide naturally continuous fields. For applications see e.g. Konyukhov and Schweizerhof (2012), Bouclier et al. (2012) and Reali and Gomez (2015) .

Basically any shell could be discretized using three-dimensional solid elements as discussed in section 3.3. A linear interpolation through the thickness is then close approximation of the assumption that plane cross sections remain plane during the deformation. Additionally a change of thickness is taken into account in this model. It is well known, see e.g. Zienkiewicz et al. (1971) for the linear case, that a pure displacement formulation leads to locking in the thin shell limit. Hence special interpolations have to be employed when three-dimensional solid formulations are used to discretize shells in order to eliminate locking, see e.g. Seifert (1996), Miehe (1998), Hauptmann and Schweizerhof (1998), Sze and Yao (2000) and Korelc et al. (2010). Higher order interpolations for such shell elements were discussed in Düster et al. (2001) and Kiendl et al. (2009).

Classically the approach based on shell theory is followed to derive the associated elements. In this approach the description of a shell is based on the definition of a middle surface. Using such parametrization, kinematics, weak form and constitutive equations can be derived from the three-dimensional continuum equations. Different approximations for the kinematical description of the shell model in thickness direction can be developed. Hence equations for thin and thick shells that additionally can model deformations in thickness direction can be developed. Depending on the number of kinematical variables these approaches are denoted 5-, 6- or 7-parameter theories. Within this line of modeling different formulations were developed to construct finite elements for shells undergoing finite deflections and rotations, see e.g. Simo et al. (1990), Simo et al. (1990B), Onate and Cervera (1993), Sansour (1995), Eberlein and Wriggers (1999), Cirak et al. (2000), Campello et al. (2003), Pimenta et al. (2004) and Gruttmann and Wagner (2005).

The degenerated concept uses the equations of a three-dimensional solid and introduces the shell kinematics at the discretization level. As in shell theory a reference midsurface is chosen, see e.g. Ramm (1976) and Bathe (1982). Within this approach no shell theory – besides the kinematical assumption – is needed for the discretization of a shell continuum. Hence this approach is conceptional simple. The introduction of stress resultants is naturally not included in this formulation. A comparison of finite elements based on classical shell theory and on the degenerated concept was investigated in Büchter and Ramm (1992), who found that both approaches lead under certain assumptions to the same finite element formulations. . In most approaches it is assumed that plane cross sections remain plane during the deformation of the shell continuum. This yields theories, named after Reissner and Mindlin, that include shear deformations which require only C^0-continuous discretizations within the finite element method.

When no further assumptions besides "plane cross sections remain plane" are introduced within the derivation of the shell equations from the non-linear continuum equations then this shell theory is called „geometrically exact". Geometrically exact shell theories were developed during the last two decades since growing computer power enabled the engineer to perform numerical simulations of complex nonlinear shell problems without assuming any approximation with respect to the size of rotations and deflections.

First investigations using geometrically exact shell theories can be found in Simo et al. (1989) and Wriggers and Gruttmann (1989). New in this work is the formulation of a singularity free parametrization of the rotations and the use of the isoparametric formulation for approximation of the shell geometry. The latter eliminates the appearance of co- and contra-variant derivatives which are replaced by standard partial derivatives. This direct approach was basis for different theoretical formulations leading to new finite elements for nonlinear shell problems, see e.g. Simo et al. (1990), Basar and Ding (1990), Wriggers and Gruttmann (1993), Wagner and Gruttmann (1994), Basar and Ding (1996) and Bischoff and Ramm (1997) a further in depth discussion can be found in Bischoff et al. (2004).

The classical Kirchhoff-Love hypothesis would, of course, be a natural assumption for kinematics of thin shells too. However this additional assumption requires C^1-continuous interpolation functions which are not easy to construct using classical approaches. In this context, the isogeometric approach that combines interpolations of the deformations with the CAD description of the shell surfaces are of interest. These discretization employ Bezier or other C^1-continuous polynomials, see e.g. Cirak et al. (2000), Onate and Cervera (1993) and lately Hughes et al. (2005). Such formulations deviate from the classical finite element concept since the C^1-continuity is not fulfilled on element but on patch level by using a patch of elements to define the interpolation functions. The formulations have a great advantage because shells are highly sensitive to geometry changes and Bezier or NURBS surfaces ideally map complex geometries into the shell model. However lately new efforts resulted in elements based on the old C^1 concept of SHEBA shells, see Argyris and Scharpf (1968), that were enhanced using a geometrically exact theory and thus can be applied to nonlinear problems, see Ivannikov et al. (2014).

When modeling with shells one has a large selection of models that are based on different mathematical models and finite element technologies. The selection of the adequate finite element depends upon the type of problem. The discussion above can help in this selection process. There exist a number of papers, like e.g. MacNeal and Harder (1985) and Korelc et al. (2010), that include test examples which can be used to check the quality of an element formulation.

The finite element formulation is essential for good results, however also the boundary conditons have to be selected with care since shells depict boundary layers. If these are present in the mathematical model they have to be resolved by the finite element mesh which can lead to a very fine mesh along the shell boundaries. If in such applications a coarse mesh is selected, stress peaks can easily be overlooked. This has led to big losses in the history since such errors resulted in fatal failures of large structures.

Users of standard software programs are advised to check the finite element included in the software - using test examples - in order to ensure good prediction of deformations and stresses in a new design.

3.6 Projection based reduction methods

Reduced order models were discussed from the general point in section 2.4. Here we will show how the projection is used to obtain a reduced model. In this approach a projection of the motion of a system onto a subspace reduces the total numbers of unknowns from N to M in a nonlinear finite element system. The motion of the system is described by a vector $\boldsymbol{u}(t) \in I\!R^N$, which is replaced by its projection onto a subspace $\boldsymbol{q}(t) \in I\!R^M$ with

$$\boldsymbol{u}(t) = \boldsymbol{\Psi}\, \boldsymbol{q}(t)\,. \tag{37}$$

The projection matrix

$$\boldsymbol{\Psi} = [\, \Psi_1, \ldots, \Psi_M \,] \tag{38}$$

contains M base vectors Ψ_i which span the subspace.

The basic formulations starts from the equation of motion (36): $\boldsymbol{M}\,\ddot{\mathbf{u}} + \boldsymbol{R}\,(\boldsymbol{u}) = \boldsymbol{P}$. By inserting the projection (37) and by pre-multiplication with the projection matrix (related to the multiplication of (36) by the test function) the reduced matrix equation

$$\boldsymbol{\Psi}^T\, \boldsymbol{M} \boldsymbol{\Psi}\, \ddot{q}(t) + \boldsymbol{\Psi}^T\, \boldsymbol{R}\,(\boldsymbol{\Psi}\, \boldsymbol{q}(t)) = \boldsymbol{\Psi}^T\, \boldsymbol{P}(t) \tag{39}$$

is obtained. Here the damping term was neglected to make the formulation more straightforward. Adding this term is just a technical matter, for details, see e.g. Spiess (2006). With the definitions $\boldsymbol{M}^* = \boldsymbol{\Psi}^T \boldsymbol{M} \boldsymbol{\Psi}$,

$\boldsymbol{R}^*(\boldsymbol{q}(t)) = \boldsymbol{\Psi}^T \boldsymbol{R}(\boldsymbol{\Psi}\,\boldsymbol{q}(t))$ and $\boldsymbol{P}^*(t) = \boldsymbol{\Psi}^T \boldsymbol{P}(t)$ the reduced set of equations with M unknowns

$$\boldsymbol{M}^*\,\ddot{q}(t) + \boldsymbol{R}^*(\boldsymbol{q}(t)) = \boldsymbol{P}^*(t) \tag{40}$$

is derived.

The problem is now to find an optimal base with a minimum number of base vectors that results in a good approximation of the nonlinear dynamic problem, see Fig. 16 for a general sketch of the deviation of the solution resulting from a projection.

There are several demands that lead to a good approximation of a nonlinear system:

- The base vectors have to fulfill the essential boundary conditions.
- The base vectors should be orthogonal: $\boldsymbol{\Psi}^T\,\boldsymbol{\Psi} = \boldsymbol{1}$.

In a nonlinear computation it is necessary to determine the error related to the projection onto the subspace. This is due to the fact that in a nonlinear system the subspace representing the solution can change. This behaviour requires a different selection of the base vectors and thus an update of the base used for the projection. The error related to the reduced projection

$$e_{RP} = \frac{\|\,\boldsymbol{R}(\boldsymbol{\Psi}\,\boldsymbol{q}(t)) - \boldsymbol{P}\,\|}{\|\,\boldsymbol{P}\,\| + \|\,\boldsymbol{M}\,\boldsymbol{\Psi}\,\ddot{q}(t)\,\|} \tag{41}$$

can be computed, see e.g. Idelsohn and Cardona (1984) and Spiess (2006). Note that the solution was obtained with (40) but the error is computed for the full system.

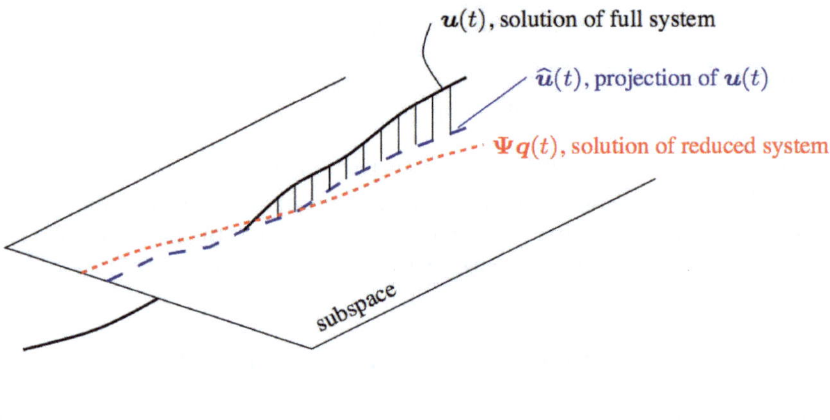

Figure 16. Approximation of a solution for a subspace.

Instead of using eigenvectors to represent the nonlinear system at a specific state $\boldsymbol{u}$, it is possible to introduce the so called Ritz vectors, see Wilson et al. (1982). These base vectors are computed from the load case, starting with the first vector

$$\boldsymbol{\psi}_1 = \boldsymbol{K}_T^{-1}\,\boldsymbol{P} \tag{42}$$

which corresponds to the static load. The further Ritz vectors are a Krylov sequence of the first vector

$$\boldsymbol{\psi}_i = \boldsymbol{K}_T^{-1}\,\boldsymbol{M}\boldsymbol{\psi}_{i-1}\,, \qquad i = 2, \ldots \tag{43}$$

These vectors have to be orthogonalized, see above, which can be achieved by a Gram-Schmidt or a QR orthonormalization.

As an example the two-dimensional T-beam, depicted in Fig. 17, is selected to show the performance of the reduction method using Ritz vectors, see also Spiess (2006). The T-beam consists of an elastic St. Venant material with $E = 1000$ and $\nu = 0.2$. The data of the geometry are given in Fig. 17 as well as the discretization using quadrilateral displacement elements with quadratic shape functions.

A time dependent point load of $F(t) = 0.1\,\sin(0.5\,t)$ is applied at the end of the beam. The system is solved using 20 Ritz vectors and compared to a solution of the unreduced system. The resulting vertical displacement at the load point is shown in Figure 18.

The full line is the solution that employs updates at given intervals. The dotted line is a Ritz approximation without updates where the Ritz vectors were computed at the beginning of the solution. The dashed, updated

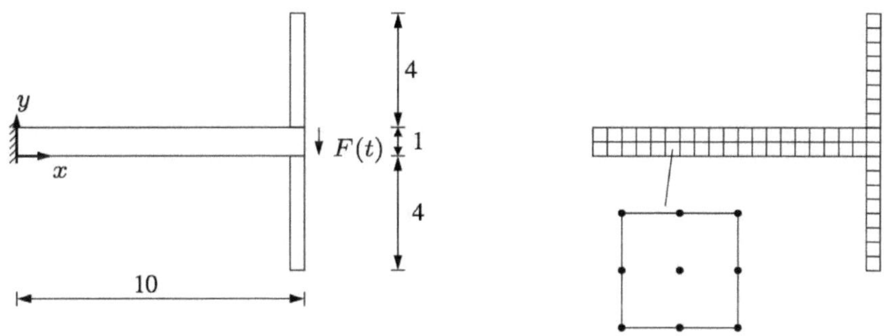

Figure 17. T-beam: geometry and discretization.

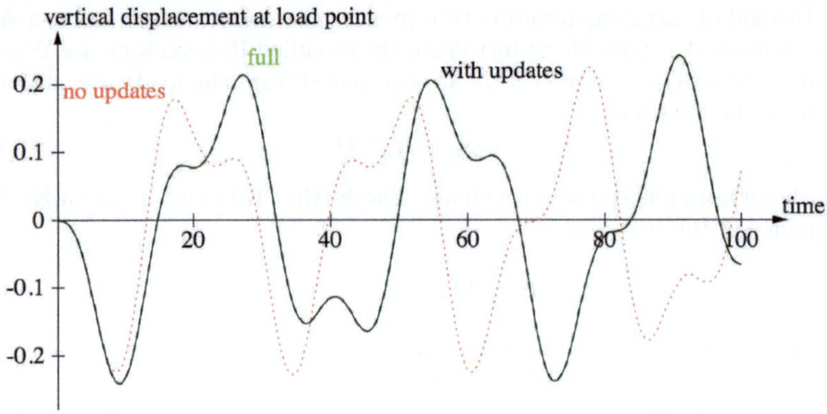

Figure 18. Vertical displacement from solution with Ritz.

reference solution cannot be distinguished from the Ritz reduction with updates and thus underlines, that the base vectors have to be updated in a nonlinear analysis. Other base vectors, like Lanczos vectors could also be used within the projection based reduction method as well. A discussion can be found in Vukazich et al. (1996).

Another technique, the Proper Orthogonal Decomposition (POD) as well as the Proper Generalized Decomposition (PGD) was developed for the analysis of data and hence often used to evaluate experimental data. The goal of POD is to identify a subspace of the solution space that includes the most relevant parts. The method determines the most energetic modes, the so-called POD vectors, as well as the fraction of energy included in each mode. These vectors are called POD vectors in the following. POD-based reduction methods were recently introduced in nonlinear finite element analysis by Krysl et al. (2001) and Meyer and Matthies (2003). The PGD method can be applied such that the method determines the most energetic modes during the analyis and thus is more efficient, see e.g. Néron and Ladevèze (2010).

The design engineer who applies these methods has to know what kind of loading is acting on the structure and what degree of nonlinearity the mathematical model is present in order to select one of the methods discussed above.

4 Conclusion

This contribution on the art of modeling in solid mechanics could only touch a few issues related to the design tasks that engineers have to face when they want to generate a good predictive model. Many examples show that a good model can save a lot of computations and time since intuition of an engineer can lead to a reduced model that has the ability to anticipate the range in which it is applied very well. Such models however can only be determined with a lot of experience, but need insight in mathematical modelling and associated analytical and numerical solution methods.

Several aspects that are also of high interest could not be discussed. Among these are contact problems, multi-physics applications and the newly developed multi-scale approaches. Many other details of solid and shell modeling could also not be touched. Hence the contribution is not complete, but hopefully gives the reader a first view on the essentials when modeling in solid mechanics.

Bibliography

Aboudi J. *Mechanics of Composite Materials - A unified micromechanical approach*. Elsevier, Amsterdam (1991).

Argyris J.H. *Energy Theorems and Structural Analysis*. Butterworth (1960).

Argyris J.H. and Scharpf D.W. The SHEBA Family of Shell Elements for the Matrix Displacement Method. *Aeron. J. Royal Aeron. Soc.*, 72:873–883 (1968).

Atluri S.N. On constitutive relations at finite strain: hypo-elasticity and elasto-plasticity with isotropic and kinematic hardening. *Computer Methods in Applied Mechanics and Engineering*, 43:137–171 (1984).

Basar Y. and Ding Y. Finite-Rotation Elements for Nonlinear Analysis of Thin Shell Structures. *International Journal of Solids & Structures*, 26:83–97 (1990).

Basar Y. and Ding Y. Shear Deformation Models for Large Strain Shell Analysis. *International Journal of Solids & Structures*, 34:1687–1708 (1996).

Bathe K.J. *Finite Element Procedures in Engineering Analysis*. Prentice-Hall, Englewood Cliffs, New Jersey (1982).

Bathe K.J. *Finite Element Procedures*. Prentice-Hall, Englewood Cliffs, New Jersey (1996).

Bathe K.J. and Bolourchi S. Large displacement analysis of three-dimensional beam structures. *International Journal for Numerical Methods in Engineering*, 14:961–986 (1979).

Bazant Z.P. and Cedolin L. *Stability of Structures: Elastic, Inelastic, Fracture, and Damage Theories*. Dover Publications (2003).

Becker A. Berechnung ebener Stabtragwerke nach der Fliešgelenktheorie II. Ordnung unter Berücksichtigung der Normal- und Querkraftinteraktion mit Hilfe der Methode der Finiten Elemente. *Technical report*, Diplomarbeit am Institut für Baumechanik und Numerische Mechanik der Universität Hannover (1985).

Belytschko T., Ong J.S.J., Liu W.K. and Kennedy J.M. Hourglass control in linear and nonlinear problems. *Computer Methods in Applied Mechanics and Engineering*, 43:251–276 (1984).

Benson D.J., Bazilevs Y., Hsu M.C. and Hughes T.J.R. A large deformation, rotation-free, isogeometric shell. *Computer Methods in Applied Mechanics and Engineering*, 200:1367–1378 (2011).

Bischoff M. and Ramm E. Shear Deformable Shell Elements for Large Strains and Rotations. *International Journal for Numerical Methods in Engineering*, 40:4427–4449 (1997).

Bischoff M., Wall W.A., Bletzinger K.U. and Ramm E. Models and Finite Elements for Thin-walled Structures. In E. Stein, R. de Borst and T.J.R. Hughes, editors, *Encyclopedia of Computational Mechanics*, pages 59–137. Wiley, Chichester (2004).

Bouclier R., Elguedj T. and Combescure A. Locking free isogeometric formulations of curved thick beams. *Computer Methods in Applied Mechanics and Engineering*, 245:144–162 (2012).

Braess D. *Finite Elements: Theory, Fast Solvers, and Applications in Solid Mechanics*. Cambridge University Press (2007).

Brenner S.C. and Scott L.R. *The Mathematical Theory of Finite Element Methods*. Springer (2002).

Büchter N. and Ramm E. Shell Theory versus Degeneration – A Comparison in Large Rotation Finite Element Analysis. *International Journal for Numerical Methods in Engineering*, 34:39–59 (1992).

Büchter N., Ramm E. and Roehl D. Three-Dimensional Extension of Non-Linear Shell Formulation Based on the Enhanced Assumed Strain Concept. *International Journal for Numerical Methods in Engineering*, 37:2551–2568 (1994).

Campello E.M.B., Pimenta P.M. and Wriggers P. A triangular finte shell element based on a fully nonlinear shell formulation. *Computational Mechanics*, 31:505–518 (2003).

Cazzani A., Malagù M. and Turco E. Isogeometric analysis of plane-curved beams. *Mathematics and Mechanics of Solids*, page 1081286514531265 (2014).

Chadwick P. *Continuum Mechanics, Concise Theory and Problems*. Dover Publications, Mineola (1999).

Chavan K. and Wriggers P. Consistent coupling of beam and shell models for thermo-elastic analysis. *International Journal for Numerical Methods in Engineering*, 59:1861–1878 (2004).

Chinesta F., Ammar A. and Cueto E. ecent advances and new challenges in the use of the proper generalized decomposition for solving multidimensional models. *Archives of Computational methods in Engineering*, 17:327–350 (2010).

Christensen R.M. A nonlinear theory of viscoelastocity for application to elastomers. *Journal of Applied Mechanics*, 47:762–768 (1980).

Cirak F., Ortiz M. and Schröder P. Subdivision surfaces: a new paradigm for thin shell finite-element analysis. *International Journal for Numerical Methods in Engineering*, 47:2039–2072 (2000).

Cottrell J.A., Hughes T.J.R. and Bazilevs Y. *Isogeometric analysis: toward integration of CAD and FEA*. John Wiley & Sons (2009).

Courant R. Variational methods for the solution of problems of equilibrium and vibrations. *Bull. Amer. Math. Soc,*, 49:1–23 (1943).

Crisfield M.A. *Non-linear Finite Element Analysis of Solids and Structures*, volume 1. Wiley, Chichester (1991).

Crisfield M.A. *Non-linear Finite Element Analysis of Solids and Structures*, volume 2. Wiley, Chichester (1997).

Desai C.S. and Siriwardane H.J. *Constitutive Laws for Engineering Materials*. Prentice-Hall, Englewood Cliffs, New Jersey (1984).

Donnell L.H. *Beams, plates and shells*. McGraw-Hill Companies (1976).

Düster A., Bröker H. and Rank E. The p-version of the finite element method for three-dimensional curved thin walled structures. *International Journal for Numerical Methods in Engineering*, 52:673–703 (2001).

Dvorkin E.N., Onate E. and Oliver J. On a nonlinear formulation for curved Timoshenko beam elements considering large displacement/rotation increments. *International Journal for Numerical Methods in Engineering*, 26:1597–1613 (1988).

Eberlein R. and Wriggers P. Finite element concepts for finite elastoplastic strains and isotropic stress response in shells: Theoretical and computational analysis. *Computer Methods in Applied Mechanics and Engineering*, 171:243–279 (1999).

Ehrlich D. and Armero F. Finite element methods for the analysis of softening plastic hinges in beams and frames. *Computational Mechanics*, 35(4):237–264 (2005).

Ekh M., Johansson A., Thorberntsson H. and Josefson B. Models for Cyclic Ratchetting PlasticityIntegration and Calibration. *Journal of Engineering Materials and Technology*, 122:49 (2000).

Fish J. and Shek K. Multiscale analysis for composite materials and structures. *Composites Science and Technology*, 60:2547–2556 (2000).

Flügge W. *Stresses in shells*. Springer Science (2013).

Gruttmann F., Sauer R. and Wagner W. A geometrical non-linear eccentric 3D-beam element with arbitrary cross-sections. *Computer Methods in Applied Mechanics and Engineering*, 160:383–400 (1998).

Gruttmann F., Sauer R. and Wagner W. Theory and numerics of three-dimensional beams with elastoplastic material behaviour. *International Journal for Numerical Methods in Engineering*, 48:1675–1702 (2000).

Gruttmann F. and Wagner W. A linear quadrilateral shell element with fast stiffness computation. *Computer Methods in Applied Mechanics and Engineering*, 194(39-41):4279–4300 (2005).

Gurson A.L. Continuum theory of ductile rupture by void nucleation and growth, part i. *Journal Engineering Material Technology*, 99:2–15 (1977).

Hain M. and Wriggers P. Computational homogenization of micro-structural damage due to frost in hardened cement paste. *Finite Elements in Analysis and Design*, 44:223–244 (2008a).

Hain M. and Wriggers P. On the numerical homogenization of hardened cement paste. *Computational Mechanics*, 42:197–212 (2008b).

Hashin D. Analysis of composite materials, a survey. *Journal of Applied Mechanics*, 50:481–505 (1983).

Hauptmann R. and Schweizerhof K. A systematic development of 'solid-shell' element formulation for linear and nonlinear analyses employing only displacement degree of freedom. *International Journal for Numerical Methods in Engineering*, 42:49–69 (1998).

Heisserer U., Hartmann S., Yosibash Z. and Düster A. On volumetric locking-free behavior of p-version finite elements under finite deformations. *Communications in Numerical Methods in Engineering. accepted for publication* (2007).

Henning A. Traglastberechnung ebener Rahmen – Theorie II. Ordnungschweizerhof und Interaktion. *Technical Report 75–12*, Institut für Statik, Braunschweig (1975).

Hofstetter G. and Mang H.A. *Computational Mechanics of Reinforced Concrete Structures*. Vieweg, Berlin (1995).

Holzapfel G.A. *Nonlinear Solid Mechanics*. Wiley, Chichester (2000).

Hughes T.J.R., Cottrell J.A. and Bazilevs Y. Isogeometric analysis: CAD, finite elements, NURBS, exact geometry, and mesh refinement. *Computer Methods in Applied Mechanics and Engineering*, 194(39-41):4135–4195 (2005).

Hughes T.R.J. *The Finite Element Method*. Prentice Hall, Englewood Cliffs, New Jersey (1987).

Hull D. and Clyne T. *An introduction to composite materials*. Cambridge university press (1996).

Ibrahimbegovic A. A finite element implementaion of geometrically nonlinear Reissner's beam theory: three-dimensional curved beam elements. *Computer Methods in Applied Mechanics and Engineering*, 122:11–26 (1995).

Idelsohn S.R. and Cardona A. Reduction methods and explicit time integration. *Advances in engineering software*, 6:36–44 (1984).

Ivannikov V., Tiago C. and Pimenta P. Tuba finite elements: application to the solution of a nonlinear kirchhoff–love shell theory. *Shell structures: theory and applications*, 3:81–84 (2014).

Jelenic G. and Saje M. A kinematically exact space finite strain beam model - finite element formulations by generalised virtual work principle. *Computer Methods in Applied Mechanics and Engineering*, 120:131–161 (1995).

Johansson G., Ekh M. and Runesson K. Computational modeling of inelastic large ratcheting strains. *International Journal of Plasticity*, 21(5):955–980 (2005).

Johnson K.L. *Contact Mechanics*. Cambridge University Press (1985).

Jones R.M. *Mechanics of composite materials*, volume 1. McGraw-Hill New York (1975).

Kahn R. Finite-Element-Berechnungen ebener Stabwerke mit Fliessgelenken und grossen Verschiebungen. *Technical Report F 87/1*, Forschungs- und Seminarberichte aus dem Bereich der Mechanik der Universität Hannover (1987).

Khan A.S. and Huang S. *Continuum theory of plasticity*. Wiley, Chichester, New York (1995).

Kiendl J., Bletzinger K.U., Linhard J. and Wüchner R. Isogeometric shell analysis with Kirchhoff–Love elements. *Computer Methods in Applied Mechanics and Engineering*, 198:3902–3914 (2009).

Kirchhoff G.R. Über das Gleichgewicht und die Bewegung einer elastischen Scheibe. *Journal für die Reine und Angewandte Mathematik* (1850).

Kirsch U., Bogomolni M. and Sheinman I. Nonlinear dynamic reanalysis for structural optimization. In van CampenD. H., L.M. D. and van den Oever W. P. J . M, editors, *Proceedings of the Fifth EUROMECH Nonlinear Dynamics Conference (ENOC)*. Eindhoven (2005).

Koiter W.T. A consistent first approximation in the general theory of thin elastic shells. In W.T. Koiter, editor, *The Theory of Thin Elastic Shells*, pages 12–33. North-Holland, Amsterdam (1960).

Kolymbas D. An outline of hypoplasticity. *Archive of Applied Mechanics*, 61:143–151 (1991).

Konyukhov A. and Schweizerhof K. Geometrically exact theory for contact interactions of 1d manifolds. algorithmic implementation with various finite element models. *Computer Methods in Applied Mechanics and Engineering*, 205:130–138 (2012).

Korelc J., Solinc U. and Wriggers P. An improved EAS brick element for finite deformation. *Computational Mechanics*, 46:641–659 (2010).

Korelc J. and Wriggers P. Consistent gradient formulation for a stable enhanced strain method for large deformations. *Engineering Computations*, 13:103–123 (1996).

Krysl P., Lall S. and Marsden J.E. Dimensional model reduction in nonlinear finite element dynamics of solids and structures. *International Journal for Numerical Methods in Engineering*, 51:479–504 (2001).

Ladeveze P., Passieux J. and Neron D. The latin multiscale computational method and the proper generalized decomposition. *Computer Methods in Applied Mechanics and Engineering*, 199:1287–1296 (2010).

Larsson R., Runesson K. and Ottosen N. Discontinuous displacement approximation for capturing plastic localization. *International Journal for Numerical Methods in Engineering*, 36:2087–2105 (1993).

Laursen T.A. *Computational Contact and Impact Mechanics*. Springer, Berlin, New York, Heidelberg (2002).

Leger P. Mode superposition methods. In M. Papadrakakis, editor, *Solving Large-scale Problems in Mechanics*, pages 225–257. Wiley (1993).

Lehmann E., Faßmann D., Loehnert S., Schaper M. and Wriggers P. Texture development and formability prediction for pre-textured cold rolled body-centred cubic steel. *International Journal of Engineering Sciences*, 68:24–37 (2013).

Lehmann E., Löhnert S. and Wriggers P. About the microstructural effects of polycrystalline materials and their macroscopic representation at finite deformation. *PAMM*, 12:275–276 (2012).

Lemaitre J. *A course on damage mechanics*. Springer, Berlin, New York (1996).

Leppin C. and Wriggers P. Numerical simulations of the behaviour of cohesionless soil. In D.R.J. Owen, E. Hinton and E. Onate, editors, *Proceedings of COMPLAS 5*. CIMNE, Barcelona (1997).

Liu W., Karpov E. and Park H. *Nano Mechanics and Materials: Theory, Multiscale Analysis and Applications*. Wiley, Chichester (2005).

Love A.E.H. *A Treatise on the Mathematical Theory of Elasticity*, volume Fourth. University Press, Cambrige (1927).

Lubliner J. A model of rubber viscoelasticity. *Mechancis Research Communications*, 12:93–99 (1985).

Lubliner J. *Plasticity Theory*. MacMillan, London (1990).

Lumpe G. Geometrisch nichtlineare berechnung von räumlichen stabwerken. *Technical Report 28*, Institut für Statik, Universität Hannover (1982).

MacNeal R. and Harder R. A proposed standard set of problems to test finite element accuracy. *Finite Elements in Analysis and Design*, 1:3–20 (1985).

Mäkinen J. Total lagrangian Reissner's geometrically exact beam element without singularities. *International Journal for Numerical Methods in Engineering*, 70:1009–1048 (2007).

Meyer M. and Matthies H. Efficient model reduction in non-linear dynamics using the Karhunen-Loeve expansion and dual-weighted-residual methods. *Computational Mechanics*, 31:179–191 (2003).

Miehe C. A theoretical and computational model for isotropic elastoplastic stress analysis in shells at large strains. *Computer Methods in Applied Mechanics and Engineering*, 155:193–233 (1998).

Miehe C. and Schröder J. Post-critical discontinous localization analysis of small-strain softening elastoplastic solids. *Archive of Applied Mechanics*, 64:267–285 (1994).

Miehe C., Schröder J. and Schotte J. Computational homogenization analysis in finite plasticity simulation of texture development in polycrystalline materials. *Computer Methods in Applied Mechanics and Engineering*, 171:387–418 (1999).

Naghdi P.M. *The Theory of Shells*, volume VIa/2 of *Handbuch der Physik, Mechanics of Solids II*. Springer, Berlin (1972).

Néron D. and Ladevèze P. Proper generalized decomposition for multiscale and multiphysics problems. *Archives of Computational Methods in Engineering*, 17:351–372 (2010).

Nickell R.E. Nonlinear dynamics by mode superposition. *Computer Methods in Applied Mechanics and Engineering*, 7:107–129 (1976).

Niemunis A., Wichtmann T. and Triantafyllidis T. A high-cycle accumulation model for sand. *Computers and Geotechnics*, 32:245–263 (2005).

Oden J.T., Prudhomme S., Hammerand D.C. and Kuczma M.S. Modeling error and adaptivity in nonlinear continuum mechanics. *Computer Methods in Applied Mechanics and Engineering*, 190:8883–6684 (2001).

Ogden R.W. *Non-Linear Elastic Deformations*. Ellis Horwood and John Wiley, Chichester (1984).

Oliver J. Continuum modeling of strong discontinuities in solid mechanics using damage models. *Computational Mechanics*, 17:49–61 (1995).

Onate E. and Cervera M. Derivation of thin plate bending elements with one degree of freedom per node: a simple three node triangle. *Engineering computations*, 10(6):543–561 (1993).

Oran C. and Kassimali A. Large deformations of framed structures under static and dynamic loads. *Computers and Structures*, 6:539–547 (1976).

Perzyna P. Fundamental problems in viscoplasticity. *Advances in Applied Mechanics*, 9:243–377 (1966).

Pflüger A. *Elementare Schalenstatik*. Springer (1981).

Pian T.H.H. Derivation of element stiffness matrices by assumed stress distributions. *AIAA-J. 2*, 7:1333–1336 (1964).

Pian T.H.H. and Sumihara K. Rational approach for assumed stress finite elements. *International Journal for Numerical Methods in Engineering*, 20:1685–1695 (1984).

Pietraszkiewicz W. Geometrically nonlinear theories of thin elastic shells. *Technical Report 55*, Mitteilungen des Instituts für Mechanik der Ruhr-Universität Bochum (1978).

Pimenta P. and Yojo T. Geometrically exact analysis of spatial frames. *Applied Mechanics Review*, 46:118–128 (1993).

Pimenta P.M., Campello E.M.B. and Wriggers P. A fully nonlinear multi-parameter shell model with thickness variation and a triangular shell finite element formulation. *Computational Mechanics*, 34:181–193 (2004).

Pimenta P.M., Campello E.M.B. and Wriggers P. An exact conserving algorithm for nonlinear dynamics with rotational dofs and general hyperelasticity. part 1: Rods. *Computational Mechanics*, 42:715–732 (2008).

Radermacher A. and Reese S. Model reduction in elastoplasticity: proper orthogonal decomposition combined with adaptive sub-structuring. *Computational Mechanics*, 54:677–687 (2014).

Ramm E. Geometrisch nichtlineare Elastostatik und Finite Elemente. *Technical Report Nr. 76-2*, Institut für Baustatik der Universität Stuttgart (1976).

Rankin C.C. and Brogan F.A. An element independent corotational procedure for the treatment of large roations. In L.H. Sobel and K. Thomas, editors, *Collapse Analysis of Structures*, pages 85–100. ASME, New York (1984).

Reali A. and Gomez H. An isogeometric collocation approach for Bernoulli–Euler beams and Kirchhoff plates. *Computer Methods in Applied Mechanics and Engineering*, 284:623–636 (2015).

Reddy J.N. A simple higher-order theory for laminated composite plates. *Journal of Applied Mechanics*, 51:745–752 (1984).

Reddy J.N. *Mechanics of laminated composite plates and shells: theory and analysis*. CRC press (2004).

Reese S. On a physically stabilized one point finite element formulation for three-dimensional finite elasto-plasticity. *Computer Methods in Applied Mechanics and Engineering*, 194:4685–4715 (2005).

Reese S. and Govindjee S. A theory of finite viscoelasticity and numerical aspects. *International Journal of Solids & Structures*, 35:3455–3482 (1998).

Reissner E. On one-dimensional finite strain beam theory, the plane problem. *Journal of Applied Mathematics and Physics*, 23:795–804 (1972).

Rice J. Solid mechanics. *http://esag.harvard.edu/rice/e0_Solid_Mechanics_94_10.pdf* (2010).

Romero I. and Armero F. An objective finite element approximation of the kinematics of geometrically exact rods and its use in the formulation of an energy-momentum conserving scheme in dynamics. *International Journal for Numerical Methods in Engineering*, 54:1683–1716 (2002).

Sansour C. A theory and finite element formulation of shells at finite deformations involving thickness change: circumventing the use of a rotation tensor. *Ingenieur-Archiv*, 65:194–216 (1995).

Schröder J. Anisotropic polyconvex energies. In J. Schröder, editor, *Polyconvex Analysis*, pages 1–53. CISM, Springer, Wien (2009). 62.

Seifert B. *Zur Theorie und Numerik Finiter Elastoplastischer Deformationen von Schalenstrukturen*. Dissertation, Institut für Baumechanik und Numerische Mechanik der Universität Hannover (1996). Bericht Nr. F 96/2.

Simo J., Oliver J. and Armero F. Analysis of strong dicontituities in rate independent softening materials. *Computational Mechanics* (1993a).

Simo J.C. A finite strain beam formulation. The three-dimensional dynamic problem. Part I. *Computer Methods in Applied Mechanics and Engineering*, 49:55–70 (1985).

Simo J.C. Numerical analysis and simulation of plasticity. In P.G. Ciarlet and J.L. Lions, editors, *Handbook of Numerical Analysis*, volume 6, pages 179–499. North-Holland (1998).

Simo J.C. and Armero F. Geometrically Non-Linear Enhanced Strain Mixed Methods and the Method of Incompatible Modes. *International Journal for Numerical Methods in Engineering*, 33:1413–1449 (1992).

Simo J.C., Armero F. and Taylor R.L. Improved Versions of Assumed Enhanced Strain Tri-Linear Elements for 3D Finite Deformation Problems. *Computer Methods in Applied Mechanics and Engineering*, 110:359–386 (1993b).

Simo J.C., Fox D.D. and Rifai M.S. On a stress resultant geometrical exact shell model. Part I: Formulation and optimal parametrization. *Computer Methods in Applied Mechanics and Engineering*, 72:267–304 (1989).

Simo J.C., Fox D.D. and Rifai M.S. On a stress resultant geometrical exact shell model. Part III: Computational Aspects of the Nonlinear Theory. *Computer Methods in Applied Mechanics and Engineering*, 79:21–70 (1990).

Simo J.C., Hjelmstad K.D. and Taylor R.L. Numerical formulations for finite deformation problems of beams accounting for the effect of transverse shear. *Computer Methods in Applied Mechanics and Engineering*, 42:301–330 (1984).

Simo J.C. and Hughes T.J.R. *Computational Inelasticity*. Springer, New York, Berlin (1998).

Simo J.C. and Pister K.S. Remarks on rate constitutive equations for finite deformation problems. *Computer Methods in Applied Mechanics and Engineering*, 46:201–215 (1984).

Simo J.C. and Rifai M.S. A class of assumed strain methods and the method of incompatible modes. *International Journal for Numerical Methods in Engineering*, 29:1595–1638 (1990).

Simo J.C., Rifai M.S. and Fox D.D. On a Stress Resultant Geometrically Exact Shell Model. Part IV. Variable Thickness Shells with Through-The-Tickness Stretching. *Computer Methods in Applied Mechanics and Engineering*, 81:91–126 (1990B).

Simo J.C. and Vu-Quoc L. Three dimensional finite strain rod model. Part II: computational aspects. *Computer Methods in Applied Mechanics and Engineering*, 58:79–116 (1986).

de Souza Neto E.A., Peric D. and Owen D.R.J. *Computational Methods for Plasticity, Theory and Applications*. Wiley, Chichester (2008).

Spiess H. Reduction methods in finite element analysis of nonlinear structural dynamics. *Technical Report F06/2*, Forschungs- und Seminarberichte aus dem Bereich der Mechanik der Universität Hannover (2006).

Stein E. and Ohnimus S. Dimensional adaptivity in linear elasticity with hierarchical test-spaces for h- and p- refinement processes. *Engineering Computations*, 12:107–119 (1996).

Szabó I. *Geschichte der mechanischen Prinzipien und ihrer wichtigsten Anwendungen*. Birkhäuser, Basel, Boston, Stuttgart (1987).

Sze K. and Yao L. A hybrid stress ans solid-shell element and its generalization for smart structure modelling. part isolid-shell element formulation. *International Journal for Numerical Methods in Engineering*, 48:545–564 (2000).

Taylor R.L., Beresford P.J. and Wilson E.L. A Non-Conforming Element for Stress Analysis. *International Journal for Numerical Methods in Engineering*, 10:1211–1219 (1976).

Temizer I. and Wriggers P. An adaptive multiscale resolution strategy for the finite deformation analysis of microheterogeneous structures. *Computer Methods in Applied Mechanics and Engineering*, 200(37):2639–2661 (2011).

Timoshenko S. *History of strength of materials: with a brief account of the history of theory of elasticity and theory of structures*. Courier Corporation (1983).

Timoshenko S. and Woinowsky-Krieger S. *Theory of plates and shells*, volume 2. McGraw-hill New York (1959).

Truesdell C. *The Elements of Continuum Mechanics*. Springer, New York (1985).

Truesdell C. and Noll W. The nonlinear field theories of mechanics. In S. Flügge, editor, *Handbuch der Physik III/3*. Springer, Berlin, Heidelberg, Wien (1965).

Turner M.J., Clough R.W., Martin H.C. and Topp L.J. Stiffness and deflection analysis of complex structures. *Journal of the Aeronautical Sciences*, 23:805–823 (1956).

Tvergaard V. and Needleman A. Analysis of the cup-cone fracture in a round tensile bar. *Archives of Mechanics*, 32:157–169 (1984).

Vlasov V.Z. *Thin-walled elastic beams*. National Technical Information Service (1984).

Vogel U. Calibrating frames, Vergleichsrechnungen an verschiedenen Rahmen. *Der Stahlbau*, 10:295–301 (1985).

Vukazich M., Mish K. and Romstad K. Nonlinear dynamic response of frames using lanczos modal analysis. *Journal of Structural Engineering*, 122:1418–1426 (1996).

Wagner W. and Gruttmann F. A Simple Finite Rotation Formulation for Composite Shell Elements. *Engineering Computations*, 11:145–176 (1994).

Wagner W. and Gruttmann F. Modeling of Shell-Beam Transitions in the Presence of Finite Rotations. *Computer Assited Mechanics and Engineering Sciences*, 9:405–418 (2002).

Washizu K. *Variational Methods in Elasticity and Plasticity*. Pergamon Press, Oxford, second edition (1975).

Wellmann C. and Wriggers P. A two-scale model of granular materials. *Computer Methods in Applied Mechanics and Engineering*, 205-208:46–58 (2012).

Wempner G. Finite elements, finite rotations and small strains of flexible shells. *International Journal of Solids & Structures*, 5:117–153 (1969).

Wilson E.L., Taylor R.L., Doherty W.P. and Ghaboussi J. Incompatible Displacements Models. In *Numerical and Computer Models in Structural Mechanics*, Fenves S. J., Perrone N., Robinson A. R. and Schnobrich W. C. (Eds.). Academic Press, New York (1973). 43–57.

Wilson E.L., Yuan M. and Dickens J.M. Dynamic analysis by direct superposition of ritz vectors. *Earthquake Engineering and Structural Dynamics*, 10:813–821 (1982).

Windels R. Traglasten von Balkenquerschnitten beim Angriff von Biegemoment, Längs- und Querkraft. *Der Stahlbau*, 39:10–16 (1970).

Wriggers P. *Computational Contact Mechanics*. 2nd ed., Springer, Berlin, Heidelberg, New York (2006).

Wriggers P. *Nonlinear Finite Elements*. Springer, Berlin, Heidelberg, New York (2008).

Wriggers P. and Gruttmann F. Large deformations of thin shells: Theory and finite-element- discretization, analytical and computational models of shells. In A.K. Noor, T. Belytschko and J.C. Simo, editors, *ASME, CED-Vol.3*, volume 135-159 (1989).

Wriggers P. and Gruttmann F. Thin shells with finite rotations formulated in biot stresses theory and finite-element-formulation. *International Journal for Numerical Methods in Engineering*, 36:2049–2071 (1993).

Wriggers P. and Reese S. A note on enhanced strain methods for large deformations. *Computer Methods in Applied Mechanics and Engineering*, 135:201–209 (1996).

Zienkiewicz O.C. and Taylor R.L. *The Finite Element Method, 4th Ed.*, volume 1. McGraw Hill, London (1989).

Zienkiewicz O.C. and Taylor R.L. *The Finite Element Method*, volume 2. Butterworth-Heinemann, Oxford, UK, 5th edition (2000).

Zienkiewicz O.C., Taylor R.L. and Too J.M. Reduced Integration Technique in General Analysis of Plates and Shells. *International Journal for Numerical Methods in Engineering*, 3:275–290 (1971).

MIX
Papier aus verantwortungsvollen Quellen
Paper from responsible sources
FSC® C105338

If you have any concerns about our products,
you can contact us on
ProductSafety@springernature.com

In case Publisher is established outside the EU,
the EU authorized representative is:
Springer Nature Customer Service Center GmbH
Europaplatz 3, 69115 Heidelberg, Germany

Printed by Libri Plureos GmbH
in Hamburg, Germany